Nanofluid Heat and Mass Transfer: Engineering Applications

Nanofluid Heat and Mass Transfer: Engineering Applications

Editor: Ricky Cole

MURPHY & MOORE
www.murphy-moorepublishing.com

Published by Murphy & Moore Publishing
1 Rockefeller Plaza,
New York City, NY 10020, USA
www.murphy-moorepublishing.com

Nanofluid Heat and Mass Transfer: Engineering Applications
Edited by Ricky Cole

International Standard Book Number: 978-1-63987-381-4 (Hardback)

Cataloging-in-Publication Data

Nanofluid heat and mass transfer : engineering applications / edited by Ricky Cole.
p. cm.
Includes bibliographical references and index.
ISBN 978-1-63987-381-4
1. Heat exchangers. 2. Heat exchangers--Fluid dynamics. 3. Heat--Transmission. 4. Heat-transfer media.
5. Nanofluids--Simulation methods. I. Cole, Ricky.
TJ263 .N36 2022
621.402 5--dc23

Contents

Preface

The fluid which contains nanometer-sized particles is termed as a nanofluid. The area of thermal engineering which is concerned with the generation, use, conversion and exchange of heat between physical systems is known as heat transfer. The net movement of mass from one location such as, stream, phase, fraction or component, to another is termed as mass transfer. Nanofluid heat and mass transfer are studied under the field of chemical engineering. The bigger the surface area of nanoparticles, the more is the heat transfer. The increased heat transfer and the micro-convection increases the dispersion of heat in the fluid at a faster rate. With the rise in temperature, the thermal conductivity of nanofluids increases significantly. Due to Brownian motion of suspended nanoparticles, mass transport is promoted by the presence of nanoparticles, inside the nanofluid. Nanofluid heat and mass transfer are studied under the field of chemical engineering. The various studies that are constantly contributing towards advancing technologies and evolution of this field are examined in detail. From theories to research to practical applications, case studies related to all contemporary topics of relevance to this field have been included herein. Those in search of information to further their knowledge will be greatly assisted by this book.

All of the data presented henceforth, was collaborated in the wake of recent advancements in the field. The aim of this book is to present the diversified developments from across the globe in a comprehensible manner. The opinions expressed in each chapter belong solely to the contributing authors. Their interpretations of the topics are the integral part of this book, which I have carefully compiled for a better understanding of the readers.

At the end, I would like to thank all those who dedicated their time and efforts for the successful completion of this book. I also wish to convey my gratitude towards my friends and family who supported me at every step.

Editor

1

Measuring Nanofluid Thermal Diffusivity and Thermal Effusivity: The Reliability of the Photopyroelectric Technique

Monir Noroozi and Azmi Zakaria

Abstract

It is important to study nanofluids to understand their extraordinary thermal properties and how the size, concentration and agglomeration of the nanoparticles affect those properties. Photopyroelectric (PPE) technique has been well established in the use of non-destructive measurement of thermal diffusivity and thermal effusivity, by using polyvinylidene fluoride (PVDF) films as sensitive pyroelectric sensors in thermally thick conditions instead of using very thick ceramic sensors. There have been two proposed practical configurations for the PPE technique, the back and the front PPE configurations, to obtain both the thermal diffusivity and effusivity, which are suitable thermal parameters of materials. This PPE technique involves the measurement of thermal waves in the sample due to absorption of optical radiation, by placing a pyroelectric sensor in thermal contact with the sample. This chapter provides a review of the back and the front PPE configurations to determine the thermal diffusivity and effusivity of nanofluids, sample preparation techniques using high-amplitude ultrasonic dispersion and data analysis for metal oxide-based nanofluid materials.

Keywords: nanofluids, thermal properties, photopyroelectric technique, thermal diffusivity, thermal effusivity

1. Introduction

Water, ethylene glycol (EG) and oil are universally used for transfer heating, but unfortunately these fluids have extremely poor thermal conductivity; thus, smaller and lighter heat exchanges could be reduced leading to reduced power and the size of the required heat transfer. As can be seen in **Figure 1**, the thermal conductivity of copper is about 700 times more than the water and about 3000 times more than the engine oil.

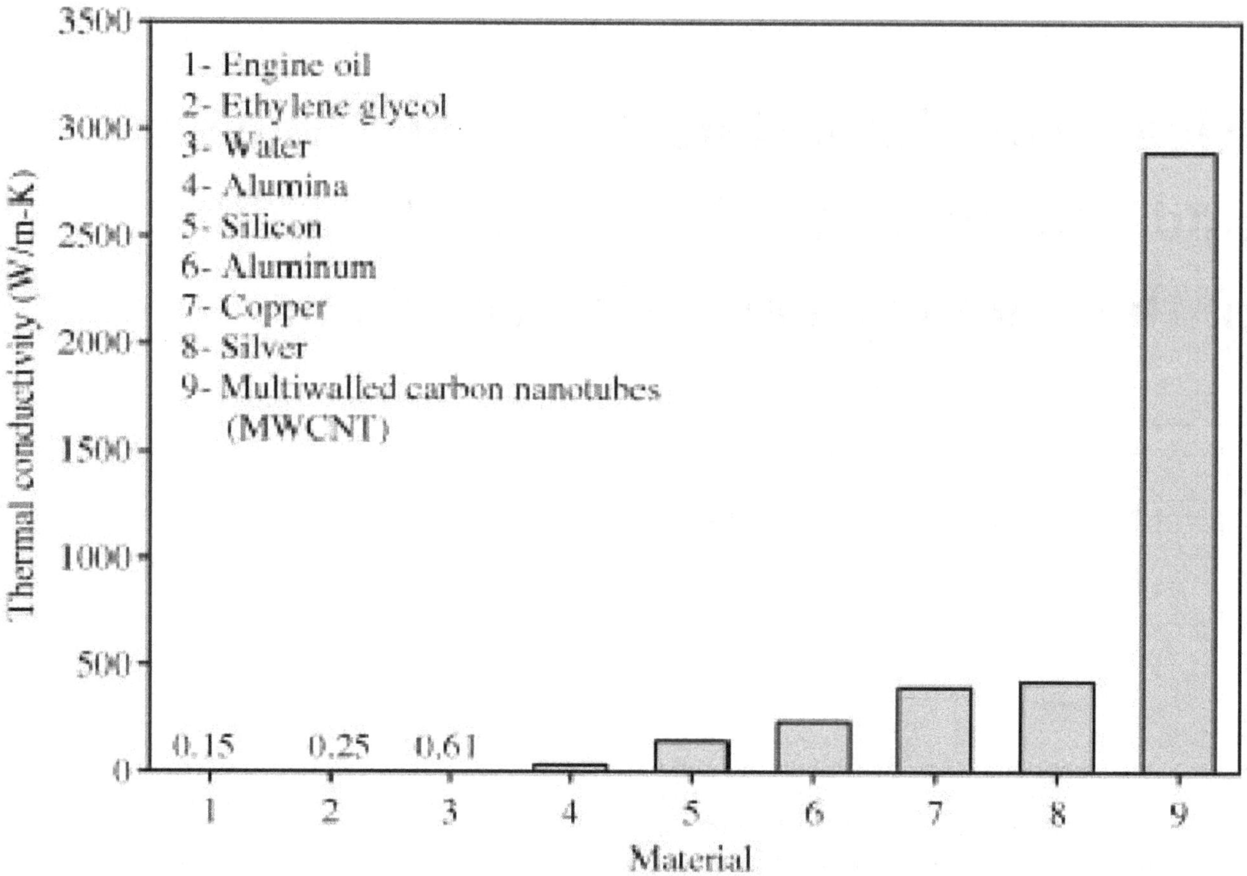

Figure 1. Thermal conductivity of the materials (solids and liquids) at room temperature [1].

The study of heat transport in solid-liquid dispersions (colloids) began as early as in the 1970s. The first thermal conductivity enhancement of nanoparticles (NPs) was reported by Masuda et al. [2]. Due to the size of the NPs, they are well suited for use in microsystems. It was observed that the thermal conductivity of an ultrafine suspension of metal oxides in water increased up to 30% for an NP volume fraction of 4.3%. The term 'nanofluid' was first coined by Choi and Eastman [3] when he reported a class of engineered fluids containing nanosized particles dispersed in ethylene glycol that had a thermal conductivity with almost a factor of 2 greater than the base fluid. Nanofluids have widespread usage in industry, including medicine applications, engineering applications, in cooling/heating systems and in micromechanical systems, due to their enhanced thermal management. Nanofluids are able to improve heat transfer and, thus, allow for smaller pumping, heaters and other elements. Nanofluids containing metals and metal oxides have been considered as the next generation of the heat transfer fluids, as they have shown an increase in their effective thermal conductivity compared to their base fluid [4]. Metal oxide nanofluids have potential applications in the main processing industries, such as the materials, chemistry, biomedicine, food and drink, oil and gas industries as they are able to enhance thermal transfer [4, 5].

The preparation of nanofluids is an important parameter in the investigation of the thermal properties of nanofluids. Nanofluids can be prepared by directly dispersed nanopowders in the base fluid, but this method could result in a large degree of NP agglomeration. Therefore, a clear understanding of the effect of concentration, dispersion/aggregation state and particle size on the thermal properties of prepared nanofluid is an essential assay step for particle validation. Data has shown that the thermal conductivity of nanofluids could significantly be increased with an increase in the volume fraction of NPs [6–8]. It is clear that four thermal parameters can be connected by two relationships, $\alpha = k/\rho C$ and ($e = k/\sqrt{\alpha}$), where C, α, k and e are the volume specific heat, thermal diffusivity, conductivity and effusivity, respectively. Although the thermophysical properties of nanofluids are intensely researched at present, most of the studies have been focused on measuring the thermal conductivity of the nanofluids.

There are only a few reports on the thermal diffusivity measurements using thermal lens spectrometry and the transient double-hot-wire method [9]. However, these methods often require high temperatures to obtain reasonable signal-to-noise (SNR) ratios, which in turn, could increase the sample temperature and thus increase the measurement error. These techniques are also disadvantageous because they require high-volume samples, long measurement times without according the effect of aggregation time and are expensive. Nowadays, highly sensitive photopyroelectric (PPE) techniques have been designed to measure the thermal properties of samples and are actually one of the several available photothermal techniques [10–12]. The thermal parameters directly resulting from the PPE experiment are usually the 'fundamental' ones: the thermal diffusivity and also the thermal effusivity. The unique features of the back and front PPE techniques to measure the thermal diffusivity and the thermal effusivity of nanofluids in high resolution are that they are not possible with other existing techniques [13]. The advantages of this method include its relatively low cost, and only a small volume of the sample is required with a short measurement time, where the concentration of the nanofluid remains constant in the measurement process, thus making this technique suitable for nanofluids. In the principle by using both the back PPE and front PPE configurations, source of information (amplitude and phase of pyroelectric signal), cavity scanning or frequency scanning, it is possible to obtain both the thermal diffusivity and thermal effusivity. Recently, a new simplified front PPE configuration was designed using a metalized polyvinylidene fluoride (PVDF) sensor in a thermally thick condition instead of using the very thick ceramic sensors of typically 300 μm [14] or 500 μm in size [15] (usually $LiTaO_3$) that have to be coated with gold with a very low chopping frequency facility.

The present chapter provides a review on the thermal properties of nanofluids measured by the PPE technique by using frequency scans of the signals employing PVDF as a pyroelectric (PE) sensor in thermally thick conditions, due to its low cost, light weight, flexibility and sensitivity. The back and front PPE configurations in 'thermally thick' conditions have been implemented to measure the thermal diffusivity and the thermal effusivity of nanofluids (containing Al_2O_3 and CuO NPs) as a function of the base fluid's particle size. To reduce agglomeration of the NPs, an ultrasonic dispersion technique was utilized for in low concentration of the NPs to produce stable nanofluids, and the effects of sonication type on the stability and thermal properties were investigated.

2. Thermal properties of the nanofluids

2.1. Synthesis and stability of nanofluids

The preparation of nanofluids is an important parameter in the investigation of the thermal properties of nanofluids. The preparation of nanofluid is not merely a simple mixing of liquid and nanopowder, and thus a good dispersion method of dispersing NPs in liquids or a direct production of stable nanofluids is crucial. A good dispersion of NP materials into liquids such as deionized water (DW), ethylene glycol (EG) or oil is needed for producing a stable nanofluid. There are primarily two methods for the synthesis of nanofluids, including the two-step process and the single-step process for the direct synthesis of nanofluids.

The two-step method is achieved by firstly synthesizing dry NPs with the preferred size and shape. In the second step, these particles are carefully mixed into the required base fluid in the desired volume fraction, typically with some additives to enhance the stability of the nanofluids. Thus, the small volume fraction of NPs and proper dispersion techniques are important for the preparation of stabile nanofluids in this technique. Many researchers have reported successful fabrication and testing of nanofluids using the two-step preparation method [16, 17]. Due to the high surface area of the NPs, they have the tendency to aggregate. A large degree of agglomeration in NPs may occur as a result of using this method. Thus, proper dispersion techniques, such as the ultrasonic dispersion technique [18] or the fragmentation process of NPs using laser irradiation, in low concentrations of NPs, are important for the production stability of nanofluids. Another technique to enhance the stability of NPs in fluids is the use of surfactants. To summarize, the optimization of thermal characteristics of nanofluids requires stabile nanofluids, which can be achieved by synthesis and dispersion processes.

2.2. Experimental investigation methods

2.2.1. Thermal diffusivity measurement techniques

Highly sensitive photothermal methods using a laser as an optical source have been widely used in the thermal diffusivity measurements of nanofluids [19–28]. The photothermal effect in a material is a consequence of the deposition of heat in the sample following absorption of a laser beam and subsequent thermal de-excitations, which results in the indirect heating of the sample. Photoacoustics, photothermal deflection, thermal lens, photothermal radiometry and photopyroelectric methods are some of the techniques commonly used powerful for thermal and optical characterization of materials using lasers. The conventional techniques such as the 'hot-wire', 'laser flash', '3ω-wire method' and 'optical (forced Rayleigh light scattering)' techniques have also been utilized by some researchers [29–33], as seen in **Table 1**.

2.2.2. Thermal effusivity measurement techniques

Very few studies have been reported on the determination of the thermal effusivity of liquids. During the last two decades, the front PPE configuration and photoacoustic techniques have been used for determining the thermal effusivity [13–15]. For the front detection configuration,

References	Particle type	Base fluid	Particle fraction	Enhancement	Method
Dadarlat et al. [19]	Fe_3O_4	Transformer oil	0–623 mg/ml	(9.06–9.84) 10^{-8} m^2/s	Thermal-wave resonator cavity
Nisha and Philip [20]	TiO_2/PVA 5–100 nm	Water	(1–5)vol%	Increases at high concentrations and high particle size	Thermal-wave interference
Philip and Nisha [21]	TiO_2/PVA	Water	(0.005–0.039) vol%	Normalized thermal diffusivity from 1 to 0.96	Thermal-wave resonator cavity
López-Muñoz et al. [22]	Urchin-like colloidal gold	Water	(0.2–1)wt%	1.02–1.05	Photopyroelectric
		EG	(0.2–1)wt%	1.06–1.11	
		Ethanol	(0.2–1)wt%	1.09–1.14	
Dadarlat et al. [23]	Fe_3O_4	Water	(8.2–81.7) mg/cm^3	High-accuracy results (within ±0.5%). Thermal diffusivity was sensitive to changes in type and NP concentration	Photopyroelectric
	$CoFe_2O_4$	Water	(6.1–24.5) mg/cm^3		
Agresti et al. [24]	Al_2O_3 20–70°C	Water, EG	1 and 2 wt%	For both water and glycol from 1.04 to 1.12	Photoacoustic
López-Muñoz et al. [22]	Gold	Water	(0.2–1)wt%	1.01–1.04	Photoacoustic
Sánchez-Ramírez et al. [25]	Au/Pd	Water	Au/Pd = 12/1, 5/1, 1/1, 1/5	The maximum diffusivity was achieved for the nanoparticles with highest Au/Pd molar ratio	Thermal lens technique
Kumar et al. [26]	Gold 30–50 nm	Water	1% wt	The value decreases with decrease in particle size	Thermal lens technique
Jiménez Pérez et al. [27]	Gold	Water, ethanol and EG	0.1 g/L	Enhanced	Thermal lens technique
Gutierrez Fuentes et al. [28]	Au/Ag nanoparticles	Water	Au/Ag = 3/1, 1/1, 1/3, 1/6	A lineal increment of the thermal diffusivity when the Ag shell thickness is increased	Thermal lens technique
Filippo et al. [29]	Ag	Deionized water	4 vol%	12%	Laser flash
Wang et al. [30]	TiO_2	Water	(1–4)vol% 20–70°C	Vary significantly with temperature An enhancement of up to 19.6% is observed at 4% and 65oc	3ω-Method
	SiO_2	Water, ethanol and EG	(1–4)vol% 20–70°C		
Faris Mohammed and Yunus [31]	Al Al_2O_3	Distilled water, ethanol and EG	Five different volume fractions of nanoparticle	Thermal diffusivity increased linearly with increasing concentration of nanoparticles in the respective base fluids	Hot-wire technique
Murshed et al. [32]	TiO2	EG	(1–5)wt%	1.1–1.3	Hot-wire technique
	Al_2O_3	EG	(1–5)wt%	1.05–1.3	
Rondino et al. [33]	Pyrolytic titania	Ethanol	0.6 vol%	0.6%	Optical technique

Table 1. Summary of experimental studies of thermal diffusivity enhancement.

two schemes were proposed, namely, the configuration with a thermally thin and the optically opaque PVDF sensor [34] and the configuration with a thermally thick and optically semi-transparent sensor using $LiTaO_3$ [15]. Balderas-López et al. [35] applied the front PE configuration to perform high precision measurements of thermal effusivity in transparent liquids in a very thermally thick regime.

Esquef et al. [36], in 2006, developed a method consisting essentially of a photoacoustic cell and a PE cell enclosed in a single compact gas analyzer for the measurement of thermal diffusivity and thermal effusivity. Concerning the front configuration, a simplified method to measure both the thermal diffusivity and thermal effusivity of sensor was proposed. For example, Streza et al., in 2009, [37]applied two PE detection configurations, 'back' and 'front', to the calorimetric studies of some liquids (liquid mixtures, magnetic material nanofluids, liquid foodstuffs, etc.). They demonstrated that if the back configuration used the phase of PE signal in the cavity scan method and the front configuration used the frequency scan, both thermal diffusivity and thermal effusivity could be measured. Dadarlat et al.[23], in 2008, measured the thermal diffusivity and thermal effusivity of Fe_3O_4 and $CoFe_2O_4$ nanofluids by using two PPE detection configurations (back and front). Their thermal diffusivity and effusivity measurements were obtained with high accuracy (within 0.5%), and the results were sensitive to changes in the relevant parameters of the nanofluid as the base fluid, concentration and type of NPs. Thus, the front PPE method [13–15, 35–39] was a suitable for accurate and simultaneous measurements of thermal diffusivity and effusivity of nanofluids.

2.3. Theoretical background: photopyroelectric technique

The photothermal method has been widely used for determining the thermal parameters of materials. This technique typically uses a modulation of laser beam for inducing a thermal-wave (TW) field in the sample. The obtained TW distribution is then detected by various photothermal methods, such as photoacoustics [40], photothermal spectroscopy [41] or PPE techniques [11, 42]. Recently, many useful applications of the photopyroelectric (PPE) effect have been reported with regard to the measurement of both thermal and optical absorption properties of a material [43]. The PPE effect has provided a calorimetric method in which a thin-film PE sensor produces a voltage proportional to its surface temperature change due to the propagation of TWs through a sample in intimate contact with the PVDF sensor. In this technique, the light modulation impinges on the front surface of a sample and the PE sensor, located in good thermal contact with the sample's backside so the PE signal can be measured by performing either a frequency or a cavity length scan. The back and front PPE configurations in 'thermally thick' conditions have been reported to measure the thermal diffusivity and thermal effusivity of a sample [39, 44]. A front PPE technique is the modification of the classical configuration of the PPE technique. In this technique, the TW is introduced to the rear of the PE detection [45]. In the back PPE technique, a very thin metal film is illuminated by a modulated laser beam, and the PE cell consisted of these two parallel walls, one the metallic foil as the TW generator and another the PE film as a PE signal sensor which was placed parallel to the TW generator surface at a fixed cavity length as a function of frequency in frequency scanning and at a given frequency as a function of cavity length in cavity scanning, respectively [46]. This experimental device has allowed the measurement of thermal properties of gas and liquid and liquid mixtures [47]. This expression is typically based on the general theory of PE detection. The experimental results can then be obtained by using the PPE technique that is designed at different configurations in the measurement of the thermal properties of the nanofluids. The following results and discussion are divided into two parts: (i) the back and (ii) front PPE configurations to measure the thermal diffusivity and thermal effusivity of the nanofluid samples.

2.3.1. *Back photopyroelectric theory*

In the back PPE technique, named the thermal-wave cavity (TWC) technique, a very thin metal film was illuminated by a modulated laser beam, and the PE cell consisted of these two parallel walls, one the metallic foil as a TW generator and another the PVDF film, as a PE signal sensor. The sample(s) converts the modulated laser beam into TWs.

The induced TWs then transmit through the intracavity medium (l) by TW transmission, and the reflection mechanism is detected by the PE sensor (p), as shown in **Figure 2**. TW's arrival at PVDF film gives rise to the surface temperature at the film ($x = 0$) [48]:

$$\theta_0 = \frac{\theta_{l_s} T_{sl} e^{-\sigma_l l_l}}{1 - R_{ls} R_{lp} e^{-2\sigma_l l_l}} \tag{1}$$

The transmitted terms of TWs, of the solid-liquid interface at ($x = -l$), are given by

$$\theta_{l_s} = \frac{Q_0 T_{gs} e^{-\sigma_s l_s}}{1 - R_{sg} R_{sl} e^{-2\sigma_s l_s}} \tag{2}$$

Hence, the surface temperature of the PVDF can be continued as

$$\theta_0 = \frac{Q_0 T_{gs} T_{sl} e^{-(\sigma_s l_s + \sigma_l l_l)}}{\left(1 - R_{sg} R_{sl} e^{-2\sigma_s l_s}\right)\left(1 - R_{ls} R_{lp} e^{-2\sigma_l l_l}\right)} \tag{3}$$

where Q_o is the TW source intensity, σ_j is the complex TW diffusion coefficient $\sigma_j = (1 + i)/\mu_j$ and T_{jk} and R_{jk} are TW transmission coefficient and TW reflection coefficient, respectively, at (j–k)

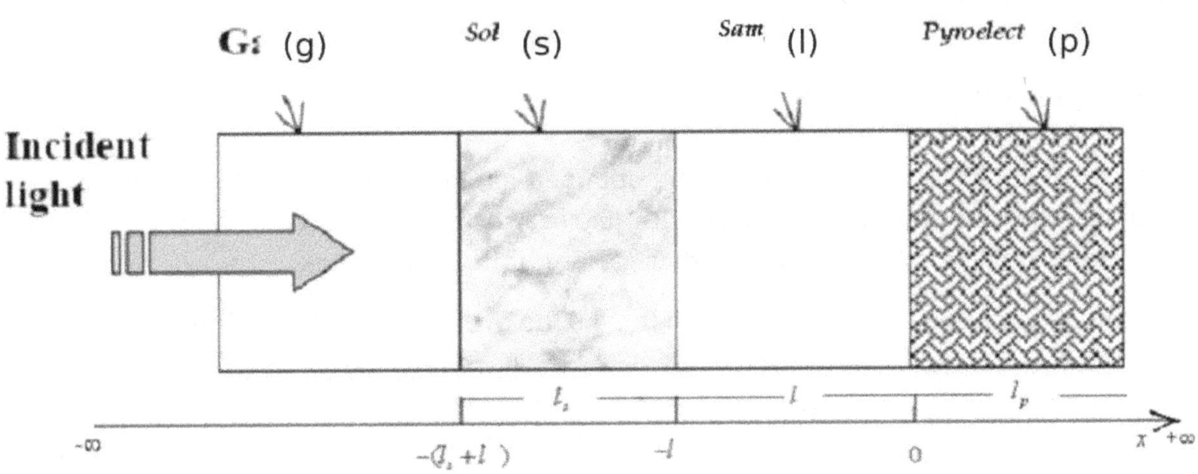

Figure 2. 1D configuration of TWC showed that the thermal waves are partially reflected and transmitted upon striking the boundaries (g, s, l, p and b) which stand for gas, solid, liquid sample, PVDF film and backing, respectively.

interface, defined as $T_{jk} = \frac{2}{1+b_{jk}}$; $R_{jk} = \frac{1-b_{jk}}{1+b_{jk}}$; $b_{jk} = \left(\frac{k_k}{k_j}\right)\left(\frac{\alpha_j}{\alpha_k}\right)^{1/2}$. The following parameters were also defined: α_j, the thermal diffusivity of j (= g, s, l, p, b); μ_j (= $(\alpha_j/\pi f)^{1/2}$), the thermal diffusion length of j at modulation frequency f; and l_j, the thickness of j. The temperature distribution in PVDF film from two parts, the PVDF film-liquid interface and the PVDF film-backing interface, can be written as

$$\theta_p(f,x) = \theta_0 \frac{T_{lp}\left(e^{-\sigma_p x} + R_{pb}e^{\left(-2\sigma_p l_p + \sigma_p x\right)}\right)}{\left(1 - R_{pb}R_{pl}e^{-2\sigma_p l_p}\right)} \tag{4}$$

The average PE voltage is given by

$$V(f,l_1) = \frac{p}{\varepsilon\varepsilon_o} < \theta_p > = \frac{Q_o T_{sl} T_{lp} p e^{-\sigma_s l_s}\left(1 - e^{-\sigma_p l_p}\right)\left(1 + R_{pb}e^{-\sigma_p l_p}\right)}{\varepsilon\varepsilon_o \sigma_p\left(1 - R_{sg}R_{sl}e^{-2\sigma_s l_s}\right)\left(1 - R_{pb}R_{pl}e^{-2\sigma_p l_p}\right)} \frac{e^{-\sigma_l l_l}}{\left(1 - R_{ls}R_{lp}e^{-2\sigma_l l_l}\right)} \tag{5}$$

If P is the PE coefficient, l_p is the thickness of the PVDF sensor, ε is the dielectric constant of the pyroelectric sensor, ε_0 is the permittivity constant of vacuum, ω is the angular frequency of modulated light and R_{jk} is the interfacial thermal coefficients. Considering that for thermally thick condition $\left|e^{-2\sigma_l l}\right| << 1$, Equation (5) can be written more simply as [49]

$$\mathrm{V}(f,L) = \mathrm{Constant}\ (f)e^{-\sigma L} \tag{6}$$

$$|\mathrm{V}(f,L)| = \mathrm{Constant}\ (f)e^{-L/\mu} \tag{6a}$$

$$\phi(f,L) = \mathrm{Constant}\ (f) - L/\mu \tag{6b}$$

The thermal diffusivity of sample can be obtained by the slope liner fitting from the plot ln (amplitude) and phase versus both cavity length (from the cavity scan) and frequency square (from the frequency scan). In frequency scanning method, the cavity was at a fixed thickness L. By plotting the phase and ln(amplitude) as a function of frequency scan, the thermal diffusivity can be determined: $\alpha = \pi L^2/\left(\frac{\phi}{\sqrt{f}}\right)^2$, $\alpha = \pi L^2/\left(\frac{\ln(|V|)}{\sqrt{f}}\right)^2$

2.3.2. *Front photopyroelectric theory*

Usually, the front PPE configuration can be obtained as shown in **Figure 3**, the sensor directly is irradiated and the sample in contact with its rear surface. Then, the cell structure, gas, PE sensor and sample, (g/p/s) becomes another variant of PPE technique.

Under this assumption, for a cell structure (*g*/*p*/*s*), the average PE voltage simplifies to [35]

$$V(f,l) = V_s \frac{\left(1 - e^{-\sigma_p l_p}\right)\left(1 - R_{ps}e^{-\sigma_p l_p}\right)}{\left(1 - R_{ps}R_{pg}e^{-2\sigma_p l_p}\right)} \tag{7}$$

where $V_s(f) = P/\varepsilon\varepsilon_o\langle\theta_s\rangle$ with thermally thick sensor and sample, the signal defined by

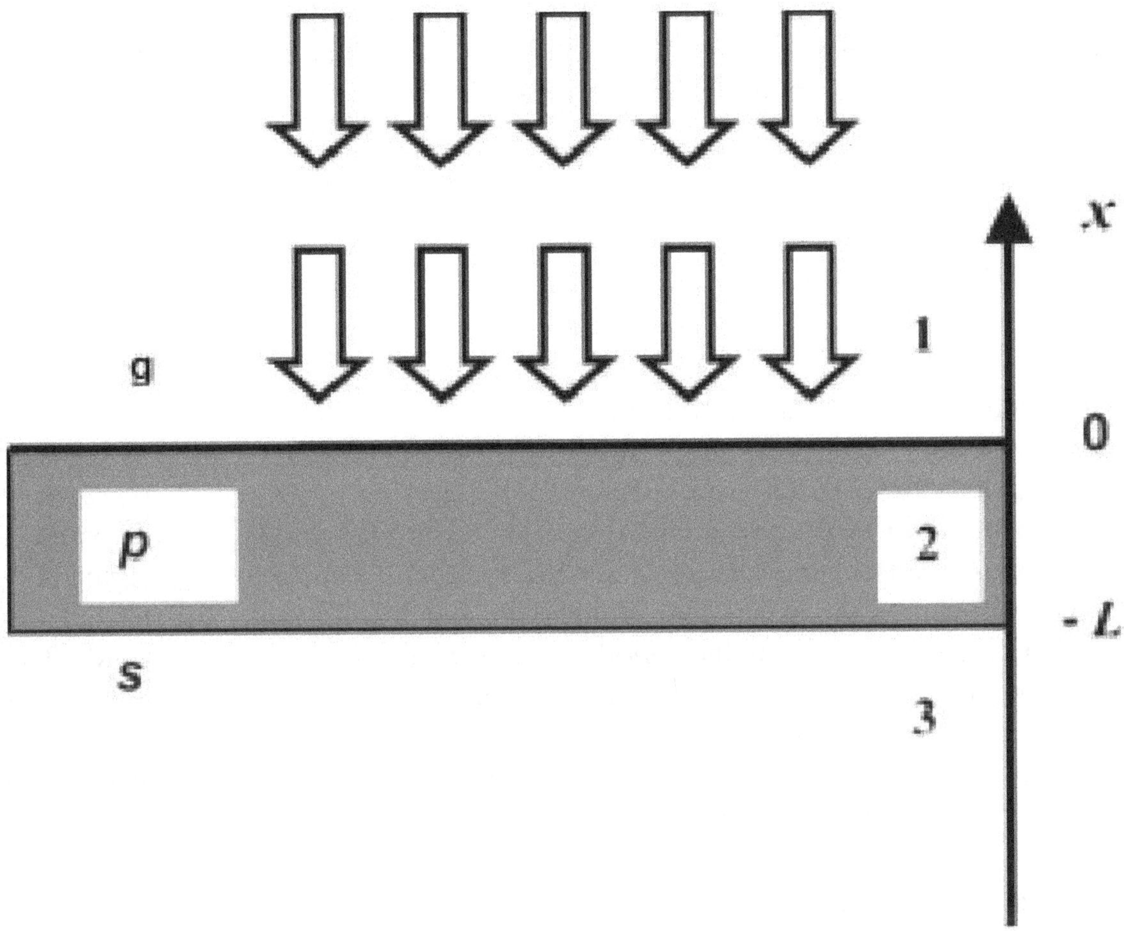

Figure 3. 1D geometry of the front PPE configuration, for a cell structure (g/p/s).

$$V(f) = V_s\left[1-(1+R_{sp})e^{-\sigma_p L_p}\right] \quad (8)$$

$$R_{sp} = (e_s-e_p)/(e_s+e_p) \quad (9)$$

where e_s and e_p are the thermal effusivity of sample and PE sensor, respectively. The normalizing signal is determined (by using air), and the normalized signal becomes

$$V_n(f) = 1-(1+R_{sp})e^{-\sigma_p L_p} \quad (10)$$

The normalized phase and amplitude of the signal are defined by

$$\theta = \arctan\left[\frac{Ae^{-L_p/\mu_p}\sin\left(L_p/\mu_p\right)}{1-Ae^{-L_p/\mu_p}\cos\left(L_p/\mu_p\right)}\right] \quad (10a)$$

$$|V_n(f)| = \left\{\left[\text{A}\sin\left(L_p/\mu_p\right)e^{-L_p/\mu_p}\right]^2 + \left[1-\text{A}\cos\left(L_p/\mu_p\right)e^{-L_p/\mu_p}\right]^2\right\}^{1/2} \quad (10b)$$

where $A = 1 + R_{sp}$, A the constant can be obtained by optimizing the fit performed on the experimental data with the normalized signal phase by using Equation (10b). It can be shown that from the phase of the normalized signal, one can obtain the thermal effusivity of the liquid sample.

3. Experimental method

3.1. Preparation of metal oxide nanofluids

Nanofluids were prepared by dispersing pre-synthesized NPs into fluids, and if necessary, in the presence of the stabilizer polyvinylpyrrolidone (PVP) to keep the NPs stable in the fluids. Nanopowders into base fluids were dispersed by stirring, and the suspensions were ultrasonicated by using probe-type or bath-type sonicator. Nanofluids were prepared using Al_2O_3 (Nanostructured & Amorphous Materials, Inc.), and copper oxide (Sigma-Aldrich) particles were dispersed in various base fluids, DW and EG. To make the desired volume concentration percentage of NPs in the nanofluids, the weights of the base fluid and NPs were measured using an electric balance (Ohaus Adventurer Balances). For example, 3.97g of Al_2O_3 NPs, which is 1 ml based on the density provided by the vendor, were added to 99g (99 ml) of DW to make 1 % volume concentration of the Al_2O_3/DW nanofluid. All nanofluids are processed by the same ultrasound power.

3.1.1. *Ultrasonication dispersion process*

Physical dispersion of powders in a liquid can be achieved by ultrasonic irradiation, either in a bath or by direct irradiation using a probe sonication method. Probe sonication has been studied to determine its effect on the particle characteristics such as the average agglomerate size and the surface charges [50]. Probe sonication is expected to provide higher power to the suspension than the ultrasonic bath as the probe is directly immersed in the suspension. The bottles containing the nanofluid were placed in the ultrasonic bath which was filled with water. The influence of the main parameter of ultrasonication such as the irradiation type (probe and bath) to dispersion and reduced size was observed in the suspension of Al_2O_3 in low concentration in water, as shown in **Figure 4**. Al_2O_3 NPs (99%, 11 nm) 0.5 wt% were dissolved in DW and magnetically stirred vigorously until a clear solution was obtained in about 1 h. The suspension was sonicated for 30 min using an ultrasonic probe (VCX 500, 25 kHz, 500 W) and labelled as sample P or using an ultrasonic bath (Powersonic, UB-405, 40 KHz, 350 W), which was labelled as sample B, respectively. As energy transferred into the liquid, the liquid become heated, and a cooling system to control the temperatures between 35 and 40°C was required. This temperature range is favourable to produce a large cavity field that greatly accelerates the integration of NPs in fluids. Unlike the bath sonication that was performed at room temperature, the tip probe sonication had higher amplitudes and, thus, a more effective creation of cavitation and heating. In the case of the ultrasonic probe, the nanoparticle/DW mixture was placed in another larger container filled with ice cubes. This was to prevent the evaporation of fluids caused by elevated temperatures. It was found the

Figure 4. Probe (VCX 500, 20 kHz, 500 W) (*left*) and bath (POWERSONIC, UB-405, 40 KHz, 350 W) (*right*) ultrasonic, respectively.

most appropriate power and conditions were obtained using the ultrasonic probe to achieve the highest dispersion and long-term stability.

3.1.2. Sample characterization

Various techniques have been applied to analyze the chemical and physical properties of the prepared nanofluids. The morphologies of the deposits were studied using an S-4700 field emission scanning electron microscope (FESEM) (Hitachi, Tokyo, Japan), operating at 5.0 kV. The size, distribution and morphology of the synthesized NPs were determined via TEM (H-7100, Hitachi, Tokyo, Japan), and the particle size distributions were determined using the UTHSCSA Image Tool software (version 3.00; UTHSCSA Dental Diagnostic Science, San Antonio, TX). In the characterization of the prepared nanofluids, the particle size and size distribution of spherical NPs in colloidal form were measured by the Nanophox particle size analyzer (Sympatec GmbH System-Partikel-Technik). This equipment is based on the principle of dynamic light scattering, which provides mean particle size as well as particle size distribution (PSD). The surface plasmon or absorption maximum in the colloidal solution spectrum provides information on the average size of the particles, and a UV-Vis spectrophotometer (Shimadzu-UV1650PC) was used to measure the absorption spectra at room temperature for wavelength range 200–800 nm.

3.2. Experimental setup of the photopyroelectric methods

The systematic experiments were to investigate the accuracy of thermal diffusivity and effusivity by the PE method using the back PPE and front PPE configurations as a special case of the different structures of the PE cell. The basic design of the analytical instrument consisted of only a laser, a TW generator and a PVDF, PE sensor. The thermal diffusivity and thermal effusivity of the nanofluids were obtained with both the back and front PPE configurations.

3.2.1. Back PPE configuration and experimental conditions

The schematic diagram of the experimental setup is shown in **Figure 5**. Here, a 52 μm PVDF film PE sensor (MSI DT1-028K/L), which is an excellent choice for signal detection due to its low cost, low weight, flexibility and sensitivity, was used in signal detection [51]. A 30 mW He-Ne laser (05-HR-991) was modulated by an optical chopper (SR540) before illumination on copper foil of 50μm thickness and 0.8cm diameter. To maximize its optical to thermal conversion efficiency, a very thin layer of carbon soot was coated on the surface of the foil. When the laser was illuminated on the copper foil, TWs were generated in this foil.

In the cell, the initiated TWs propagated across the fluid and reached the PE sensor. Since the PVDF film is very flexible and any film wrap can cause a change of signal, it was fixed with silicon glue to a Perspex substrate. On its top side, a plastic ring of 1 cm diameter was glued to it to act as the sample container. A small volume of the liquid sample, <0.1 cm^3, was simply filled in the inner side of the ring, with a sample depth or thickness of around 1 mm. The PE signal generated by PVDF sensor was analyzed by using a lock-in amplifier (SR-530) to produce the PE amplitude and phase. The electromagnetic noise was reduced by eliminating all the ground loops via proper grounding.

The typical PE signal was measured with respect to time to investigate the steady state of the signal. The sensitivity of the back PPE technique was tested by maintaining the cavity length at about 100 μm, and the PE signal was recorded over 300 s. The experiment was carried out with

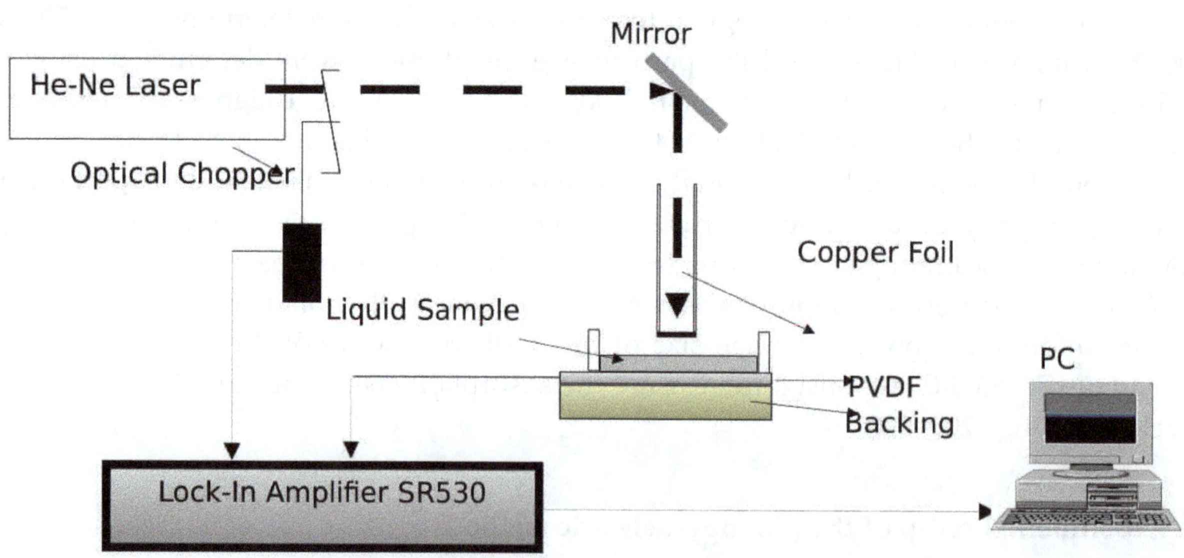

Figure 5. Schematic diagram of back PPE configuration [51].

a single drop of DW. In **Figure 6**, it can be observed that the PE signal was quite stable around 1.49 $\times 10^{-3}$ V with a standard deviation of 5.12 $\times 10^{-5}$ V.

A frequency scan was carried out as it was important to choose the optimal value of frequency for thermophysical measurements of nanofluids. **Figure 7** displays the frequency behaviour of the signal amplitude obtained from the distilled water as a reference sample with known thermal properties. It can be observed that at frequencies above 7 Hz, the effect of thermally thick regime become obvious. The amplitude of the PE signal of the sample decreased exponentially to zero with increasing modulation of the frequency in the thermally thick regime. Therefore, the frequency range between 7 and 30 Hz was used for the frequency scan, which is shown in **Figure 5**. The noise level in the present setup was about 75 μV. The ln(amplitude) of the PE signal as a function of $f^{1/2}$ in this useful frequency range was linear. The thermal diffusivity was calculated from the slope of the linear part of the logarithmic amplitude of the signal curves by using Equation (6b).

3.2.2. Front photopyroelectric configuration

In the new section design, a simplified front PPE configuration was setup using the similar PE sensor. The metalized PVDF sensor was used as an optically opaque sensor and in a thermally thick regime for both the sensor and sample, instead of a very thick sensor (usually $LiTaO_3$) in

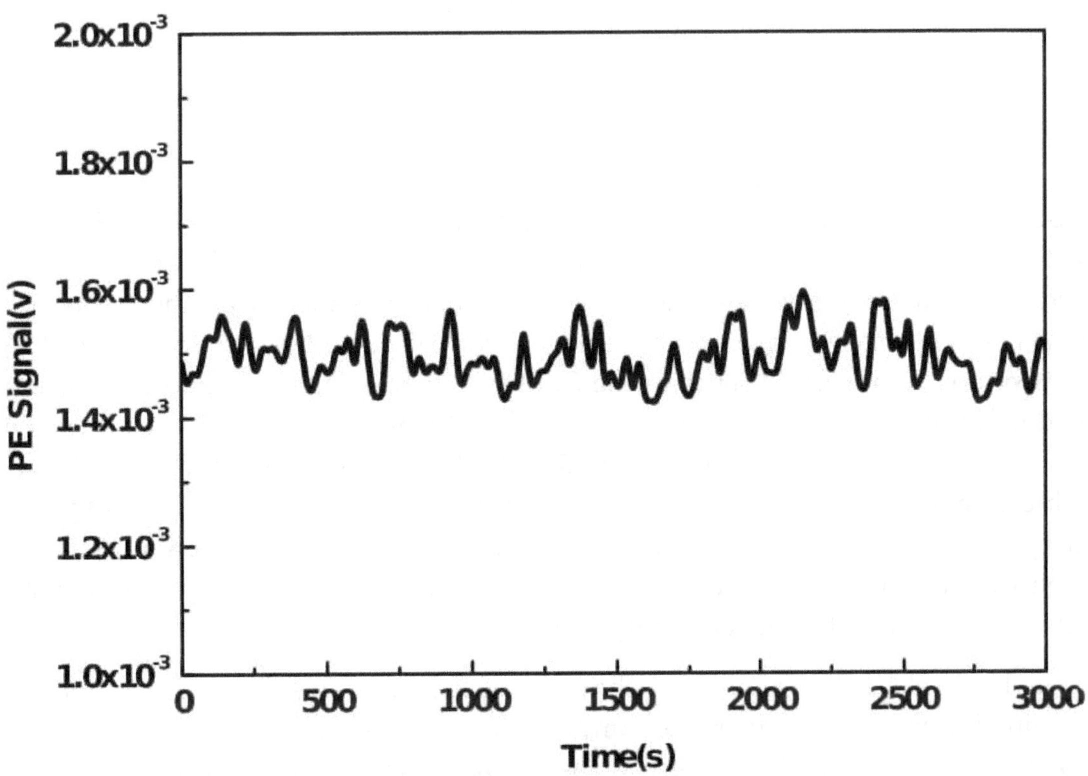

Figure 6. PVDF signals recorded versus time for distilled water; the baseline is a steady-state signal in various times.

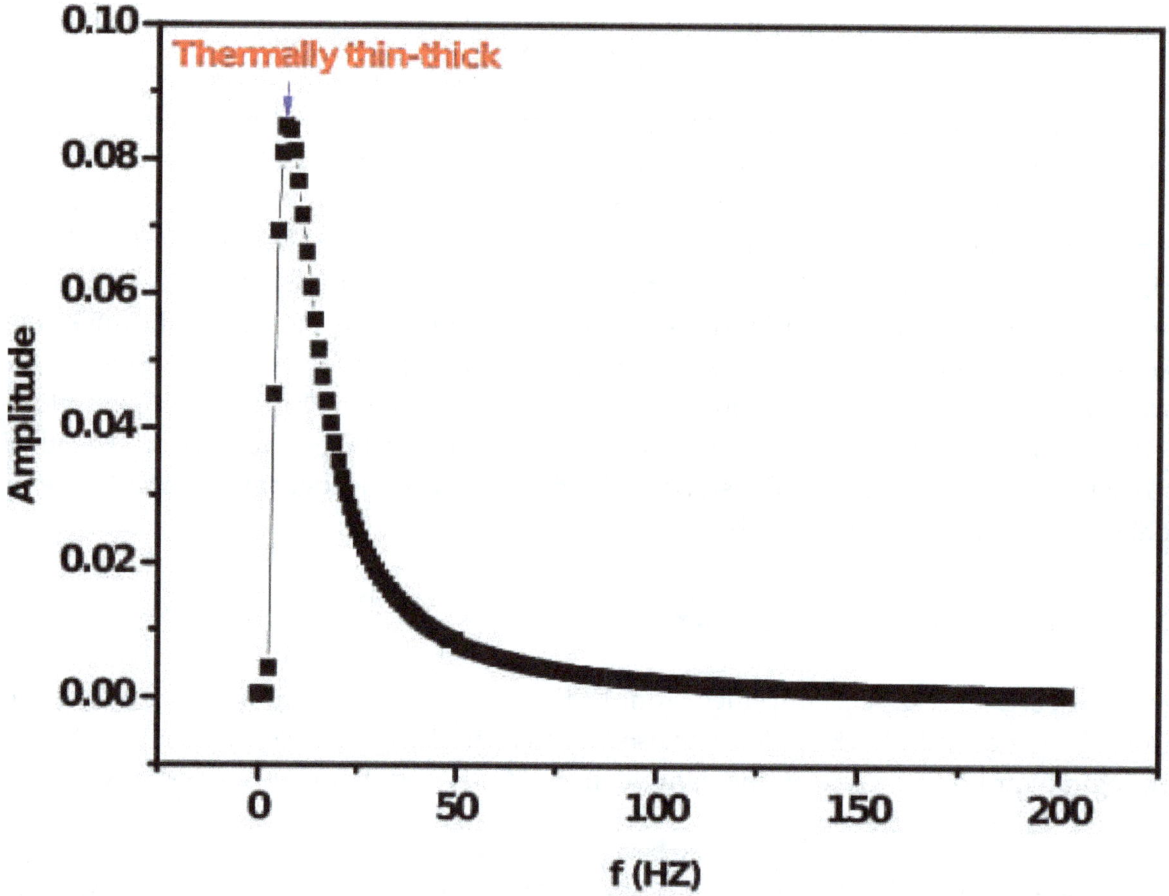

Figure 7. Frequency behaviour of the amplitude of signal obtained from the distilled water [52].

the conventional front PPE configuration [53, 57]. The radiation from the similar He-Ne laser was modulated by the mechanical chopper, and the signal from the PVDF sensor was processed with the lock-in amplifier. The liquid sample was simply filled into a plastic ring and glued on the rear side of the sensor, and the overall thickness was about 1 mm. As the sample thickness decreased, the contribution from the reflected TW power increased. A schematic view of the experimental setup of the front PPE is presented in **Figure 8**. The scan was performed in thermally thick conditions in a frequency range of 7 to 30Hz with 1Hz steps. The S/N ratio of the experiment was more than 750. The LabVIEW software was used to capture the amplitude and phase data, and the data were analyzed using Microcal Origin 8. The following procedure describes the steps from the recorded experimental data up to obtaining thermal effusivity of the nanofluid by fitting the normalized phase of the PE signal versus frequency scan to obtain thermal effusivity (e_p or e_s).

Figure 9 (a–b) displays the frequency behaviour of the normalized amplitude and phase of signal obtained from DW as a reference sample of known thermal effusivity, 1600Ws$^{1/2}$m^{-2}K^{-1} [53], to determine the thermal effusivity of the PVDF sensor. In **Figure 9 (a–b)**, the frequency range between 7 and 30 Hz was the best choice for fitting to find the parameters. However, here, the phase was used instead of the amplitude because it produced more accurate results as it did not change with source intensity fluctuations.

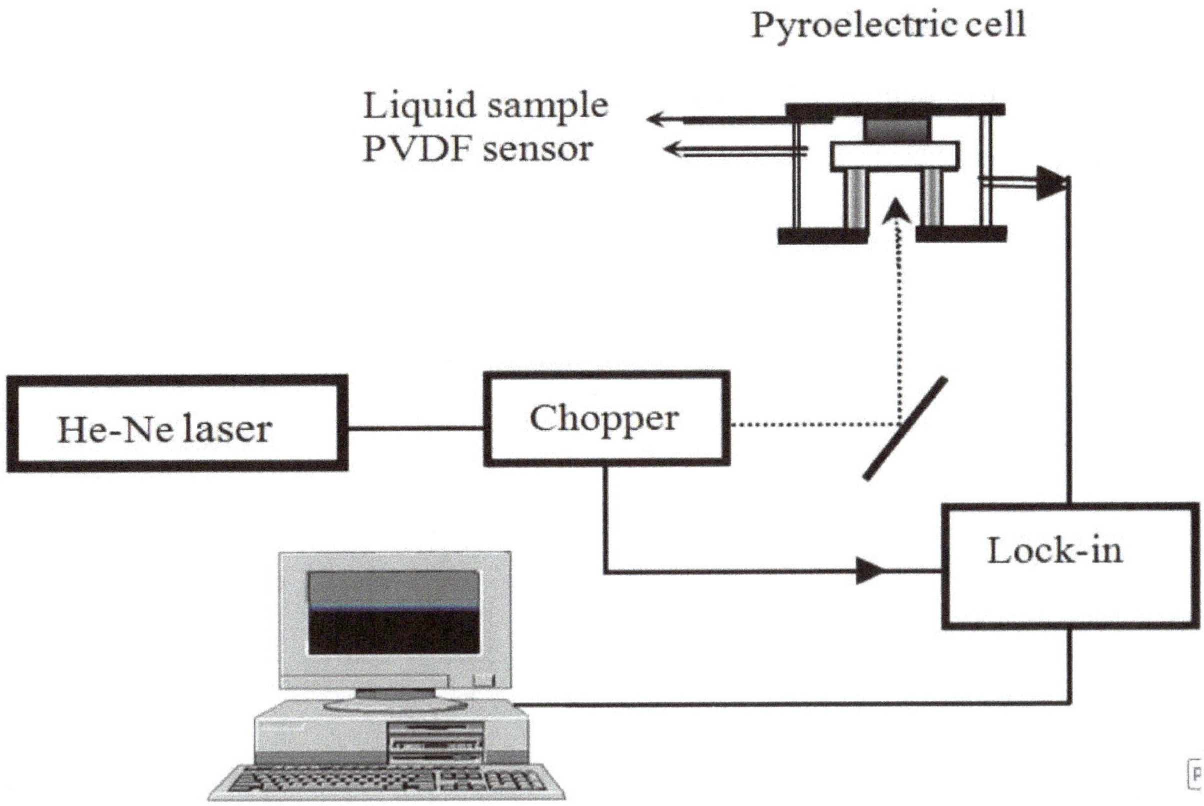

Figure 8. Schematic view of experimental setup of front PPE configuration [54].

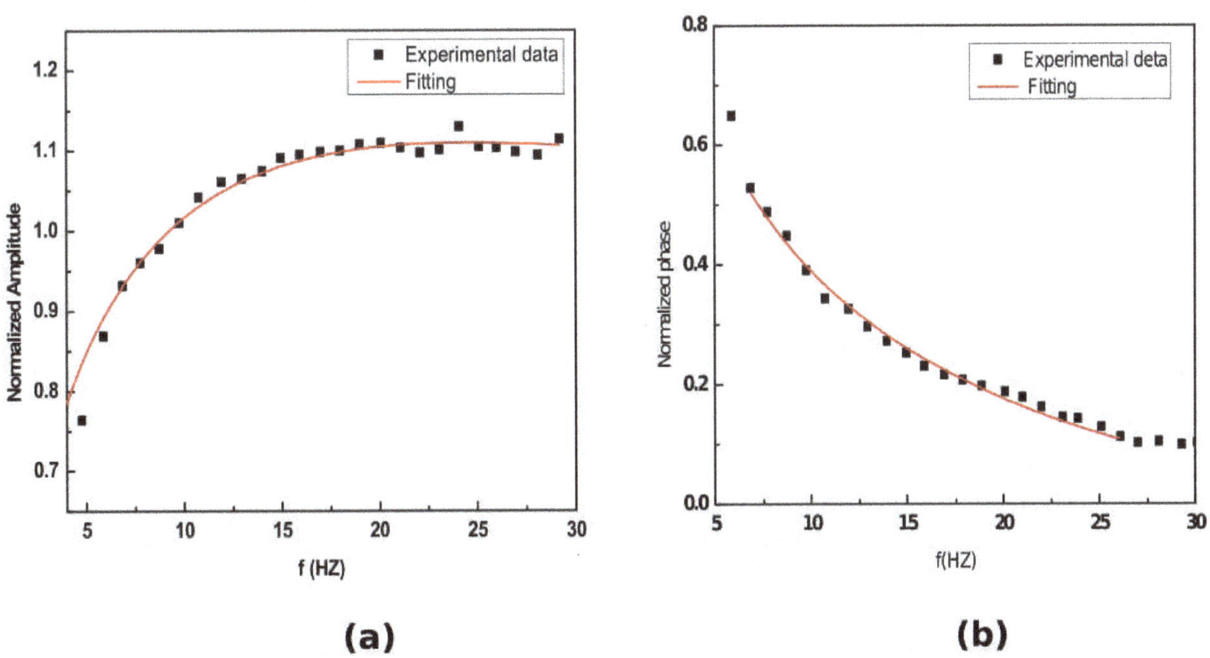

Figure 9. Frequency behaviour of the normalized (a) amplitude and (b) phase measured for the PVDF sensor with water as substrate. Solid lines are the best fit of amplitude to Equation (10b) and phase to Equation (10a), respectively [54].

4. Results and discussion

4.1. Effect of ultrasonication on the thermal diffusivity of Al2O3 nanofluids

4.1.1. Sample preparation and characterization

In the study, the influence of ultrasonication on the thermal diffusivity of low concentration of Al_2O_3 nanofluids in two sizes of NPs, size A (11 nm) and size B (30 nm) in water were investigated. Each nanofluid sample 0.125%, 0.25% and 0.5 wt% was dissolved in DW and magnetically stirred vigorously until a clear solution was observed after about 1 h. Two different ultrasonic systems were chosen to disperse the NPs in DW for 30 min using the bath sonicator, called sample B, or the probe sonicator, called sample P, respectively. The total

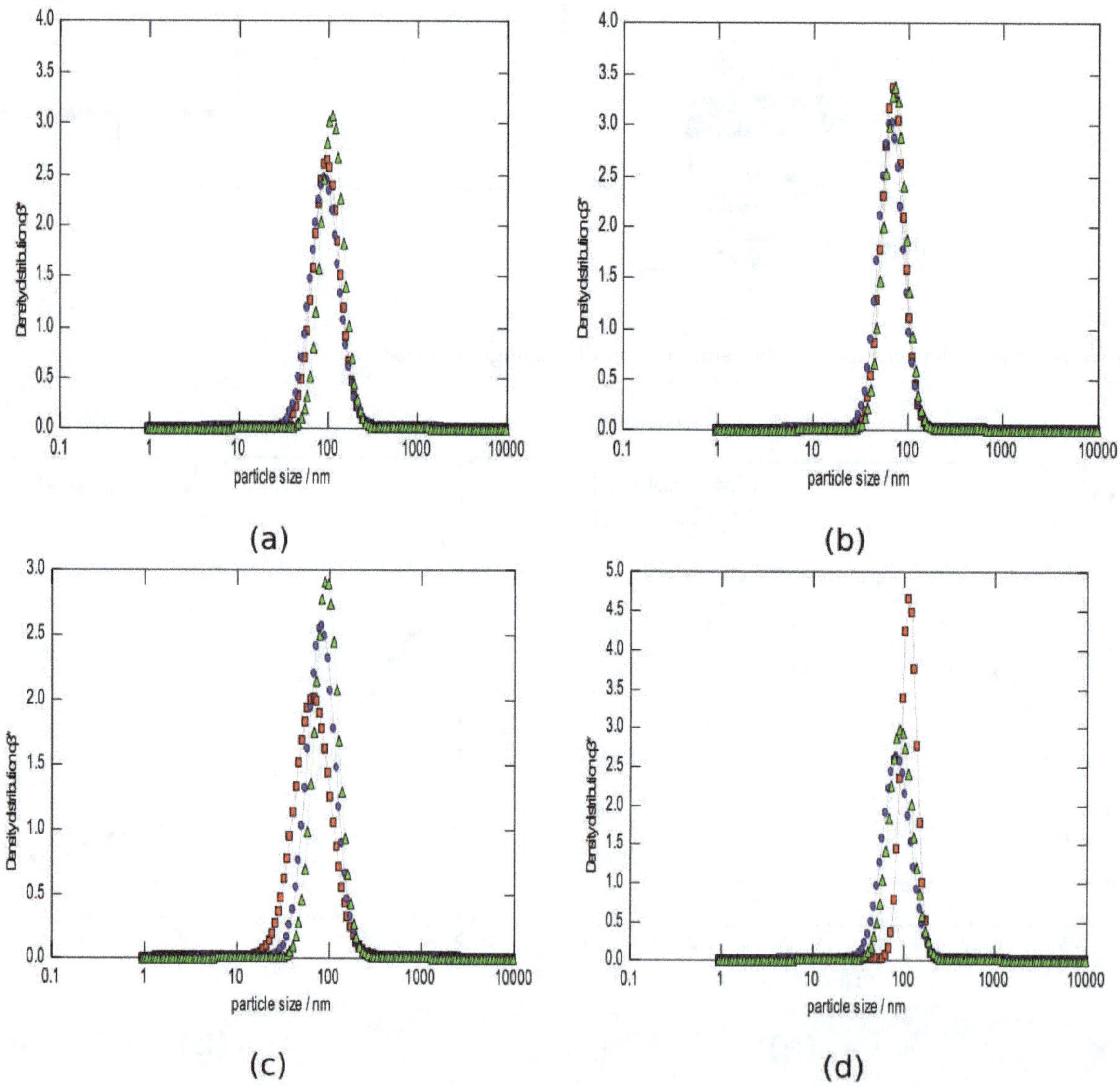

Figure 10. Particle size distributions determined using the Nanophox analyzer of Al_2O_3 particles in the nanofluids after three measurements at 15 min intervals, for NPs of size A (a,b) and B (c,d) prepared using the bath (a,c) and probe (b,d) sonicators. PDS just after sonication (□), after 15 min (◉) and after 30 min (Δ) [51].

amount of energy delivered to the sample was constant for both sonicators. After each ultrasonication, the mean particle size was measured using the Nanophox particle size analyzer (Sympatec GmbH, D-38678), and an average was taken from at least three measurements. The morphology of the alumina clusters was characterized by TEM.

Figure 10 shows the particle size distribution (PSD) and the hydrodynamic diameters of the Al_2O_3 NPs in the nanofluids. It can be seen that the NP agglomerates were only slightly broken up by the bath sonicator (**Figure 10 a, c**); however, the large agglomerates were completely broken down by the probe sonicator (**Figure 10 b, d**). The smallest mean PSD was recorded for samples with small particle size, A, prepared using probe sonication. There was no significant change in the mean particle size for the three measurements (**Figure 10 b**). However, in all cases, the NPs agglomerated in water were not completely broken up using sonication, whether by using the bath or probe sonicators.

The UV-Vis absorption spectra of the Al_2O_3 NPs prepared in DW, using bath- and probe-type ultrasonicator for the dispersion of the particles, are shown in **Figure 11**. The increase of absorption behaviour of the sample prepared using the ultrasonic probe could be attributed

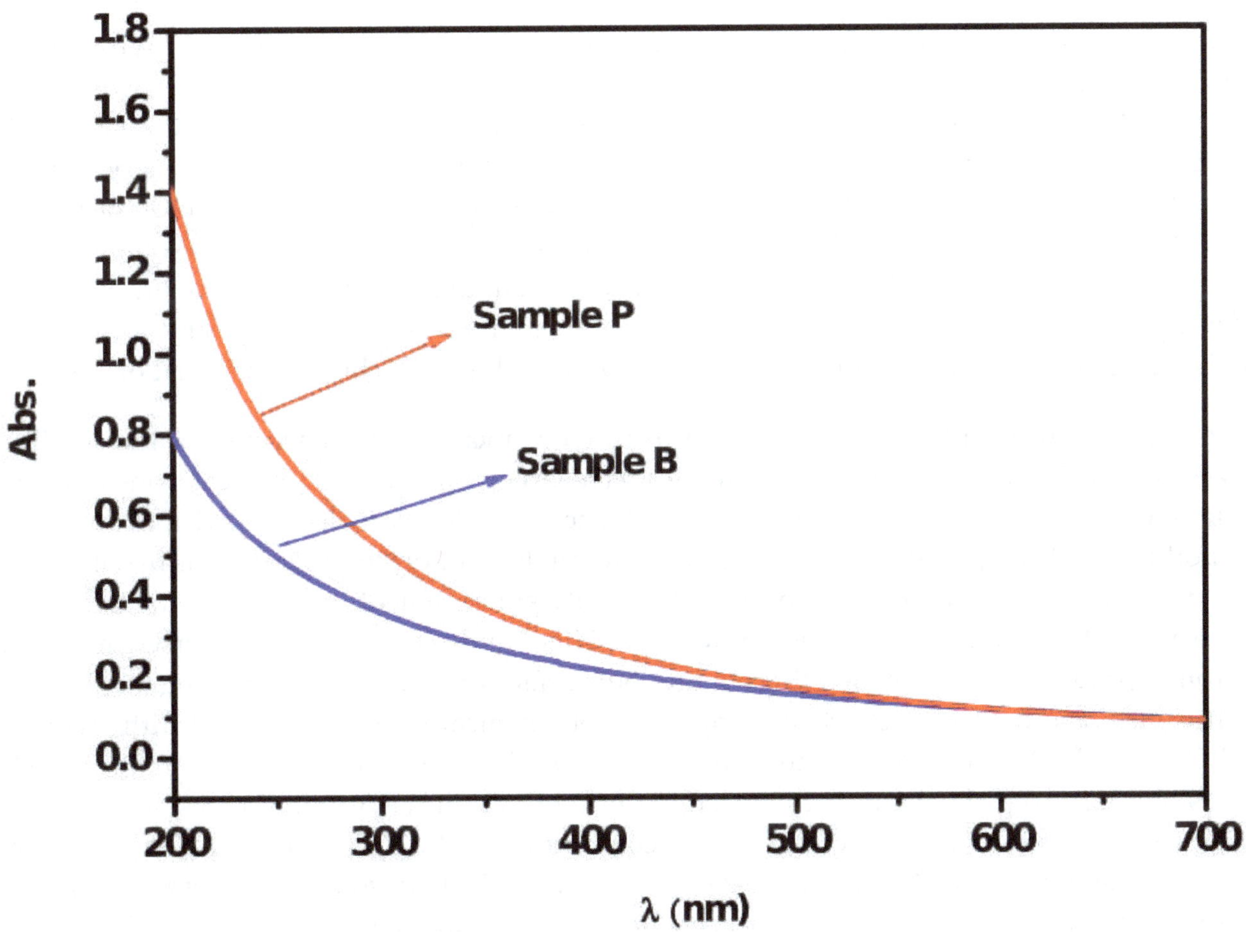

Figure 11. UV-Vis absorption spectra of the Al_2O_3 nanofluids, in the treatment by bath (sample B) and probe (sample P) sonication, respectively.

to the increase in quantity of Al_2O_3 NPs assembled within the fluid, as proven by the Nanophox results. This indicated that for the sample prepared using the ultrasonic probe, the absorption of nanofluids was at a maximum; therefore, the stability of the nanofluid was high, and the agglomeration between particles was reduced [50].

The effect of ultrasonic irradiation on the synthesized Al_2O_3 nanofluids was analyzed by TEM. **Figure 12** shows the TEM images of Al_2O_3 NPs of two sizes A and B prepared in DW without sonication (a,b) and prepared using the bath (c,d) and the probe (e,f) sonicators, for NPs of size A (a,c,e) and B (b,d,f), respectively. It can be seen that most of the NPs were spherical and were connected to each other to form a porous structure. The size of the NPs was well distributed in both ultrasonic sonicators, as shown in **Figure 12** (c,d). However, the probe sonicator was more effective in reducing particle sizes to below 100 nm, as shown in **Figure 12** (e,f). As previously mentioned, in all nanofluids, the measured particle sizes were larger than the nominal particle sizes claimed by the vendor. This indicated that the oxide NPs agglomerated in water and the hard aggregates could not be broken down into individual NPs under these operating conditions or even with very high-energy input [18].

4.1.2. Enhancement of thermal diffusivity

Before measuring thermal diffusivity of the nanofluids, the PPE setup was tested with DW as the base fluid. The recorded α value was $(1.431 \pm 0.030)\times10^{-3}$ cm^2/s, which differed by less than 2% from the values reported in literature [49]. The thermal diffusivity of the Al_2O_3 nanofluids prepared using different sonication techniques at different concentrations of NPs of sizes A and B was obtained. **Figure 13** shows the typical behaviour of the (a) amplitude of the PE signal versus the frequency and (b) the plot of ln(amplitude) of PE signal versus square root of frequency. The thermal diffusivity can be calculated from the fitting slope of the linear part of the signal curves. The thermal diffusivity data are summarized in **Tables 2** and **3**. The data indicated that the thermal diffusivity of the Al_2O_3 nanofluids was higher than that of water.

The data also proved that the thermal diffusivity enhancement was greater for the smaller-sized NPs. This was because smaller particles have larger surface area (the heat transfer area), thus increasing the thermal diffusivity [55]. Hence, smaller particles helped form a stable nanofluid, and the probe sonicator had a substantial effect on the thermal diffusivity. At a given particle concentration, the thermal diffusivity enhancement was greater for the probe than the bath sonicator. This was because the NPs were more widely dispersed in water through probe sonication, generating a larger NP surface area and thus increasing the thermal diffusivity. The beneficial effect of using the probe sonicator on the thermal diffusivity of Al_2O_3 nanofluids was more pronounced at high particle concentrations and small particle sizes. For example, the greatest enhancement of thermal diffusivity of 6% was achieved for the probe sonicator with NPs of size A at a concentration of 0.5 wt%. The smallest enhancement was about ≈1% for NPs of size B at 0.125 wt% with the bath sonicator. These findings are possibly attributable to the rapid particle clustering at a high concentration, which necessitates using a more powerful sonication tool to break up large agglomerates into smaller-sized particles.

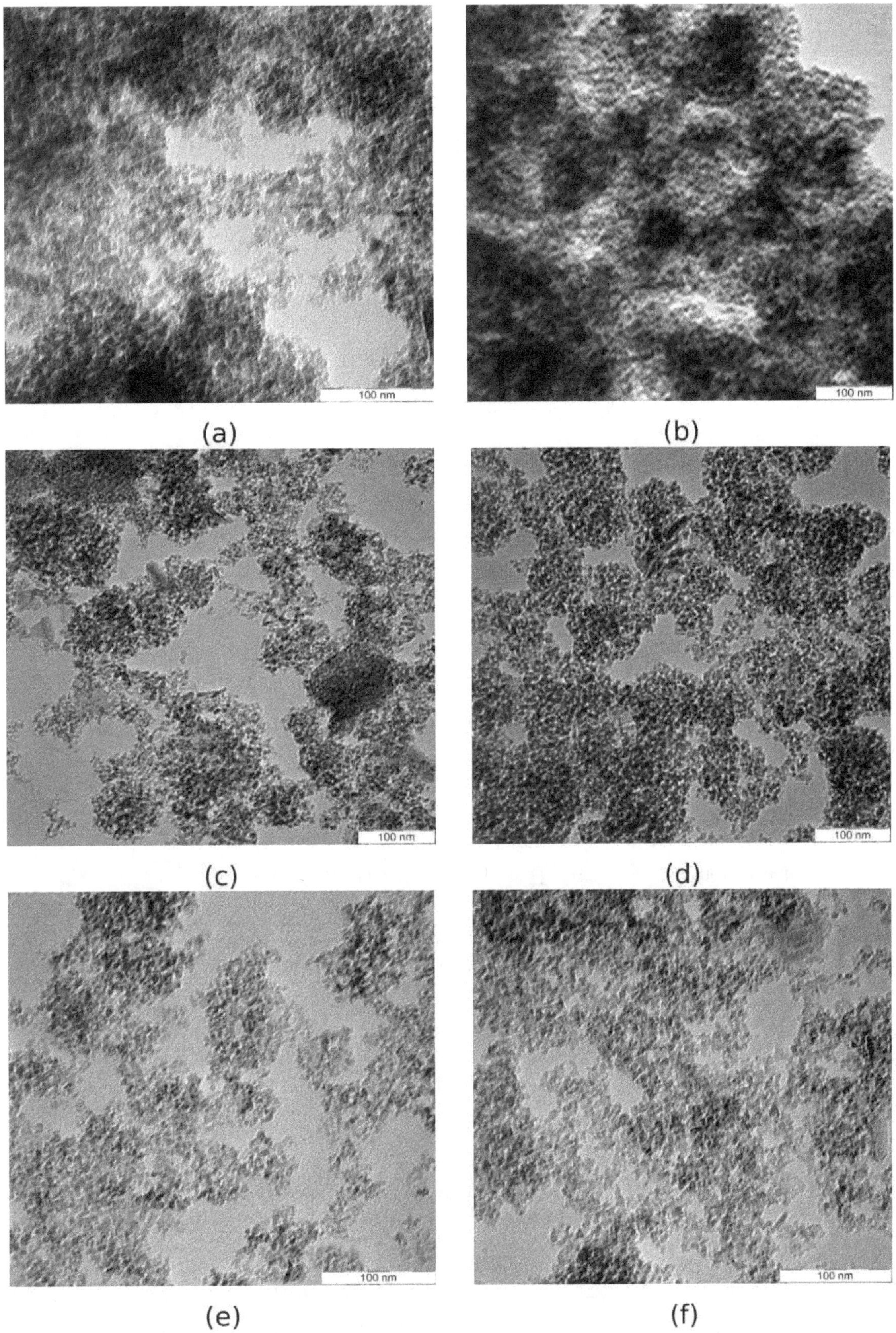

Figure 12. TEM images of Al_2O_3 NPs prepared in DW without (a,b) and with (c,d) the bath sonicator and (e,f) probe sonicators, for NPs of size 11 nm (a,c,e) and 30 nm (b,d,f) [51].

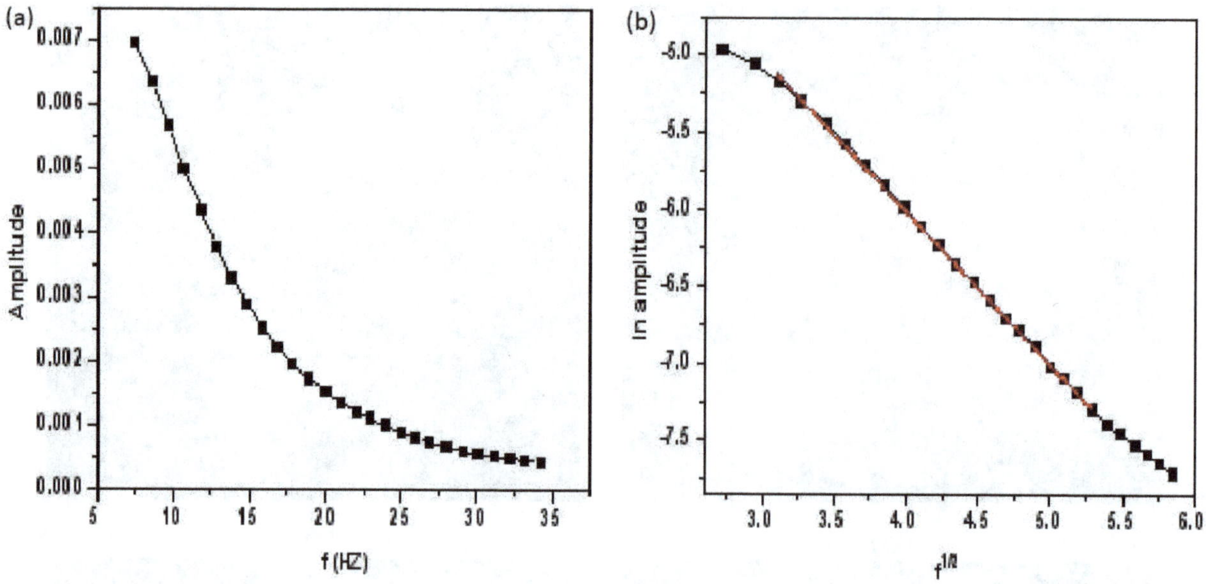

Figure 13. (a) Amplitude of the PE signal as a function of the chopping frequency *f* and (b) natural log of the amplitude of the PE signal as a function of the square root of the chopping frequency and its fitting by using Equation (6b), for one of the samples [51].

	Bath		Probe	
Concentration wt%	**Thermal diffusivity** $(cm^2/s)10^{-3}$	**Thermal diffusivity enhancement %**	**Thermal diffusivity** $(cm^2/s)10^{-3}$	**Thermal diffusivity enhancement %**
0.125	1.476 ± 0.002	3.1	1.482 ± 0.004	3.5
0.25	1.483 ± 0.003	3.5	1.494 ± 0.002	4.3
0.5	1.492 ± 0.004	4.2	1.515 ± 0.003	5.8

Table 2. Thermal diffusivity of Al_2O_3 nanofluids, NP type A (11 nm), prepared by using different sonication techniques at different NP concentrations [51].

	Bath		Probe	
Concentration wt%	**Thermal diffusivity** $(cm^2/s)10^{-3}$	**Thermal diffusivity enhancement %**	**Thermal diffusivity** $(cm^2/s)10^{-3}$	**Thermal diffusivity enhancement %**
0.125	1.446 ± 0.003	0.9	1.448 ± 0.001	1.1
0.25	1.461 ± 0.002	2.1	1.473 ± 0.002	2.9
0.5	1.478 ± 0.004	3.2	1.498 ± 0.003	4.6

Table 3. Thermal diffusivity of Al_2O_3 nanofluids, NP type B (30 nm), prepared by using different sonication techniques at different NP concentrations [51].

4.2. Effect of base fluids on thermal effusivity of nanofluids

4.2.1. Sample preparation and characterization

The thermal effusivity of Al_2O_3 (11 nm) and CuO (50 nm) NPs dispersed in three different base fluids, DW, EG and olive oil, in the presence of the stabilizer polyvinylpyrrolidone (PVP) was investigated. In each nanofluid, sample 0.125 wt% was dissolved in each base fluid and magnetically stirred vigorously until a clear solution were observed after about 1 h. The solution was then sonicated by probe sonicator for 30 min to ensure a uniform dispersion of NPs in the fluids. TEM was employed to obtain the morphology of the CuO and Al_2O_3 particles and to determine the average particle size.

Figure 14 shows the TEM images and their corresponding size distributions of (a) CuO and (b) Al_2O_3 nanofluids prepared in water. It can be seen that most of the NPs were well dispersed and some agglomerates were present. The CuO and Al_2O_3 NP sizes were about 52.3 ± 4.2 nm and 7.5 ± 2.5 nm, respectively. These commercial NPs determined from TEM images were

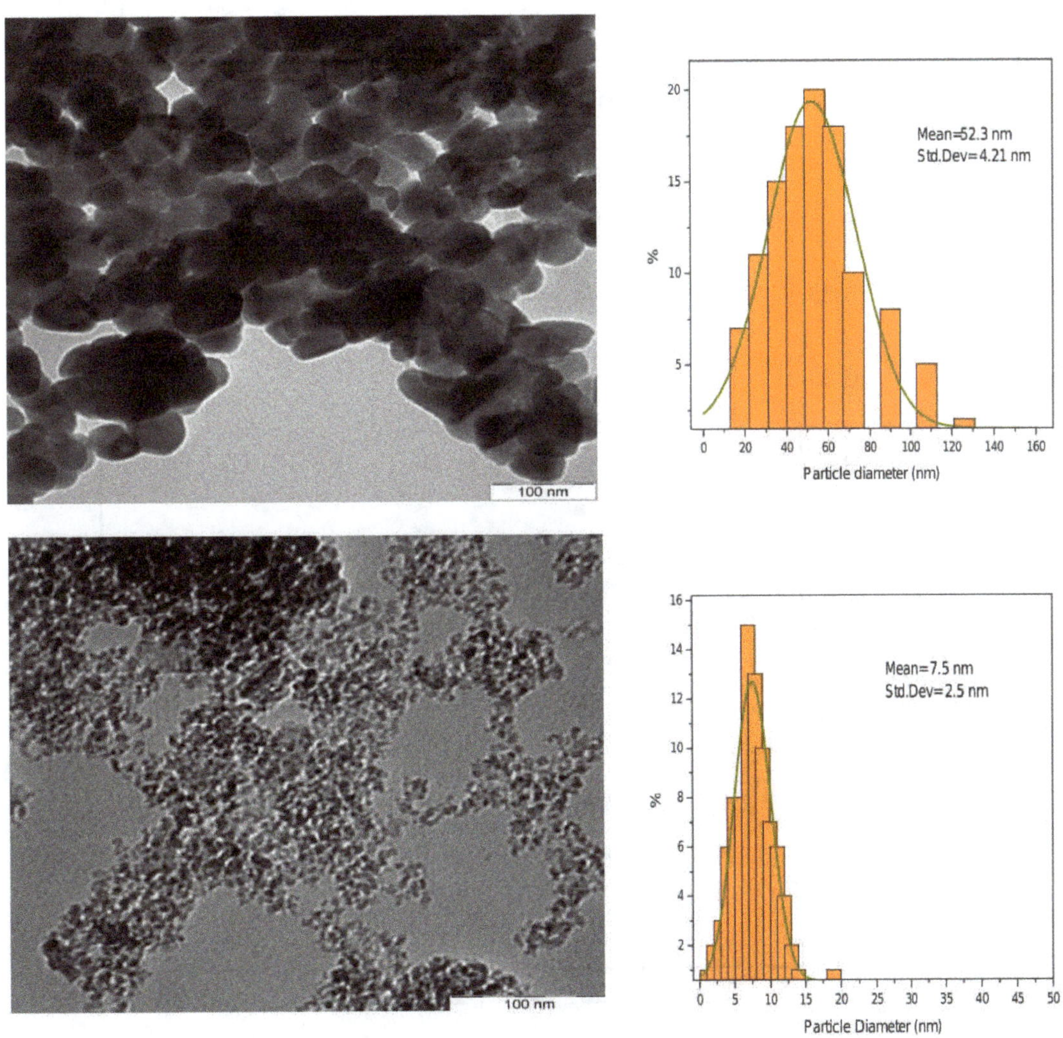

Figure 14. TEM images and their size distributions of (a) CuO particles and (b) Al_2O_3 nanofluids prepared in water [56].

±

±

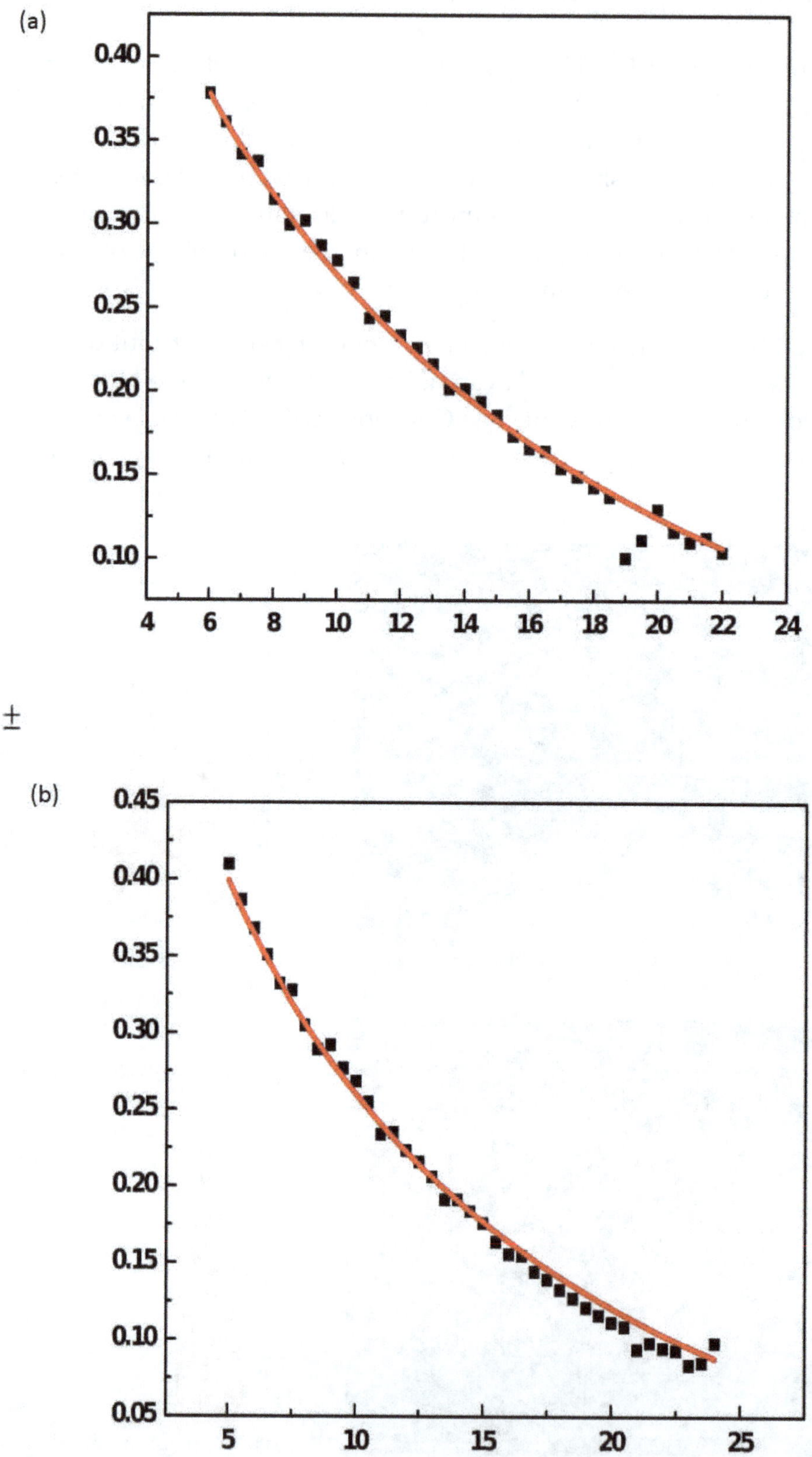

Figure 15. The experimental data and the best fit of the PE normalized phase versus modulation frequency in (a) Al_2O_3/olive oil and (b) CuO/olive oil, obtained by using Equation (10a) [56].

slightly different from those reported by the vendors. This indicated that some of particles in each sample were aggregated with some uniform size distribution as reported by them.

4.2.2. Thermal effusivity measurements

Figure 15 shows the PE-normalized phase versus modulation frequency in (a) Al_2O_3/olive oil and (b) CuO/olive oil. It was observed that from this fit the values of A from Equation (9) were obtained at (1.112 ± 0.005) and (1.175 ± 0.006), corresponding to the values of thermal effusivity of Al_2O_3/olive oil (0.614 ± 0.003) × 10^3 $Ws^{1/2}m^{-2}K^{-1}$ and CuO/olive oil (0.697 ± 0.003) × 10^3 $Ws^{1/2}m^{-2}K^{-1}$, respectively, obtained by using Equation (10a). The values of thermal effusivity measured for all nanofluids and their comparison with pure solvents are summarized in **Table 3** and **Figure 16**. The comparisons indicated that the thermal effusivity of the various base fluids mixed with NPs in the presence of PVP were reduced as compared to pure fluids, possibly due to the effect of the surfactant that inhibited the thermal effusivity of the nanofluids [56]. The results also showed that the base fluids had more influence on effusivity than the NPs. The relative standard deviation for measuring the thermal effusivity of

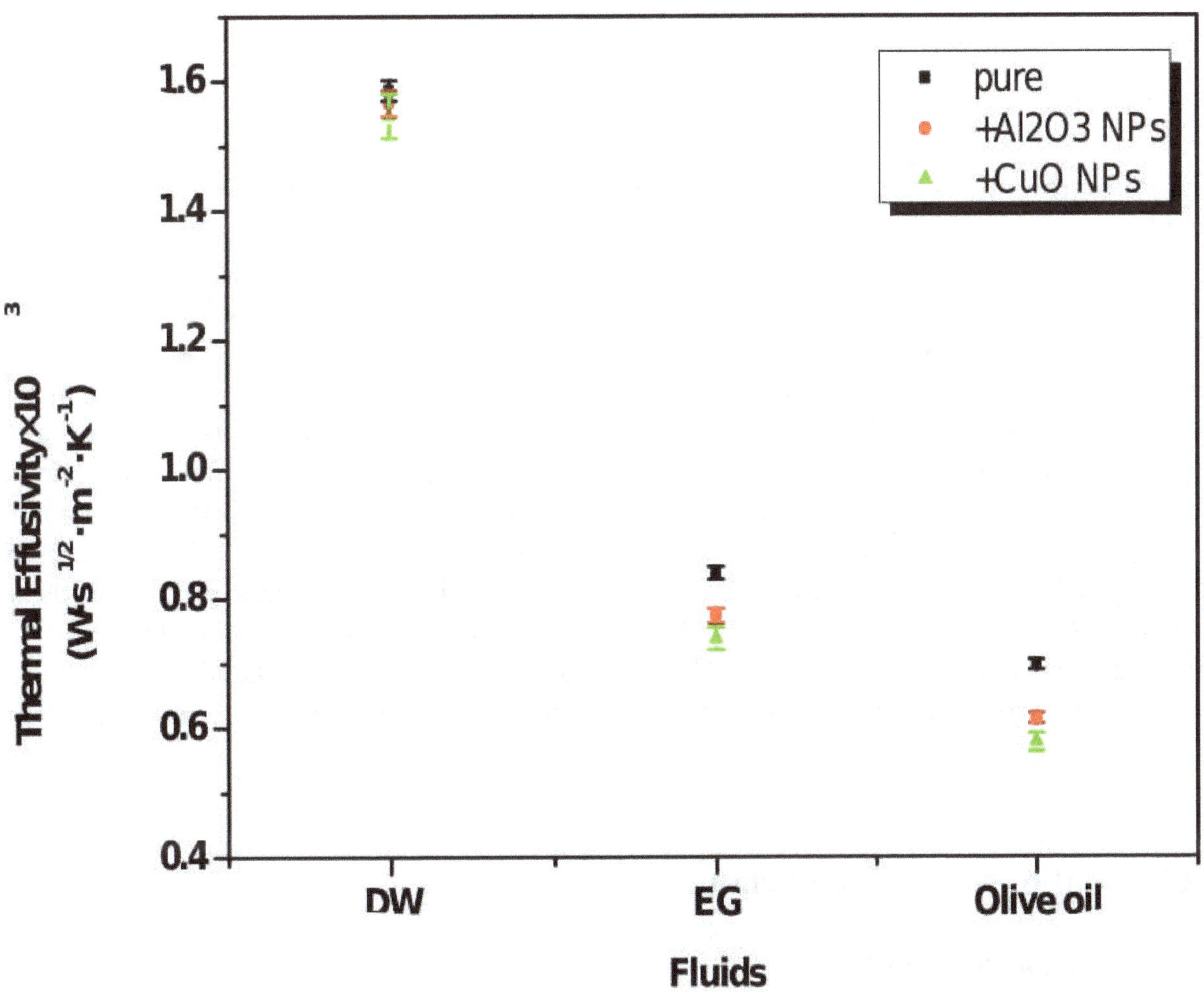

Figure 16. Thermal effusivity of Al_2O_3 and CuO nanofluids with (DW, EG and olive oil) and their pure solvents [56].

nanofluids was below 2%, as shown in **Table 4**. Therefore, the front PPE technique is a promising high-accuracy alternative for this measurement.

NPs	Base fluid	Fitting parameter(A)	Thermal effusivity × 10^3 ($Ws^{1/2}$ m^{-2} K^{-1}) measurement	Relative error%	Thermal effusivity × 10^3 ($Ws^{1/2}$ m^{-2} K^{-1}) literature
Al_2O_3	Water	1.523 ± 0.014	1.566 ± 0.015	0.95	–
Al_2O_3	EG	1.223 ± 0.009	0.773 ± 0.006	0.77	–
Al_2O_3	Olive	1.112 ± 0.005	0.614 ± 0.003	0.48	–
CuO	Water	1.519 ± 0.028	1.547 ± 0.029	1.87	–
CuO	EG	1.202 ± 0.021	0.738 ± 0.012	1.75	–
CuO	Olive	1.081 ± 0.018	0.577 ± 0.009	1.56	–
–	Water	1.528 ± 0.011	1.586 ± 0.011	0.69	1.579 [10]
–	EG	1.263 ± 0.008	0.839 ± 0.005	0.59	0.810 [11]
–	Olive	1.175 ± 0.006	0.697 ± 0.003	0.43	0.621 [11]

Table 4. Experimental thermal effusivity of Al_2O_3 and CuO nanofluids and their pure solvents and their literature values [56].

5. Conclusions

The PPE technique is a sensitive method to measure the thermal properties of nanofluids in small volumes. Following this, the back PPE configuration was used to obtain the influence of ultrasonic irradiation modes (either bath or probe sonication) such as the cluster size of Al_2O_3 nanofluids in low concentrations on the thermal diffusivity. The ultrasonic bath proved to be almost ineffective in size reduction, as most of the Al_2O_3 particles were spherical and were connected to each other to form a porous structure ranging in size from 1 μm to larger, and the probe sonication effectively reduced the particle size to below 100 nm. This showed that the oxide NPs in water were agglomerated and some hard aggregates could not be broken into individual NPs under these operating conditions or even at very high-energy inputs. The proposed front PPE technique, with a metalized PVDF sensor in a thermally thick regime, was applied to measure thermal effusivity by utilizing the phase signal of nanofluids that contained Al_2O_3 and CuO NPs dispersed in different solvents, water, ethylene glycol and olive oil. As expected, the relative standard deviation of this measurement, 2%, confirmed that this method was also suitable for measuring the thermal effusivity of nanofluid with a high degree of accuracy.

Abbreviations and Nomenclature

Q_o	TW source intensity
ω	Angular frequency of modulated light
f	Modulation frequency

ε_0	Permittivity constant of vacuum
T_{jk}	TW transmission coefficient
R_{jk}	TW reflection coefficient
σ	Complex TW diffusion
μ	Thermal diffusion length
k	Thermal conductivity
α	Thermal diffusivity
e	Thermal effusivity
L_s	Sample thickness
DW	Deionized water
NPs	Nanoparticles
PPE	Photopyroelectric
PE	Pyroelectric
PVDF	Polyvinylidene fluoride
PSD	Particle size distribution
SNR	Signal-to-noise ratio
TW	Thermal wave
TWC	Thermal-wave cavity
EG	Ethylene glycol

Author details

Monir Noroozi* and Azmi Zakaria

*Address all correspondence to: monir.noroozi@gmail.com

Physics Department, Faculty of Science, Universiti Putra Malaysia, Malaysia

References

[1] Murshed SMS, Leong KC, Yang C. Thermophysical and electrokinetic properties of nanofluids—a critical review. Applied Thermal Engineering, 2008; **28**: 2109–2125.

[2] Masuda H, Ebata A, Teramae K, Hishinuma N. Alteration of thermal conductivity and viscosity of liquid by dispersing ultra-fine particles [dispersion of Al_2O_3, SiO_2 and TiO_2 ultra-fine particles. Netsu Bussei. 1993; **4**: 227–233

[3] Choi SUS, Eastman JA. Enhancing thermal conductivity of fluids with nanoparticles. osti. gov, International mechanical engineering congress and exhibition, San Francisco, CA (United States), 1995; 12–17 Nov 1995

[4] Wang XQ, Mujumdar AS. Heat transfer characteristics of nanofluids: a review. International Journal of Thermal Sciences 2007; **46**: 1–19.

[5] Sheikholeslami M, Hayat T, Alsaedi A. MHD free convection of Al_2O_3–water nanofluid considering thermal radiation: a numerical study. International Journal of Heat and Mass Transfer, 2016; **96**: 513–524.

[6] Eastman JA, Choi SUS, Li S, Yu W, & Thompson LJ. Anomalously increased effective thermal conductivities of ethylene glycol-based nanofluids containing copper nanoparticles. Applied Physics Letters. 2001; **78**: 718–720.

[7] Evans W, Prasher R, Fish J, Meakin P, Phelan P, Keblinski P. Effect of aggregation and interfacial thermal resistance on thermal conductivity of nanocomposites and colloidal nanofluids. International Journal of Heat and Mass Transfer. 2008; **51**: 1431–1438.

[8] Sheikholeslami M, Ellahi R. Three dimensional mesoscopic simulation of magnetic field effect on natural convection of nanofluid. International Journal of Heat and Mass Transfer, 2015;**89**: 799–808.

[9] Murshed SMS, Leong KC, Yang C. Determination of the effective thermal diffusivity of nanofluids by the double hot-wire technique. Journal of Physics D: Applied Physics 2006; **39**: 5316.

[10] Shen J, and Mandelis A. Thermal-wave resonator cavity. Review of Scientific Instruments 1995; **66**: 4999–5005

[11] Chirtoc M, Mihilescu G. Theory of the photopyroelectric method for investigation of optical and thermal materials properties. Physical Review B 1989; **40**: 9606–9617.

[12] Kwan CH, Matvienko A, Mandelis A. Optimally accurate thermal-wave cavity photopyroelectric measurements of pressure-dependent thermophysical properties of air: theory and experiments. Review of Scientific Instruments 2007; **78**: 104902–104910.

[13] Chirtoc M, Bentefour EH, Glorieux C., Thoen J. Development of the front-detection photopyroelectric (FPPE) configuration for thermophysical study of glass-forming liquids. Thermochimica Acta. 2001; **377**: 105–112.

[14] Dădârlat D, Neamtu C. Detection of molecular associations in liquids by photopyroelectric measurements of thermal effusivity. Measurement Science and Technology 2006; **17**: 3250.

[15] Longuemart S, Quiroz AG, Dadarlat D, Sahraoui AH, Kolinsky C, Buisine JM. An application of the front photopyroelectric technique for measuring the thermal effusivity of some foods. Instrumentation Science & Technology 2002; **30**: 157–165.

[16] Wei Y, Xie H. A review on nanofluids: preparation, stability mechanisms, and applications. Journal of Nanomaterials, 2012; **2012**: 17 pages

[17] Taylor R, Phelan P, Otanicar T, Adrian R, Prasher R. Nanofluid optical property characterization: towards efficient direct absorption solar collectors. Nanoscale Research Letters, 2011; **6**: 225.

[18] Nguyen VS, Rouxel D, Hadji R, Vincent B, Fort Y. Effect of ultrasonication and dispersion stability on the cluster size of alumina nanoscale particles in aqueous solutions. Ultrasonics Sonochemistry, 2011; **18**: 382–388.

[19] Dadarlat D, Longuemart S, Turcu R, Streza M, Vekas L, Hadj Sahraoui A. Photopyroelectric calorimetry of Fe_3O_4 magnetic nanofluids: effect of type of surfactant and magnetic field. International Journal of Thermophysics, 2014;**35**: 2032–2043.

[20] Nisha MR, Philip J. Dependence of particle size on the effective thermal diffusivity and conductivity of nanofluids: role of base fluid properties. Heat and Mass Transfer, 2012; **48**:1783–1790.

[21] Philip J, Nisha MR. Thermal diffusion in dilute nanofluids investigated by photothermal interferometry. Journal of Physics: Conference Series, 2010; **214**: 012035.

[22] López-Muñoz GA, Balderas-López JA, Ortega-Lopez J, Pescador-Rojas JA, Salazar JS. Thermal diffusivity measurement for urchin-like gold nanofluids with different solvents, sizes and concentrations/shapes. Nanoscale Research Letters, 2012; **7**: 1–7.

[23] Dadarlat D, Neamtu C, Streza M, Turcu R, Craciunescu I, Bica D, et al. High accuracy photopyroelectric investigation of dynamic thermal parameters of Fe_3O_4 and $CoFe_2O_4$ magnetic nanofluids. Journal of Nanoparticle Research, 2008; **10**: 1329–1336.

[24] Agresti F, Ferrario A, Boldrini S, Miozzo A, Montagner F, Barison S, et al. Temperature controlled photoacoustic device for thermal diffusivity measurements of liquids and nanofluids. Thermochimica Acta, 2015; **619**:48–52.

[25] Sánchez-Ramírez JF, Pérez JLJ, Orea AC, Fuentes RG, Bautista-Hernández A, Pal U. Thermal diffusivity of nanofluids containing Au/Pd bimetallic nanoparticles of different compositions. Journal of Nanoscience and Nanotechnology, 2006; **6**: 685–690

[26] Kumar BR, Basheer NS, Kurian A, George SD. Effect of particle size on the thermo-optic properties of gold nanofluids—a thermal lens study. AIP Conference Proceedings, 2014; **1576**: 118–121.

[27] Jiménez Pérez LJ, Gutierrez Fuentes R, Sanchez Ramirez FJ, Cruz-Orea A. Study of gold nanoparticles effect on thermal diffusivity of nanofluids based on various solvents by using thermal lens spectroscopy. The European Physical Journal Special Topics, 2008; **153**: 159–161.

[28] Gutierrez Fuentes R, Pescador Rojas JA, Jiménez-Pérez JL, Sanchez Ramirez JF, Cruz-Orea A, Mendoza-Alvarez J G. Study of thermal diffusivity of nanofluids with bimetallic nanoparticles with Au(core)/Ag(shell) structure. Applied Surface Science 2008; **255**: 781–783.

[29] Filippo A, Simona B, Simone B, Cesare P, Laura C, Laura F, et al. Tuning the thermal diffusivity of silver based nanofluids by controlling nanoparticle aggregation. Nanotechnology, 2013; **24**: 365601.

[30] Wang ZL, Tang DW, Liu S, Zheng XH, Araki N. Thermal-conductivity and thermal-diffusivity measurements of nanofluids by 3ω method and mechanism analysis of heat transport. International Journal of Thermophysics, 2007;**28**: 1255–1268.

[31] Faris Mohammed A, Yunus WMM. Study of the effect of volume fraction concentration and particle materials on thermal conductivity and thermal diffusivity of nanofluids. Japanese Journal of Applied Physics, 2011; **50**: 085201.

[32] Murshed SS, de Castro CN, Lourenço MJV, Lopes MM, Santos FJV. Experimental investigation of thermal conductivity and thermal diffusivity of ethylene glycol-based nanofluids. Proceedings of the International Conference on Mechanical Engineering 2011 (ICME2011), 18–20 December 2011, Dhaka, Bangladesh

[33] Rondino F, D'Amato R, Terranova G, Borsella E, Falconieri M. Thermal diffusivity enhancement in nanofluids based on pyrolytic titania nanopowders: importance of aggregate morphology. Journal of Raman Spectroscopy2014; **45**: 528–532

[34] Dadarlat D, Frandas A. Inverse photopyroelectric detection of phase transitions. Applied Physics A, 1993; **57**: 235–238.

[35] Balderas-López, JA, & Mandelis A. New photopyroelectric technique for precise measurements of the thermal effusivity of transparent liquids. International Journal of Thermophysics. 2003; **24**: 463–471

[36] Esquef IA, Siqueira APL, da Silva MG, Vargas H, Miranda LCM. Photothermal gas analyzer for simultaneous measurements of thermal diffusivity and thermal effusivity. Analytical Chemistry, 2006; **78**: 5218–5221

[37] Streza M, Pop MN, Kovacs K, Simon V, Longuemart S, Dadarlat D. Thermal effusivity investigations of solid materials by using the thermal-wave-resonator-cavity (TWRC) configuration. Theory and mathematical simulations. Laser Physics, 2009; **19**: 1340–1344.

[38] Balderas-López JA, Jaime-Fonseca MR, Díaz-Reyes J, Gómez-Gómez YM, Bautista-Ramírez ME, Muñoz-Diosdado A, et al. Photopyroelectric technique, in the thermally thin regime, for thermal effusivity measurements of liquids. Brazilian Journal of Physics, 2016; **46**: 105–110.

[39] Gutiérrez-Juárez G, Ivanov R, Pichardo-Molina JP, Vargas-Luna M, Alvarado-Gil JJ, & Camacho A. Metrological aspects of auto-normalized front photopyroelectric method to measure thermal effusivity in liquids. International Journal of Thermophysics 2008; **29**: 2102–2115.

[40] George NA, Vallabhan C P G, Nampoori VPN, George AK, Radhakrishnan P. Use of an open photoacoustic cell for the thermal characterisation of liquid crystals. Applied Physics B 2001; **73**: 145–149.

[41] De Albuquerque JE, Balogh DT, Faria RM. Quantitative depth profile study of polyaniline films by photothermal spectroscopies. Applied Physics A 2007; **86**: 395–401

[42] Coufal H and Mandelis A. Pyroelectric sensors for the photothermal analysis of condensed phases. Ferroelectrics 1991; **118**: 379–409

[43] Caerels J, Glorieux C, Thoen J. Absolute values of specific heat capacity and thermal conductivity of liquids from different modes of operation of a simple photopyroelectric setup. Review of Scientific Instruments 1998; **69**: 2452–2458.

[44] Delenclos S, Chirtoc M, Sahraoui AH, Kolinsky C, Buisine JM. Assessment of calibration procedures for accurate determination of thermal parameters of liquids and their

temperature dependence using the photopyroelectric method. Review of Scientific Instruments. 2002; **73**: 2773–2780.

[45] Longuemart S, Sahraoui AH, Dadarlat D, Daoudi A, Laux V, & Buisine JM. Investigations of the thermal parameters of ferroelectric liquid crystals using the pyroelectric effect in the S C * phase. Europhysics Letters (EPL). 2003; **63**: 453.

[46] Shen J, Mandelis A, and Helen T. Signal generation mechanisms, intracavity-gas thermal-diffusivity temperature dependence, and absolute infrared emissivity measurements in a thermal-wave resonant cavity. Review of Scientific Instruments 1997; **69**: 197–203

[47] Streza M, Dadarlat D, Socaciu C, Bele C, Dulf F, Simon V. Photopyroelectric detection of vegetable oils' adulteration. Food Biophysics 2009; **4**: 147–150.

[48] Matvienko A, Mandelis A. Theoretical analysis of PPE measurements in liquids using a thermal-wave cavity. The European Physical Journal Special Topics. 2008; **153**: 127–129.

[49] Balderas-Lopez JA, & Mandelis A. Simple, accurate, and precise measurements of thermal diffusivity in liquids using a thermal-wave cavity. Review of Scientific Instruments. 2001; **72**: 2649–2652.

[50] Veeradate P, Voranuch T, Piyapong A, Pichet L. Preparation and characterization of alumina nanoparticles in deionized water using laser ablation technique. Journal of Nanomaterials. 2012; **2012**:6 pages

[51] Noroozi M, Radiman S, & Zakaria A. Influence of sonication on the stability and thermal properties of Al_2O_3 nanofluids. Journal of Nanomaterials 2014; **2014**: 10.

[52] Noroozi M, Radiman S, Zakaria A. & Soltaninejad S. Fabrication, characterization, and thermal property evaluation of silver nanofluids. Nanoscale Research Letters 2014; **9**: 1–10.

[53] Dadarlat D & Neamtu C. High performance photopyroelectric calorimetry of liquids. Acta Chimica Slovenica 2009; **56**: 225–236.

[54] Noroozi M, Zakaria A, Husin MS, Moksin MM, Wahab ZA. Investigating thermal parameters of PVDF sensor in the front pyroelectric configuration. International Journal of Thermophysics 2013; **34**: 2136–2143.

[55] Zhu D, Li X, Wang N, Wang X., Gao J, Li H. Dispersion behavior and thermal conductivity characteristics of Al_2O_3–H_2O nanofluids. Current Applied Physics. 2009; **9**: 131–139.

[56] Noroozi M, Zakaria A, Moksin M.M, Wahab ZA. An Investigation on the thermal effusivity of nanofluids containing $Al_2O_3$and CuO nanoparticles. International Journal of Molecular Sciences 2012; **13**: 10350–10358.

[57] Menon PC, Rajesh RN, Glorieux C. High accuracy, self-calibrating photopyroelectric device for the absolute determination of thermal conductivity and thermal effusivity of liquids. Review of Scientific Instruments, 2009: **80**: 054904–054909.

2

Enhancing Heat Transfer in Internal Combustion Engine by Applying Nanofluids

Wenzheng Cui, Zhaojie Shen, Jianguo Yang and Shaohua Wu

Abstract

Nanofluids exhibit novel properties including significant heat transfer properties that make them potentially useful in internal combustion engine cooling. However, although there is a substantial number of mechanisms proposed, modeling works related to their enhanced thermal conductivity, systematic mechanisms, or models that are suitable for nanofluids are still lacked. With molecular dynamics simulations, thermal conductivities of nanofluids with various nanoparticles have been calculated. Influence rule of various factors for thermal conductivity of nanofluids has been studied. Through defining the ratio of thermal conductivity enhancement by nanoparticle volume fraction, *K*, the impacts of nanoparticle properties for thermal conductivity are further evaluated. Furthermore, the ratio of energetic atoms in nanoparticles, *E*, is proposed to be an effective criterion for judging the impact of nanoparticles for the thermal conductivity of nanofluids. Mechanisms of heat conduction enhancement are investigated by MD simulations. Altered microstructure and movements of nanoparticles in the base fluid are proposed to be the main reasons for thermal conductivity enhancement in nanofluids. Both the static and dynamic mechanisms for heat conduction enhancement in nanofluids have been considered to establish a prediction model for thermal conductivity. The prediction results of the present model are in good agreement with experimental results.

Keywords: nanofluids, internal combustion engine, heat transfer, mechanism

1. Introduction

In recent years, nanofluids (NFs) have received much attention due to their strengthening heat transfer properties, which possess important application in heat transfer. The concept of nanofluids was first proposed by Choi in 1995, which indicates the fluids containing nanometer-sized particles, called nanoparticles (NPs) [1]. These fluids are engineered colloidal suspensions of nanoparticles in a base fluid. Numerous experimental studies discovered that nanofluids exhibit thermal properties superior to those of base fluid or conventional solid-liquid suspensions. Most of the thermal properties of nanofluids measured greatly exceed the values predicted by classical macroscopic theories and models. Nanofluids possess significantly increased thermal conductivity and improved convective heat transfer coefficient. Therefore, they are potentially useful in many enhanced heat transfer application, including engine cooling, vehicle thermal management, and power battery. Researchers are working to explain the significant high thermal properties of nanofluids [2–4]. However, although there is a substantial number of mechanisms proposed, and modeling works related to their enhanced thermal conductivity, systematic mechanisms, or models that are suitable for nanofluids are still lacked.

Regarding the excellent thermal properties of nanofluids, researchers are interested in the application of nanofluids in internal combustion engine, and began the study of applying nanofluids in internal combustion engine. In 1999, Wambsganss in Argonne national laboratory proposed the idea of applying nanofluids in car engine to improve the vehicle thermal management performance [5]. Choi indicates in a report that in Argonne national laboratory a research program of enhanced heat transfer by nanofluids is launched aiming at the cooling and heat transfer problems in the heavy-duty engine [6]. The results show that due to the excellent heat transfer performance of nanofluids, the size and weight of the engine can be reduced by 10%. Choi pointed out that the application of nanofluids in engine is one of the best methods of improving heat transfer performance of the cooling system. Saripella et al. studied the heat transfer performance of nanocoolant (nanofluids) in Volvo truck engine, and the results indicate that with nanofluids the temperatures of combustion chamber components and coolant are lowered [7]. Lockwood et al. in Valvoline Company reported the application of nanofluids in the cooling for internal combustion engine [8]. The experiments found that adding 1% vol. carbon nanotube in engine oil could increase the thermal conductivity by 150%. Wallner et al. in Delphi Company found that applying nanofluids can efficiently improve the efficiency of internal combustion engine and decrease the size and weight of the cooling system [9]. Huminic et al. studied the performance of nanofluids in a car radiator with a numerical method and found that the convective heat transfer performance is distinctly better than that of single-phase fluids [10]. Furthermore, the heat transfer properties of nanofluids are influenced by many factors, including the volume concentrations, temperatures, and fluid velocities. Vajjha et al. reported their research on the flow and heat transfer properties of Al_2O_3 and CuO nanofluids when applying them in the car radiator [11]. Their results reveal that nanofluids possess improved convective heat transfer properties and the increase rate is increased by increased volume concentrations. Leong et al. found that the heat transfer coefficient and heat transfer rate in the cooling system of internal combustion engine are improved by using nanofluids [12]. Peyghambarzadeh et al. experimentally verified that the application of nanofluids improves heat transfer efficiency of the car radiator by 45% when using Al_2O_3-H_2O nanofluids [13].

The authors have focused on the application of nanofluids in internal combustion engine (ICE) for heat transfer enhancement. In order to apply nanofluids in ICE, the mechanisms of heat transfer enhancement and the rules of enhanced heat transfer by nanofluids should be clarified first. The original cause of heat transfer enhancement is due to the adding of nanoscale particles. Therefore, we have attempted to use molecular dynamics (MD) simulations to study these microscopic mechanisms [14]. By using MD simulation, we have calculated thermal conductivity of nanofluids via the Green-Kubo equation and proposed an effective criterion for predicting the enhancement of apparent thermal conductivity. Furthermore, possible mechanisms of heat conduction increase in nanofluids are studied by MD simulation, including: (1) the micromotions of nanoparticles, (2) changed microstructure of base fluid by adding nanoparticles, and (3) the influence of absorption layer of base fluid at the surface of nanoparticles. On the basis of the microscopic mechanisms found by MD simulations, we have also proposed a revised thermal conductivity model, which considered both the static and dynamic mechanisms. The revised model is verified by experimental data, which has been proved to be quite accurate for predicting thermal conductivity of common types of nanofluids.

2. Influence rule and criterion for nanofluids' thermal conductivity

2.1. Simulation results of thermal conductivity

MD simulation is used to calculate thermal conductivity of nanofluids via the Green-Kubo equation [14]. A series of influencing factors for the thermal conductivity of nanofluids have been considered, including: nanoparticles' volume concentrations, sizes, materials, and shapes [15]. In this work, inert liquid Ar is chosen as the base fluid because of the mature and credible potential function. The NP materials include Cu, Ag, Fe, and Au. We have chosen these types of nanoparticles because they are commonly reported in the literatures on nanofluids. The MD simulation results reveal that the thermal conductivity of nanofluids can be obviously increased by adding nanoparticles, as shown in **Table 1**. However, the contributions of several influencing factors for thermal conductivity of nanofluids are different.

Figure 1 shows the MD simulation results of thermal conductivities for nanofluids containing spherical nanoparticles. In this case, the nanoparticle volume fractions, nanoparticle diameter, and thermal conductivity of nanoparticles are considered. For the influencing factors that have been considered in this work, the influencing rules are regular. The thermal conductivity of nanofluids is increased with increased volume fraction of NPs, decreased NP sizes, and higher thermal conductivity of NPs. For instance, the ratios of thermal conductivity enhancement for Ag, Cu, Fe, and Au nanofluids are 1.41, 1.15, 1.11, and 1.08 sequentially when the other conditions are the same.

Diameters (nm)	Volume fractions (%)	Cu	Ag	Au	Fe
2	0.5	1.112	1.326	1.062	1.036
	1	1.149	1.405	1.112	1.059
	2	1.359	1.498	1.205	1.079
	3	1.414	1.570	1.322	1.096
4	0.5	1.095	1.287	1.055	1.041
	1	1.134	1.325	1.096	1.065
	2	1.303	1.396	1.175	1.085
	3	1.376	1.462	1.202	1.112
6	0.5	1.065	1.244	1.049	1.045
	1	1.097	1.265	1.086	1.079
	2	1.266	1.326	1.152	1.098
	3	1.336	1.369	1.256	1.132

Table 1. Ratios of nanofluids' thermal conductivity enhancement with different influence factors.

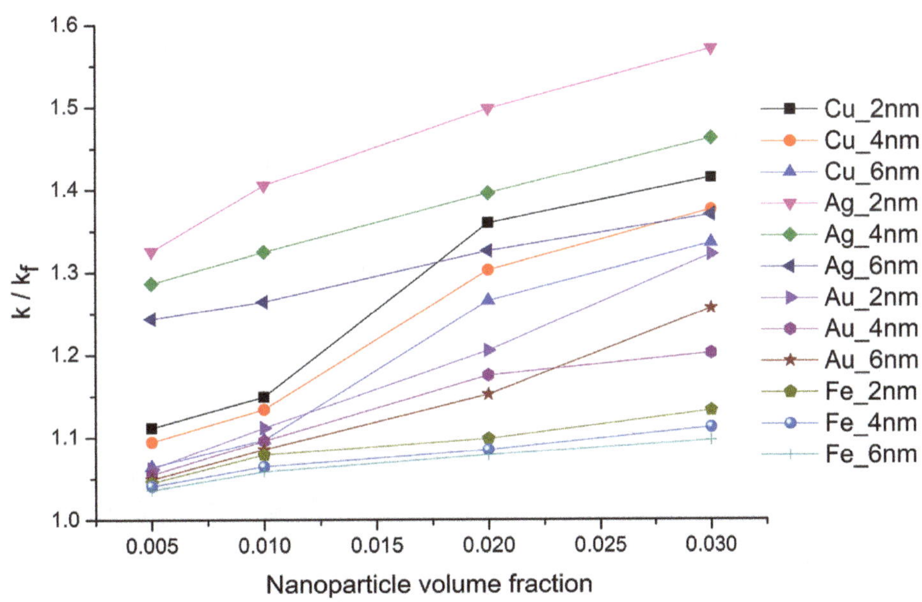

Figure 1. MD simulation results for thermal conductivity of nanofluids.

To examine the influencing rules for thermal conductivity of nanofluids in depth, the ratio of thermal conductivity enhancement by nanoparticle volume fraction, *K*, is defined:

$$K = \frac{k / k_f - 1}{V_{np} / V}, \tag{1}$$

where V_{np} and V are the volume of nanoparticles and nanofluids, respectively; k and k_f represent the thermal conductivities of nanofluids and base fluid, respectively.

The physical significance of K is to evaluate the ratio of thermal conductivity increase and nanoparticle volume fraction. In other words, K could be used to evaluate the impact of nanoparticle properties for thermal conductivity. With K, the contributions from volume fraction and thermal conductivity of nanoparticles could be compared, as shown in **Figures 2** and **3**. From the figures i,t could be easily found that the contributions of nanoparticle materials are in the order of Ag, Cu, Au, and Fe. Furthermore, through comparing K values, it is also found that the influence of nanoparticle materials is weakened when the volume fraction or nanoparticle size is increased. With the help of K value the other influencing factors could be further evaluated, please refer to reference [15].

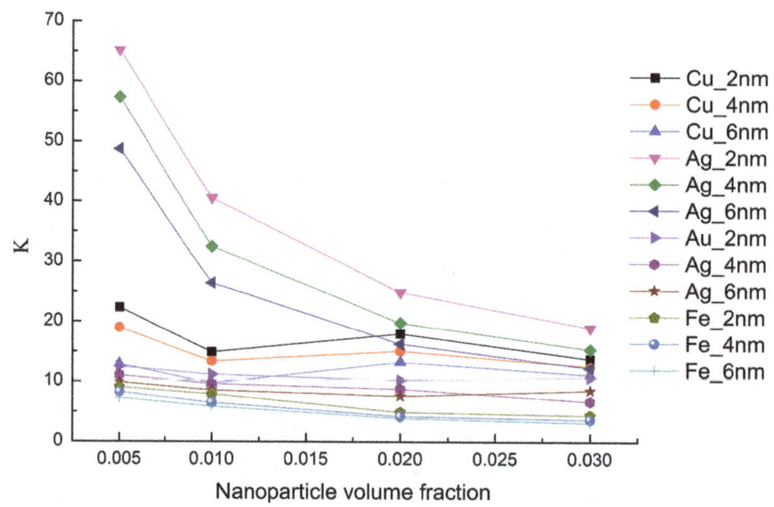

Figure 2. Comparison of K values against nanoparticle volume fraction for various nanofluids.

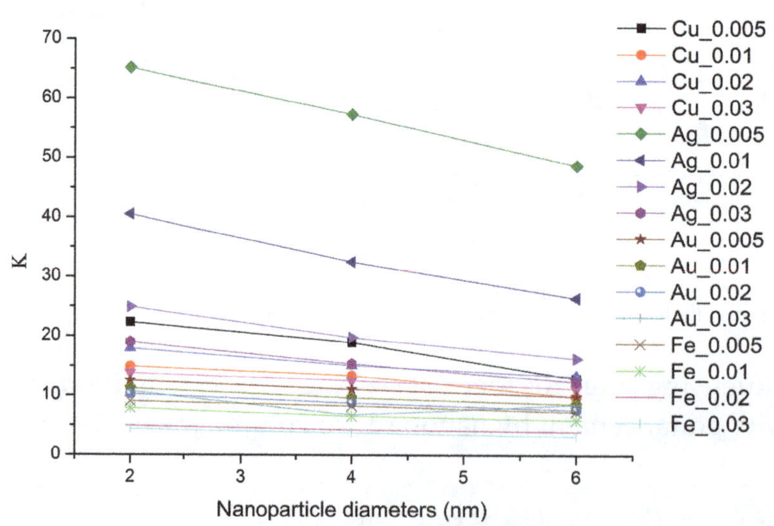

Figure 3. Comparison of K values against nanoparticle size for various nanofluids.

2.2. Criterion for the increased thermal conductivity

It is found that the nanoparticles containing higher atomic potential energy (energetic atoms) are better for thermal conductivity enhancement of nanofluids [15]. The ratio of energetic atoms in a nanoparticle E is proposed as a criterion for enhanced thermal conductivity of nanofluids, which is written as:

$$E = \frac{N_E}{N}, \tag{2}$$

where N and N_E are the quantity of atoms and energetic atoms in a nanoparticle, respectively.

If we set a standard for delimiting the energetic atoms in a nanoparticle, E can be calculated according to Eq. (2). **Figure 4** illustrates E of different types of nanoparticles. The ratio of energetic atoms in an Ag nanoparticle is the largest, and that of Fe nanoparticle is the lowest. The larger E value in a nanoparticle is better for thermal conductivity enhancement in nanofluids. **Figure 5** shows the influence of nanoparticle shapes (surface area to volume ratio S/V) for the E value. It is found that nanoparticle with larger S/V value possesses larger E value. **Figure 6** illustrates that the E value of spherical Ag nanoparticle is larger than that of Cu nanoparticle with larger S/V value. Therefore, the thermal conductivity of nanofluids with spherical Ag nanoparticles is higher than that of nanofluids with nonspherical Cu nanoparticles.

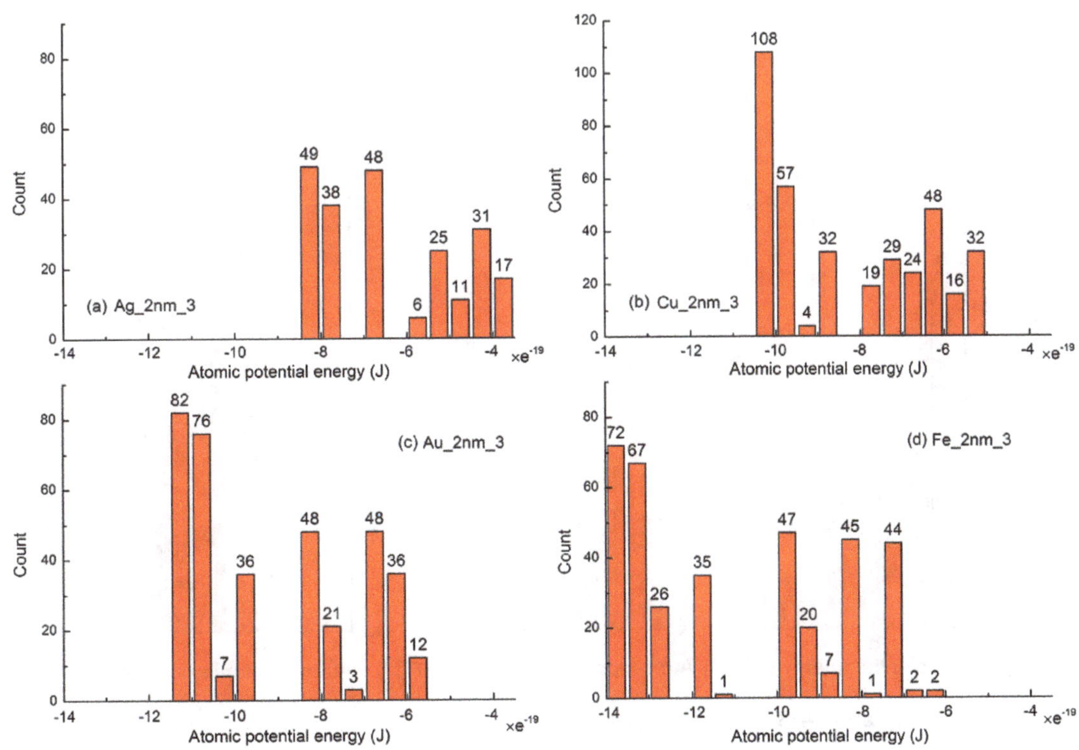

Figure 4. Atomic potential energy distributions of various nanoparticles.

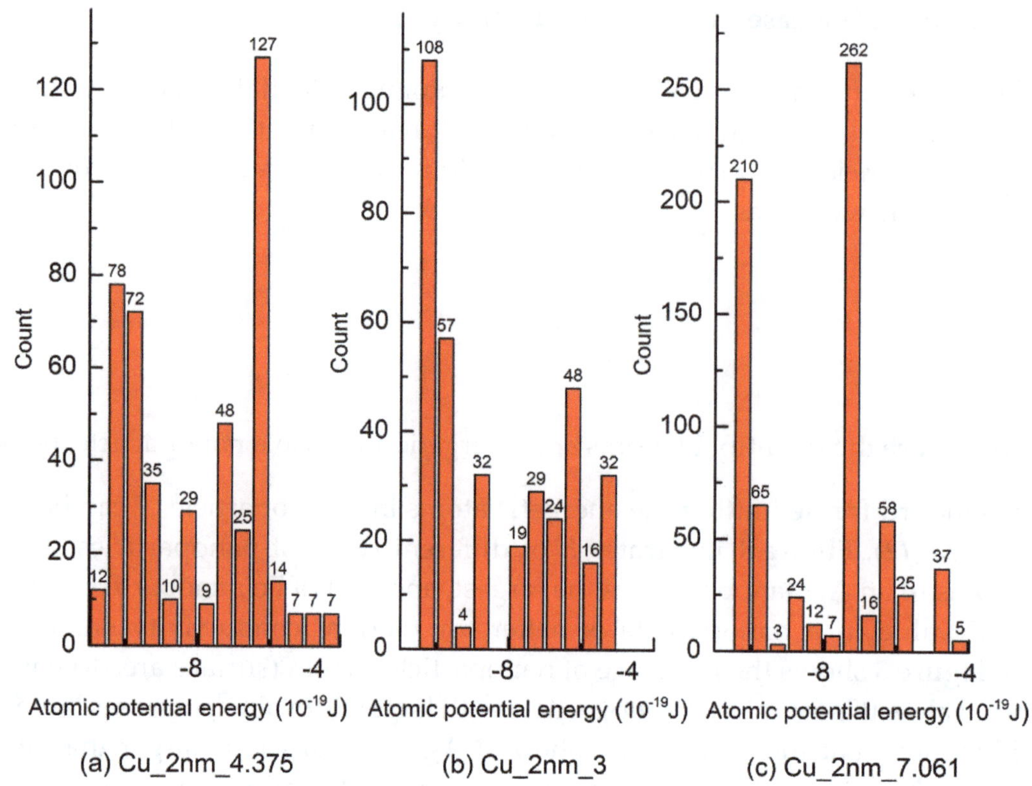

Figure 5. Comparison of atomic potential energy distributions for nanoparticles with different shapes.

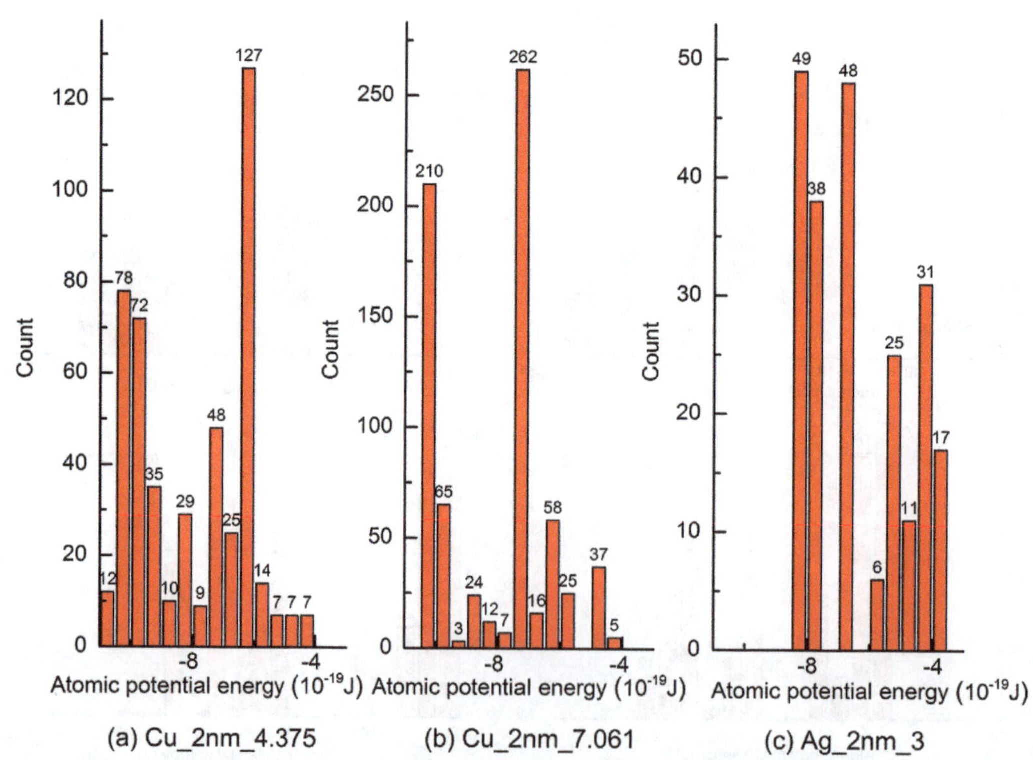

Figure 6. Comparison of atomic potential energy distributions for nanoparticles with different materials.

3. Proposed mechanisms of heat conduction in nanofluids

3.1. Altered microstructure of nanofluids

In order to analyze the microscopic structure characteristics of nanofluids, number density distribution, radial distribution function (RDF), coordination number, and potentials of mean force (PMF) should be considered [16].

Number density distribution represents the distribution of liquid atoms around a centered nanoparticle. **Figure 7** illustrates the number density distributions of base fluid atoms in different types of nanofluids. It is found that at the positions of 0.25 and 0.5 nm all the curves show the first and second peak values. But for different types of nanoparticles, the first peak values of curves are different. The order of first peak values is of the same order of thermal conductivity of bulk materials of nanoparticles.

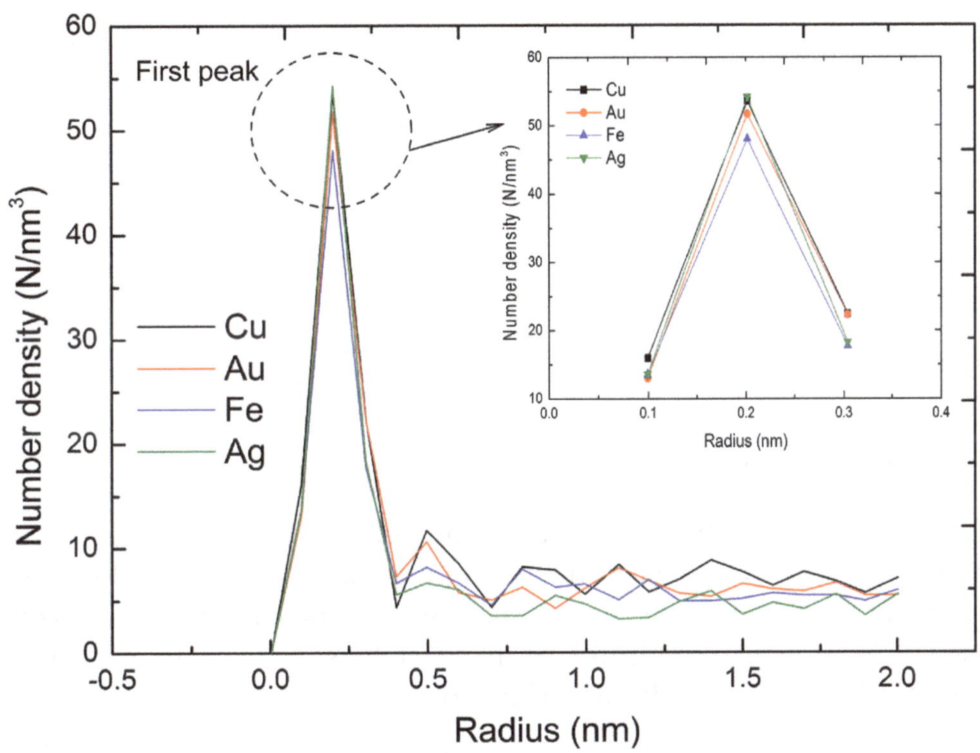

Figure 7. Statistical result of absorption layers around nanoparticles of different materials.

RDF represents the probability of finding an atom of a specified type near the central atom. Through RDF, the microscopic structure of fluid could be examined. **Figure 8** illustrates the RDF of Cu nanofluids with a 2 nm-diameter nanoparticle. In the figure, the RDF curve of "Ar-Ar" represents the chance of finding an Ar atom near the central Ar atom. It could be found that the Ar-Ar RDF shows typical characteristics of the liquid: "short-range order and long-range disorder." In the figure, the "total" RDF represents the chance of finding an atom of any type near the central atom, which represents the microscopic structure of nanofluid. It is found that the first curvilinear peak in the RDF of nanofluids is larger than that of base fluid, which means the probability of finding an atom is higher than that in a single-phase base fluid. It could also be found that there are several diminutive curvilinear peaks in the RDF of nanofluids, which is due to the adding of nanoparticles in base fluid. In general, the microscopic structure of nanofluid exerts a mixed up structure characteristics of liquid and solid. Both the liquid characteristic of "short-range order and long-range disorder" and the solid characteristic of "long-range order" have been found in the microscopic structure of nanofluids. Therefore, the microscopic structure of nanofluids is always ordered.

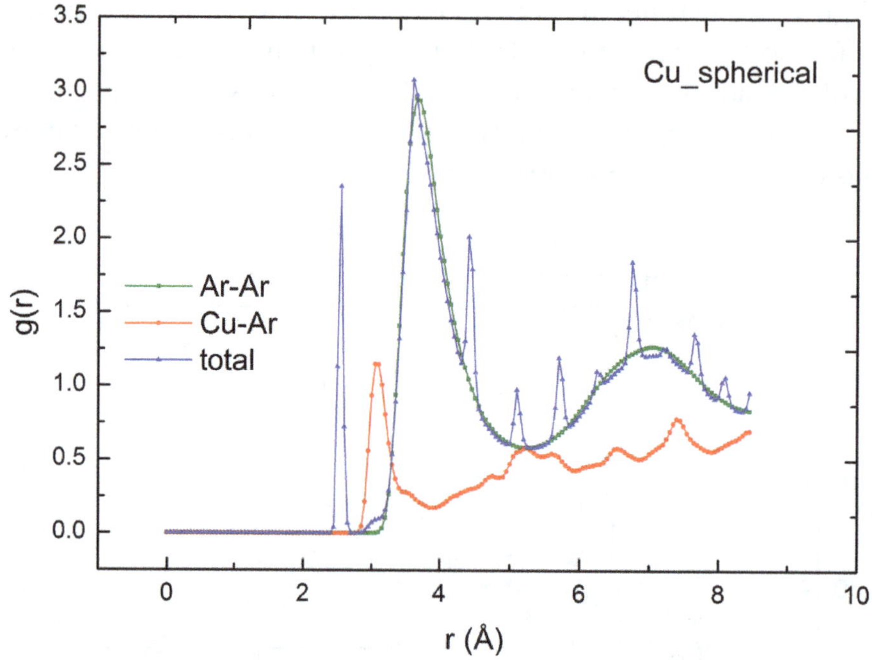

Figure 8. Radial distribution functions of nanofluids with spherical copper nanoparticle.

Coordinate number indicates the average adjacent atomic number for a certain atom within an interval of *r*. By comparing the coordinate number curve of Cu nanofluids with a 2 nm-diameter spherical nanoparticle, it is found that the coordinate number curves of Cu-Ar and Ar-Ar cross at 0.35 nm, which confirms that the local density of Ar atoms near a Cu atom is larger (**Figure 9**). Therefore, the existence of absorption layer is verified.

PMF reflects the combining capacity between particles in pairs. The value of PMF could be used to investigate the combining capacity between different particle pairs. **Figure 10** shows the PMF curve for Cu-Ar nanofluids. The contact minimum (CM), separated minimum (SM), and the layer barrier (LB) could easily be found in the PMF curve of nanofluids. But the positions and values of CM, SM, and LB are different for disparate atom pairs. The cis-trans direction of energy barrier between molecular layers of liquid Ar is different. When an atom is approaching the central particle, then it needs to conquer the energy barrier between the first and second molecular layers. But the atom is harder to leave the central particle. An atom of base fluid needs to conquer greater energy barrier to reenter the base fluid. The PMF of nanofluids is different from that of base fluid. At 0.3 nm, there is a huge energy barrier in the PMF of nanofluids, which indicates the surrounding atom needs to conquer two energy barriers to get close to the central atom. The cis-trans direction of the first energy barrier is nearly the same, but the cis-trans direction of the second energy barrier is obviously different. Once a base fluid atom enters the adjacent area of the central atom, then it is very hard to reenter the base fluid because of the large energy barrier.

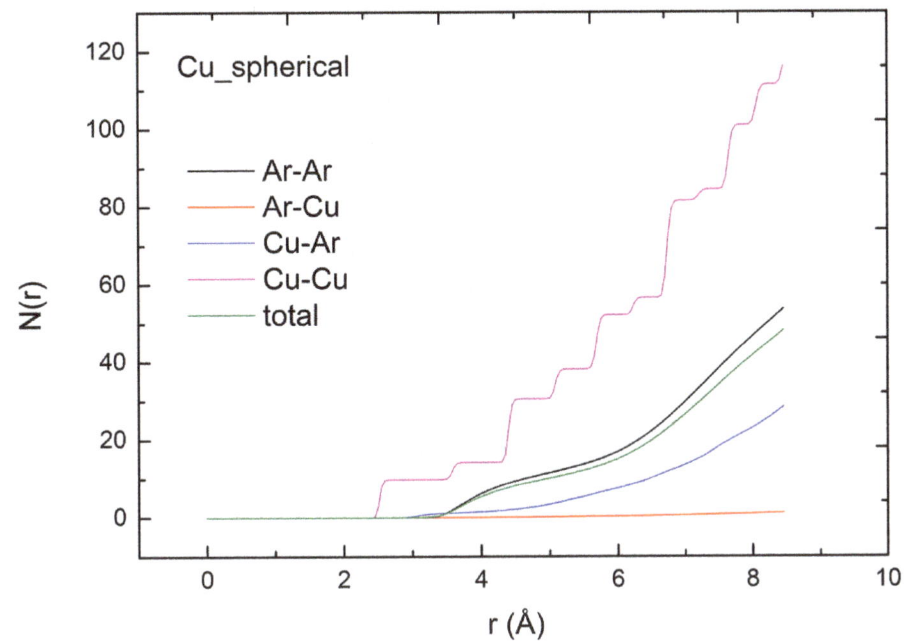

Figure 9. The coordination number of nanofluids with spherical copper nanoparticle.

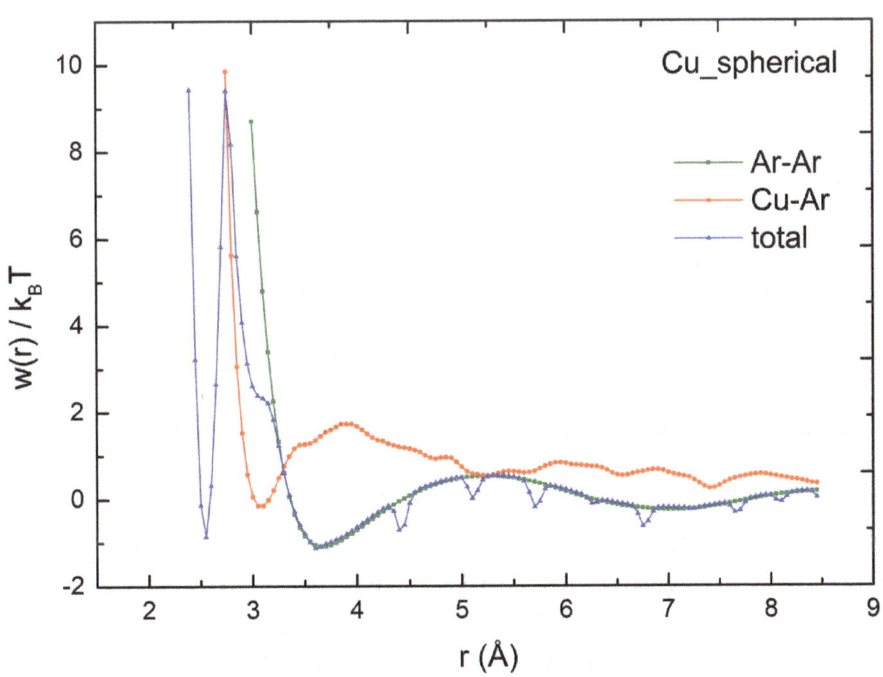

Figure 10. PMF of nanofluids with spherical copper nanoparticle.

3.2. Movements of nanoparticles in the base fluid

Through MD simulation, the nanoparticles are observed to move chaotically at high speed in the base fluid. Through MD simulation, the instantaneous velocity and position coordinates of each atom could be obtained [17]. The translational and rotational velocity of nanoparticles could be acquired by defining a group for the Cu atoms within the nanoparticle. With commands provided by LAMMPS the time-averaged translational and rotational velocity of the atom group could be calculated and output derived. For the case of imposed shearing velocity v = 50 m/s on the fluid, the translational velocity components of nanoparticles are statistically analyzed, as shown in **Figure 11**. Along x-directions, the average translational velocity components are -2 m/s~2 m/s, and the instantaneous peak value can reach 5 m/s. The translational velocity components of nanoparticles oscillate sharply, which demonstrate the chaotic movements of nanoparticles are mainly caused by their Brownian motion.

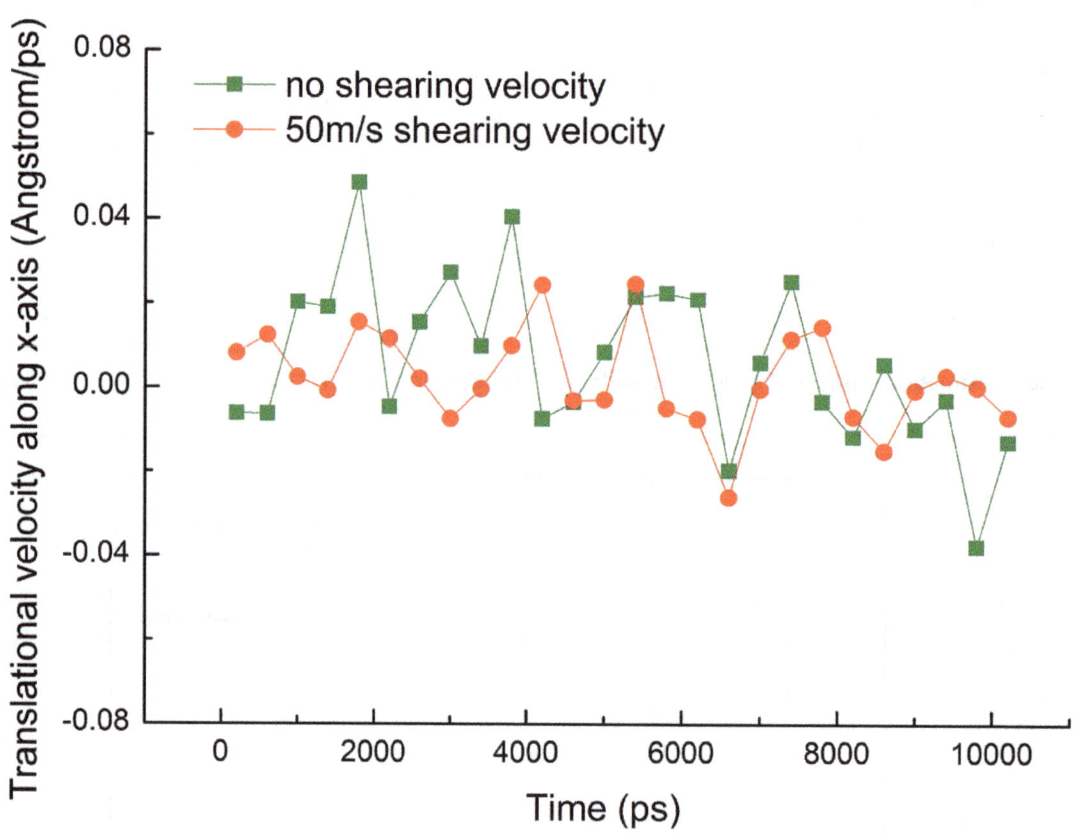

Figure 11. Translational velocity component of nanoparticle along x-axis in shearing flow.

Rotation of nanoparticles is also statistically analyzed. For the case of imposed shearing velocity 50 m/s, the angular velocity component along the x-axis of nanoparticle is shown in **Figure 12**. The peak angular velocity of nanoparticles can reach 6×10^9 rad/s, meanwhile the rotational directions of nanoparticles change randomly because of their nanoscale size. It is found that imposing shearing velocity affects little for rotation of nanoparticles. Imposed shearing velocity or not, the angular velocity components of nanoparticles are of the order of magnitude of 10^9 rad/s.

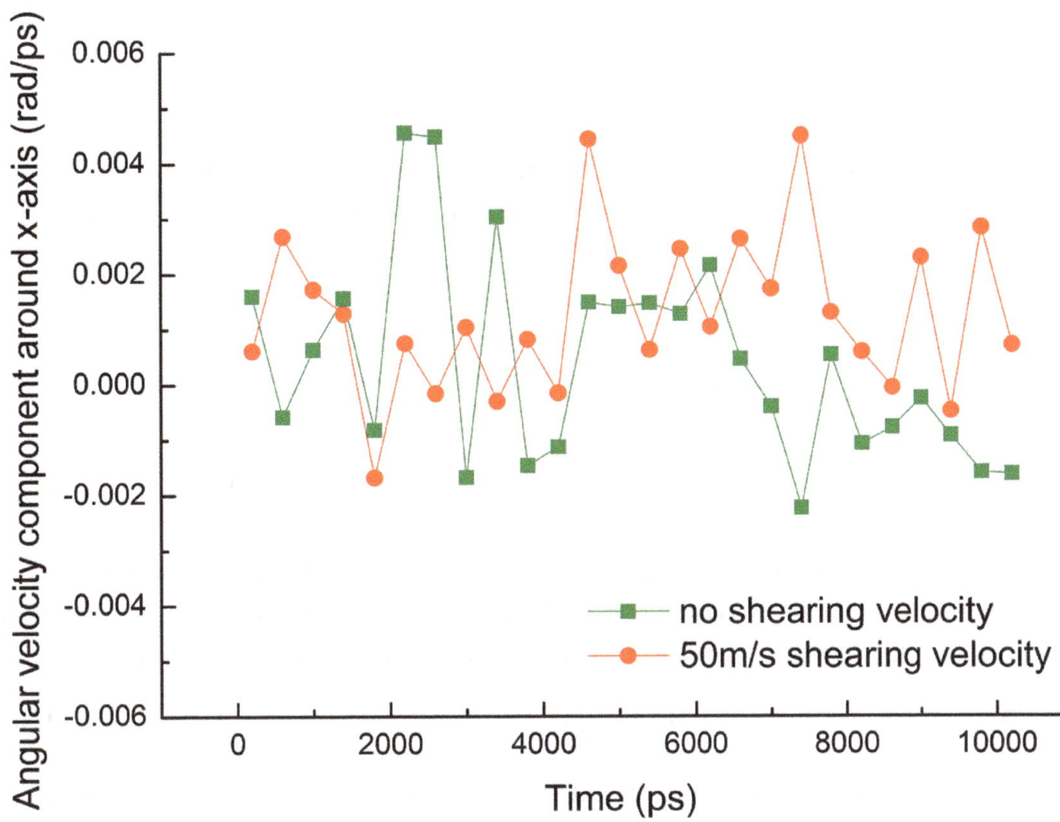

Figure 12. Angular velocity component of nanoparticle around *x*-axis.

4. Modeling thermal conductivity of nanofluids

Jeffrey applied Green's function method and relaxed the requirement of uniform configuration for particles. The formula, which is suitable for predicting suspensions with nonuniformly distributing nanoparticles and relatively large volume concentration, is written as [18],

$$\frac{k}{k_f} = 1 + 3\beta\phi_p + 3\beta(\beta + \Sigma)\phi_p^{\ 2}, \tag{3}$$

where Σ is a convergent series, which depends on the specific value of thermal conductivity of nanoparticle and base fluid k_m/k_f. β is a coefficient, which is determined by k_m/k_f:

$$\beta = \frac{\alpha - 1}{\alpha + 2}, \tag{4}$$

where α is the specific value of thermal conductivity of nanoparticle and base fluid k_m/k_f:

$$\alpha = \frac{k_m}{k_f}. \tag{5}$$

Through considering the above-mentioned mechanisms of thermal conductivity enhancement in nanofluids, a revised model for predicting thermal conductivity of nanofluids that takes into account both the static and dynamic mechanisms is proposed, which is written as[19]:

$$\frac{k_{nf}}{k_f} = \frac{k_S}{k_f} + \frac{k_D}{k_f} = 1 + 3\beta\left(1 + \frac{2t_{ab}}{d_p}\right)^3 \phi_p + c\left(1 + \frac{2t_{ab}}{d_p}\right)^6 \phi_p^{\ 2} + \frac{\phi_p\left(\rho c_p\right)_p}{2k_f}\sqrt{\frac{k_B T}{3\pi r_{cl}\mu_f}} \tag{6}$$

In the equation, k_{nf} is thermal conductivity of nanofluids; k_f is thermal conductivity of base fluid; k_S is the static part of thermal conductivity model for nanofluids, which takes into account the static mechanisms of heat conduction enhancement; k_D is the static part of thermal conductivity model for nanofluids; $\beta = (\alpha - 1)/(\alpha + 2)$ and $\alpha = \frac{k_{\mathrm{cl}}}{k_f}$, where k_{cl} is thermal conductivity of nanoparticle cluster; t_{ab} is the thickness of absorption layer formed by absorbed liquid molecules which can be estimated by Langmuir monolayer equation; d_p is the average diameter of nanoparticles; Φ_p is the initial volume concentration of nanoparticles in nanofluids; $c = 3\beta^2 + 3\beta\Sigma$ is a coefficient in Jeffrey Model; ρ is the density of nanoparticle bulk material; c_p is the specific heat of nanoparticle bulk material; k_B is the Boltzmann constant and $k_B = 1.381 \times 10^{-23} J/K$; T is thermodynamic temperature; r_{cl} is the radius of the nanoparticle cluster; and μ_f is the viscosity of base fluid.

Compared with existing prediction models for thermal conductivity of nanofluids, the present model takes into account the static and dynamic mechanisms of strengthened heat conduction in nanofluids simultaneously and possesses more definite physics meaning. In addition, parameters used in the current model are more precise that ensures the veracity of prediction result. For instance, the thermal conductivity of nanoparticles k_p is distinguished with that of bulk material of nanoparticles k_m and used in the prediction model as an independent parameter. And the thermal conductivity of nanoparticles-cluster k_{cl} is introduced as an independent parameter to include the heat conduction of absorption layer, which further improves the prediction accuracy of the present model.

Through comparing the prediction results of the present model and existing experimental data, the present prediction model is proved to be quite effective for predicting thermal conductivity of common nanofluids, as shown in **Table 2**. For various types of nanofluids (with different materials including: metal, metallic oxide, and nonmetallic oxide, different volume fractions, or different nanoparticle diameters), the present model gives good predictions.

References	NF	NP size (nm)	Volume fraction (%)	Thermal conductivity ratio (k_{nf}/k_f)	Present model	Maximum error
Masuda [20]	Al_2O_3/Water	13	1.3–4.3	1.107–1.318	1.097–1.322	0.9%
Teng et al. [21]	Al_2O_3/Water	20	0.001–0.005	1.018–1.065	1.010–1.043	2.1%
Chon et al. [22]	Al_2O_3/Water	13	1	1.081	1.072	0.9%
Xie et al. [23]	Al_2O_3/EG	25	1.7–5	1.097–1.294	1.103–1.303	0.8%
Vajjha [24]	Al_2O_3/ (60:40% EG/W)	53	1–4	1.069–1.159	1.048–1.195	3.1%
Wang et al. [25]	Al_2O_3/EG	28	5–8	1.246–1.404	1.276–1.445	2.9%
Eastman et al. [26]	CuO/EG	35	1–4	1.050–1.227	1.061–1.245	3.7%
Lee et al. [27]	CuO/EG	24	1–4	1.060–1.242	1.047–1.212	3.2%
Xuan and Li [28]	Cu/Water	100	1–5	1.078–1.434	1.090–1.459	3.1%
Li et al. [29]	Cu/Water	20	1–3	1.120–1.289	1.096–1.293	2.1%
Eastman et al. [26]	Cu/EG	10	0.33–0.55	1.041–1.101	1.061–1.122	2%
Hwang et al. [30]	CuO/Water	33	1	1.05	1.08	2.9%
Zhang et al. [31]	SiO_2/Water	7	0.5–3	1.016–1.084	1.014–1.088	0.4%
Hwang et al. [30]	SiO_2/Water	12	1	1.03	1.03	0

Table 2. Comparison between the prediction results and experimental data.

5. Concluding remarks

Thermal conductivities of nanofluids with various nanoparticles have been calculated through MD simulations. Influence rule of various factors for thermal conductivity of nanofluids has been studied. Through defining the ratio of thermal conductivity enhancement by nanoparticle volume fraction, *K*, the impacts of nanoparticle properties for thermal conductivity are further evaluated. Furthermore, the ratio of energetic atoms in nanoparticles, *E*, is proposed to be an effective criterion for judging the impact of nanoparticles for the thermal conductivity of nanofluids.

Mechanisms of heat conduction enhancement are investigated by MD simulations. Altered microstructure and movements of nanoparticles in the base fluid are proposed to be the main reasons for thermal conductivity enhancement in nanofluids. Number density distribution, radial distribution function (RDF), coordination number, and potentials of mean force (PMF) are used to analyze the microscopic structure characteristics of nanofluids. Through MD simulation, the average translational and rotational velocities of nanoparticles are obtained.

Both the static and dynamic mechanisms for heat conduction enhancement in nanofluids have been considered to establish a prediction model for thermal conductivity. The parameters in the model have definite physical meaning and are more precise. The prediction results of the present model are in good agreement with experimental results.

Acknowledgements

The authors have received funds from the National Natural Science Foundation of China (Grant Nos. 51506038, 51606052), Shandong Provincial Natural Science Foundation, China (Grant No. ZR2015EQ003), and China Postdoctoral Science Foundation (Grant Nos. 2016T90284, 2015M571411). We acknowledge the reviewers' comments and suggestions very much.

Nomenclature

E Ratio of energetic atoms in a nanoparticle

K Ratio of thermal conductivity enhancement by nanoparticle volume fraction

N Quantity of atoms in a nanoparticle

N_E Quantity of energetic atoms in a nanoparticle

S/V Surface area to volume ratio

T Thermodynamic temperature

V Volume of nanofluids

V_{np} Volume of nanoparticles

c_p Specific heat of nanoparticle bulk material

d_p Average diameter of nanoparticles

k Thermal conductivity

kB Boltzmann constant

k_f Thermal conductivity of base fluids

k_m Thermal conductivity of bulk material

k_p Thermal conductivity of nanoparticles

k_{cl} Thermal conductivity of nanoparticles' cluster

r_{cl} Radius of the nanoparticle cluster

t_{ab} Thickness of absorption layer

α Specific value of thermal conductivity of nanoparticle and base fluid

μ_f Viscosity of base fluid

ρ Density of nanoparticle bulk material

Φ_p Volume concentration of nanoparticles in nanofluids

Author details

Wenzheng Cui[1*], Zhaojie Shen[1], Jianguo Yang[1] and Shaohua Wu[2]

*Address all correspondence to: cuiwenzheng@hit.edu.cn

1 School of Automotive Engineering, Harbin Institute of Technology, Weihai, China

2 School of Energy Science and Engineering, Harbin Institute of Technology, Harbin, China

References

[1] Lee S, Choi SUS, Li S, Eastman JA. Measuring thermal conductivity of fluids containing oxide nanoparticles. Journal of Heat Transfer. 1999;121(2):280–289. DOI: 10.1115/1.2825978

[2] Wen DS, Ding YL. Experimental investigation into convective heat transfer of nanofluids at the entrance region under laminar flow conditions. International Journal of Heat and Mass Transfer. 2004;47(24):5181--5188. DOI: 10.1016/j.ijheatmasstransfer.2004.07.012

[3] Hussein AM, Sharma KV, Bakar RA, Kadirgama K. A review of forced convection heat transfer enhancement and hydrodynamic characteristics of a nanofluid. Renewable and Sustainable Energy Reviews. 2014;29:734–743. DOI: 10.1016/j.rser.2013.08.014

[4] Sundar LS, Singh MK. Convective heat transfer and friction factor correlations of nanofluid in a tube and with inserts: A review. Renewable and Sustainable Energy Reviews. 2013;20:23–35. DOI: 10.1016/j.rser.2012.11.041

[5] Wambsganss MW. Thermal management concepts for higher-efficiency heavy vehicles. In: SAE Technical Paper ; 1999. Paper number: 1999-01-2240. DOI: 10.4271/1999-01-2240

[6] Choi SUS. Nanofluids for improved efficiency in cooling systems. In: Heavy Vehicle Systems Review; April 18–20,2006; Chicago. 2006.

[7] Saripella SK, Yu W, Routbort JL, France DM. Effects of nanofluid coolant in a class 8 truck engine. In: SAE Technical Paper; 2007. Paper number: 2007-01-2141. DOI: 10.4271/2007-01-2141

[8] Lockwood FE, Zhang ZG, Forbus TR, Choi SUS, Yang Y, Grulke EA. The current development of nanofluid research. In: SAE Technical Paper; 2005. Paper number: 2007-01-2141. DOI: 10.4271/2007-01-2141

[9] Wallner E, Sarma DHR, Myers B, Shah S, Ihms D, Chengalva S, Parker R, Eesley G, Dykstra C. Nanotechnology application in future automobiles. In: SAE Technical Paper; 2010. Paper number: 2010-01-1149. DOI: 10.4271/2010-01-1149

[10] Huminic G, Huminic A. The cooling performances evaluation of nanofluids in a compact heat exchanger. In: SAE Technical Paper; 2012. Paper number: 2012-01-1045. DOI: 10.4271/2012-01-1045

[11] Vajjha RS, Das DK, Namburu PK. Numerical study of fluid dynamic and heat transfer performance of Al2O3 and CuO nanofluids in the flat tubes of a radiator. International Journal of Heat and Fluid Flow. 2010;31(4):613–621. DOI: 10.1016/j.ijheatfluidflow.2010.02.016

[12] Leong KY, Saidur R, Kazi SN, Mamun AH. Performance investigation of an automotive car radiator operated with nanofluid based coolant (Nanofluid as a coolant in a radiator). Applied Thermal Engineering. 2010;30(17–18):2685–2692. DOI: 10.1016/j.applthermaleng.2010.07.019

[13] Peyghambarzadeh SM, Hashemabadi SH, Jamnani MS, Hoseini SM. Improving the cooling performance of automobile radiator with Al2O3/water nanofluid. Applied Thermal Engineering. 2011;31(10):1833–1838. DOI: 10.1016/j.applthermaleng.2011.02.029

[14] Allen MP, Tildesley DJ. Computer simulation of liquids. 1st ed. New York: Oxford University Press; 1987. 385 p. DOI: 10.1063/1.2810937

[15] Cui WZ, Shen ZJ, Yang JG, Wu SH, Bai ML. Influence of nanoparticle properties on the thermal conductivity of nanofluids by molecular dynamics simulation. RSC Advances. 2014;4(98):55580–55589. DOI: 10.1039/C4RA07736A

[16] Cui WZ, Shen ZJ, Yang JG, Wu SH. Molecular dynamics simulation on the microstructure of absorption layer at the liquid–solid interface in nanofluids. International Communications in Heat and Mass Transfer. 2016;71:75–85. DOI: 10.1016/j.icheatmasstransfer.2015.12.023

[17] Cui WZ, Shen ZJ, Yang JG, Wu SH. Effect of chaotic movements of nanoparticles for nanofluid heat transfer augmentation by molecular dynamics simulation. Applied Thermal Engineering. 2015;76:261–271. DOI: 10.1016/j.applthermaleng.2014.11.030

[18] Jeffrey DJ. Conduction through a random suspension of spheres. Proceedings of the Royal Society of London. Series A, Mathematical and Physical Sciences. 1973;335(1602):355–367. DOI: 10.1098/rspa.1973.0130

[19] Cui WZ, Shen ZJ, Yang JG, Wu SH. A modified prediction model for thermal conductivity of spherical nanoparticle suspensions (nanofluids) by introducing static and dynamic mechanisms. Industrial and Engineering Chemistry Research. 2014;53(46):18071–18080. DOI: 10.1021/ie503296g

[20] Masuda H, Ebata A, Teramae K, Hishinuma N. Alteration of thermal conductivity and viscosity of liquid by dispersing ultra-fine particles. Dispersion of Al2O3, SiO2 and TiO2 ultra-fine particles. Netsu Bussei. 1993;7(4):227–233. DOI: 10.2963/jjtp.7.227

[21] Teng TP, Hung YH, Teng TC, Mo HE, Hsu HG. The effect of alumina/water nanofluid particle size on thermal conductivity. Applied Thermal Engineering. 2010;30(14–15): 2213–2218. DOI: 10.1016/j.applthermaleng.2010.05.036

[22] Chon CH, Kihm KD, Lee SP, Choi SUS. Empirical correlation finding the role of temperature and particle size for nanofluid (Al2O3)(Al2O3) thermal conductivity enhancement. Applied Physics Letters. 2005;87:153107. DOI: 10.1063/1.2093936

[23] Xie HQ, Wang JC, Xi TG, Liu Y, Ai F, Wu QR. Thermal conductivity enhancement of suspensions containing nanosized alumina particles. Journal of Applied Physics. 2002;91:4568. DOI: 10.1063/1.1454184

[24] Vajjha RS, Das DK. Experimental determination of thermal conductivity of three nanofluids and development of new correlations. International Journal of Heat and Mass Transfer. 2009;52(21–22):4675–4682. DOI: 10.1016/j.ijheatmasstransfer.2009.06.027

[25] Wang XW, Xu XF, Choi SUS. Thermal conductivity of nanoparticle – fluid mixture. Journal of Thermophysics and Heat Transfer. 1999;13(4):474–480. DOI: 10.2514/2.6486

[26] Eastman JA, Choi SUS, Li S, Yu W, Thompson LJ. Anomalously increased effective thermal conductivities of ethylene glycol-based nanofluids containing copper nanoparticles. Applied Physics Letters. 2001;78(6):718. DOI: 10.1063/1.1341218

[27] Lee S, Choi S, Li S, Eastman J. Measuring thermal conductivity of fluids containing oxide nanoparticles. Journal of Heat Transfer. 1999;121(2):280–289. DOI: 10.1115/1.2825978

[28] Xuan YM, Li Q. Heat transfer enhancement of nanofluids. International Journal of Heat and Fluid Flow. 2000;21(1):58–64. DOI: 10.1016/S0142-727X(99)00067-3

[29] Li Q, Xuan YM, Yu F. Experimental investigation of submerged single jet impingement using Cu–water nanofluid. Applied Thermal Engineering. 2012;36:426–433. DOI: 10.1016/j.applthermaleng.2011.10.059

[30] Hwang Y,Park HS, Lee JK, Jung WH. Thermal conductivity and lubrication characteristics of nanofluids. Current Applied Physics. 2006;6(Supplement 1): e67–e71. DOI: 10.1016/j.cap.2006.01.014

[31] Zhang SB, Luo ZY, Wang T, Shou CH, Ni MJ, Cen KF. Experimental study on the convective heat transfer of CuO-water nanofluid in a turbulent flow. Journal of Enhanced Heat Transfer. 2010;17(2):1–14. DOI: 10.1615/JEnhHeatTransf.v17.i2.60

3

Nanofluid with Colloidal Magnetic Fe_3O_4 Nanoparticles and its Applications in Electrical Engineering

Lucian Pîslaru-Dănescu, Gabriela Telipan, Floriana D. Stoian, Sorin Holotescu and Oana Maria Marinică

Abstract

In this study, we propose a new type of a cooling agent based on magnetic nanofluid for the purpose of replacing the classical cooling fluids in electrical power transformers. The magnetite (Fe_3O_4) nanoparticles were synthesized by the co-precipitation method from an aqueous medium of salts $FeCl_3x6H_2O$ and $FeSO_4x7H_2O$ in the molar ratio Fe^{3+}/Fe^{2+} = 2:1, by alkalization with 10% aqueous solution of NaOH at 80°C, for 1 h. The size of the magnetite nanoparticles, as measured by X-ray diffraction method, was 14 nm and by scanning electron microscopy (SEM), they are between 10 and 30 nm. Magnetite powder was placed in oleic acid as a surfactant to prevent agglomeration of nanoparticles. The resulting mixture was dispersed in transformer oil UTR 40, with the role of carrier liquid. The magnetic, rheological, thermal and electrical characteristic properties of the obtained Fe_3O_4 transformer oil-based nanofluid were determined. A mathematical model and numerical simulation results are very useful for investigating the heat transfer performances of the magnetic nanofluid. Based on this study, it was tested the cooling performance of this magnetic nanofluid for two types of electrical power transformers as compared to classical methods. We also presented a microactuator based on the same magnetic nanofluid.

Keywords: colloidal magnetic Fe_3O_4 nanoparticles, X-ray diffraction, SEM, electrical transformer, magnetic nanofluid coolant, heat transfer, magnetic properties, microactuator, mass transfer, pulse width modulation, mathematical model, numerical simulation

1. Introduction

Magnetic nanofluids, known also as ferrofluids or magnetic liquids, are stable colloidal suspensions of superparamagnetic nanoparticles such as γ-Fe_2O_3, α-Fe_2O_3, Fe_3O_4, $CoFe_2O_4$,

$Mn_{1-x}Zn_xFe_2O_4$, in a carrier liquid (an organic solvent or water) [1–5]. In order to prevent the aggregation of magnetic nanoparticles and to attain a stable magnetic nanofluid, the nanoparticles are coated with a surfactant during the preparation process [1, 4–8]. The characterization of magnetic nanoparticles and of the obtained magnetic nanofluid is carried out by various techniques in order to determine their structural, magnetic, rheological and magneto-rheological properties [1–11]. The development of the synthesis methods leads to the possibility of tailoring the magnetic nanofluids and consequently, their physical properties, such that it fulfills the requirements of a certain application [1, 5, 12–24].

Besides the well-known application of ferrofluids in sealing and lubrication, recent developments envisaged their potential in fields like actuation [22, 23], medicine [1, 16], biotechnology [14, 24], as cooling fluids [25–29] or as liquid core in power transformers [30, 31]. The applications in medicine and biotechnology require from the magnetic nanofluid to be biocompatible. Therefore, water-based magnetic nanofluids are the candidates for magnetic hyperthermia for cancer treatment and targeted drug delivery, as well magnetic separation for purification of cells, proteins or else. The applications envisaged in electrical engineering require from the proposed magnetic nanofluid to also have good thermal and insulating properties. These conditions can be fulfilled by the transformer oil-based magnetic nanofluids.

This chapter is addressing the application of a Fe_3O_4 magnetic nanofluid based on transformer oil, as cooling and insulating fluid of a power transformer. Thus, the preparation procedure and the characterization of structural, magnetic, rheological, thermal and electrical properties are presented. The mathematical model applied to the problem is introduced, and the numerical results for two power transformers are discussed. The use of this Fe_3O_4 magnetic nanofluid in a micro-actuation application is also presented.

2. The synthesis and complex characterization of nanofluid with colloidal magnetic Fe_3O_4 nanoparticles

2.1. Nanofluid synthesis

The materials used for the synthesis of nanofluid with colloidal magnetic Fe_3O_4 nanoparticles we can mention: hexahydrated ferric chloride ($FeCl_3x6H_2O$) of 99% purity obtained from Merck Germany; ferrous sulfate sheptahydrate ($FeSO_4x7H_2O$) of 98% purity purchased from Chimopar, Romania; sodium hydroxide (NaOH) of 99% purity and oleic acid ($C_{18}H_{34}O_2$) of 99% purity provided by Riedel de Haen. Other chemicals were of analytic grade. The reagents were used without further purification. All solutions were prepared with deionized water.

The transformer oil-based ferrofluid with Fe_3O_4 nanoparticles was synthesized by chemical co-precipitation method [23], using $FeCl_3 \times 6H_2O$ and $FeSO_4 \times 7H_2O$ with a molar ratio of $Fe^{2+}/Fe^{3+} = 1:2$, dissolved in 300 ml of water and treated with NaOH 10%. The mixture was stirred at 80°C for 1 h. The resulting black color precipitate of Fe_3O_4 was washed with deionized water and magnetically decanted until a pH of 7 was reached. Then, 10 ml of HCl 0.1 N was added to the Fe_3O_4 precipitate for peptization. The mixture was washed and decanted again until pH 7, dried at 90°C and treated with acetone for water removal. A small part of the obtained powder was analyzed [X-ray diffraction, scanning electron microscopy (SEM) and elemental analysis EDX] for structural properties. The remaining part of the powder was treated with 2 ml of oleic acid as surfactant, and with 5 ml of toluene and heated at 90°C for toluene removal. Finally, the mixture was dispersed in 50 ml of UTR 40 transformer oil, under strong stirring for 20 h in order to obtain the magnetic nanofluid. **Figure 1** schematically presents the transformer oil-based magnetic nanofluid of Fe_3O_4 synthesis.

2.2. Nanofluid characterization

2.2.1. Powder characterization of Fe_3O_4 nanoparticles

The Fe_3O_4 powder was structurally characterized by X-ray diffraction using a diffractometer D8 ADVANCE type X Bruker-AXS in conditions: Cu-Kα radiation (γ = 1.5406 Å), 40 KV/40 mA,

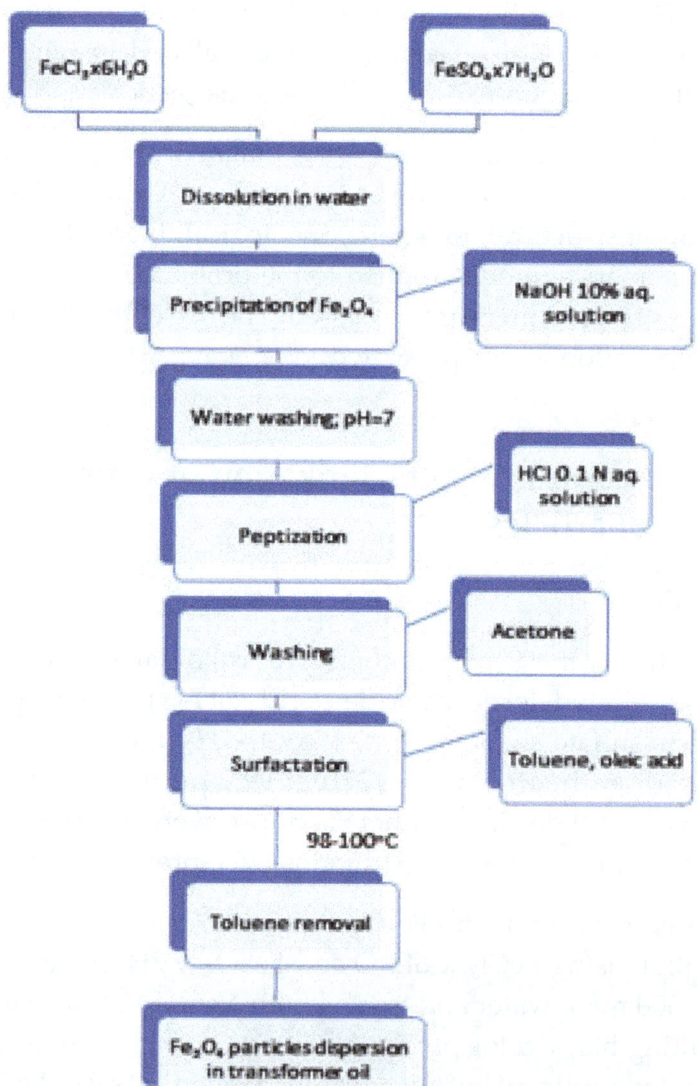

Figure 1. Transformer oil-based magnetic nanofluid of Fe_3O_4 synthesis.

filter k_β of Ni, in the 2θ range of 25–70°, using a step of 0.04° and measuring time on point of 1 s.

The XRD pattern of the powder is presented in **Figure 2** and shows the peaks corresponding to the Fe_3O_4 highlighted by "hkl" Miller indices (220), (311), (400), (422), (511) and (440), [2, 4], which denote a spinel structure with lattice parameter a = 0.83778 nm, in accord with the literature data (JCPDS file no. 19-629). According to the (311) peak, the medium size of the crystallites determined by Scherrer formula (1) is 14 nm.

$$D = \frac{0.9\lambda}{B^{*}\cos\theta}, \tag{1}$$

where D is the crystallites medium size, γ is the wavelength of this X-ray (γ = 0.154059 nm), B^* is the full width at half maximum (FWHM) and θ is the half diffraction angle of crystal orientation peak.

The morphology of the sample was studied by SEM using a Carl Zeiss SMT FESEM-FIB Auriger type scanner. The elemental analysis (energy-dispersive X-ray spectroscopy EDX) was performed with an energy dispersive probe of Inca Energy 250 type Oxford Instruments LTD England coupled to SEM.

The topology of the Fe_3O_4 powder analyzed by scanning electron microscopy evidenced two types of surface: smooth surface and rough surface (fracture). A crystalline structure of the material was found for the first type of surface, which is composed of crystallites having

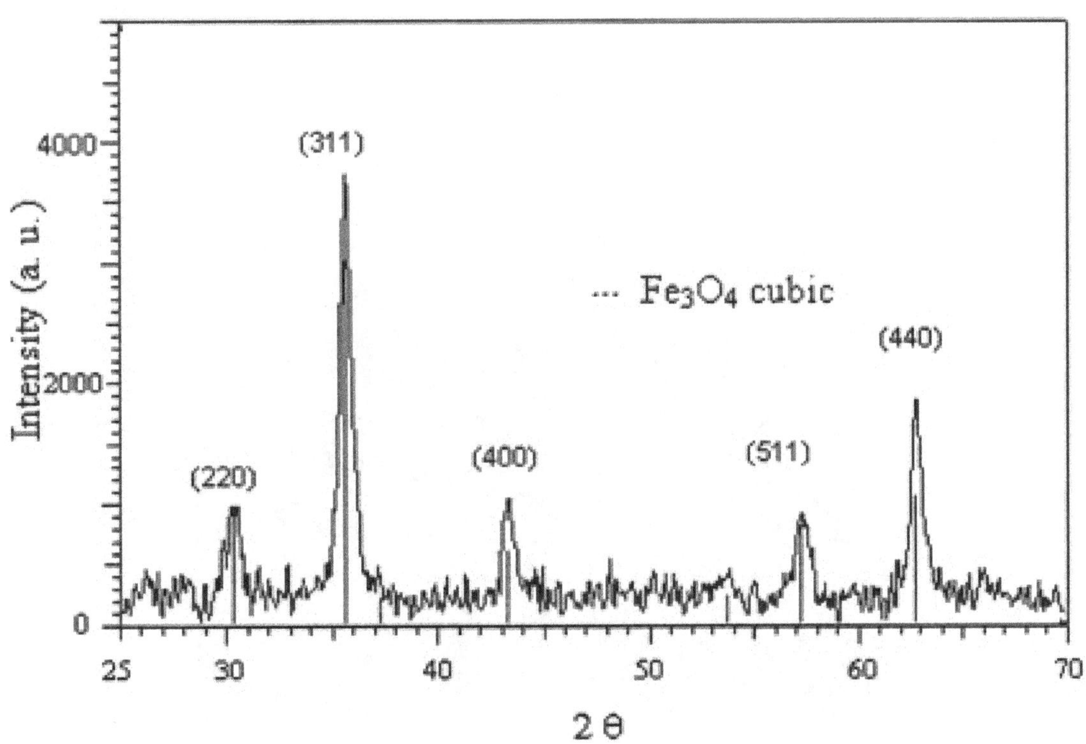

Figure 2. The Fe_3O_4 XRD pattern of the Fe_3O_4 nanoparticles.

average sizes between 10 and 30 nm (**Figure 3**) in good agreement with the diffraction analysis. For the second type of surface, a structure of acicular type agglomerates was found (**Figure 4**).

The elemental analysis confirms the presence of Fe_3O_4 (**Figure 5**) and shows that the resulting black powder contains 70.14%Fe, 24.96% O, 4.16% C and 0.74% Cl (**Table 1**). The presence of the Fe_3O_4 is exhibited by elemental Fe-peaks of about 6.45 and 0.75 keV. The high percentage of oxygen is related to its existence in the iron oxide.

The SEM image (**Figure 5a**) and the elemental energy dispersive X-ray analysis (**Figure 5b**) confirms the data determined by X-ray diffraction. The content of C and Cl represents the little impurities.

Figure 3. The SEM image for the first structure.

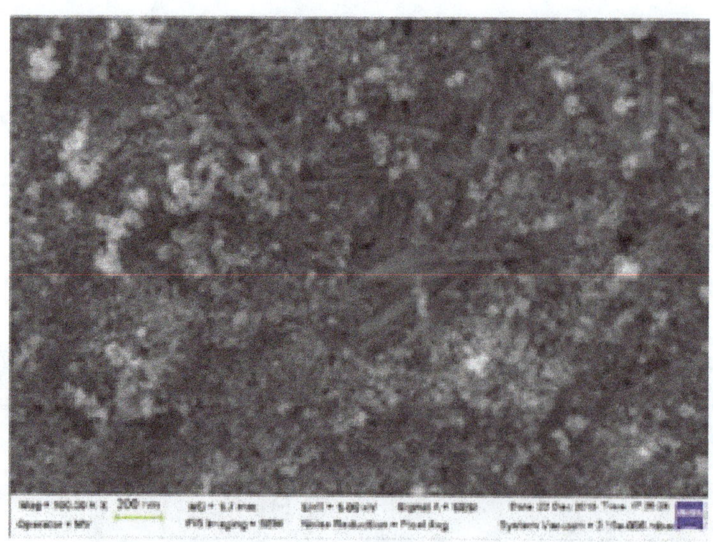

Figure 4. The SEM image for the second structure.

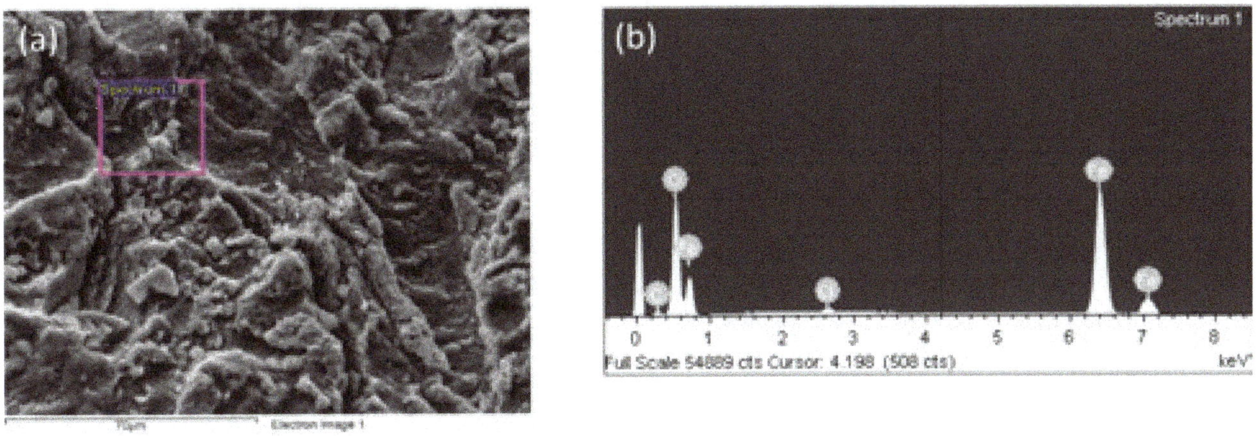

Figure 5. The elemental analysis for Fe_3O_4 powder: (a) the SEM image and (b) the elemental energy dispersive X-ray analysis.

Element	Weight (%)	Atomic (%)
C K	4.16	10.89
O K	24.96	49.01
Cl K	0.74	0.66
Fe K	70.14	39.45
Totals	100.00	100.00

Table 1. Analysis data for Fe_3O_4 nanoparticles, energy dispersive.

2.2.2. Characteristic properties of the nanofluid with colloidal magnetic Fe_3O_4 nanoparticles used as a cooling fluid for power transformers

2.2.2.1. Magnetic properties

The full magnetization curve and the hysteresis loop of the transformer oil-based magnetic nanofluid (MNF/UTR 40), with a solid volume fraction of the dispersed magnetite particles of 1.67%, were measured at room temperature (25°C), using a vibrating sample magnetometer—VSM 880—ADE Technologies USA, in the magnetic field range of 0–950 kA/m.

The magnetization M measured at the maximum value of the applied magnetic field, approx. 900 kA/m, is considered to be the nominal magnetization of the investigated sample. Also, the absence of hysteresis loop area indicates a specific behavior of soft magnetic material (**Figure 6**) with the magnetic characteristics shown in **Table 2**.

In **Table 2**, M_r represents the remnant magnetization and H_c is the coercive magnetic field; $\rho_{24^\circ C}$ is the density of the magnetic fluid at 24°C and $\varphi_{Fe_3O_4}$ is the solid volume fraction of the dispersed magnetite.

According to Shliomis [32], the magnetic behavior of a diluted magnetic nanofluid ($\varphi_{Fe_3O_4} < 5$) under the action of an external magnetic field is well described by the single-particle model,

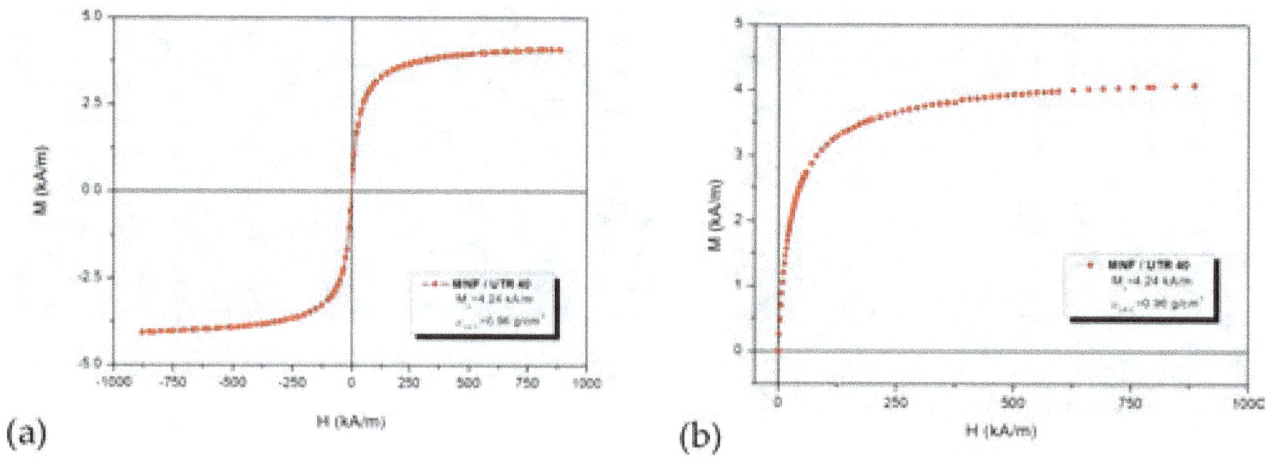

Figure 6. Hysteresis loop and full magnetization curve of the transformer oil-based magnetic fluid sample shows specific behavior of soft magnetic materials: (a) hysteresis loop for UTR 40-based MNF sample and (b) full magnetization curve for UTR 40-based MNF sample.

Sample	M (Gs)	M (kA/m)	M_r (kA/m)	H_c (kA/m)	$\rho_{24°C}$ (g/cm^3)	$\varphi_{Fe_3O_4}$ (%)
MNF/UTR 40	50	3.98	–	–	0.96	1.67

Table 2. Physical properties of MNF/UTR 40 sample.

which states that the energy of dipolar interactions is lower than the thermal energy. In this case, the equilibrium static magnetization is a superposition of Langevin functions,

$$M = \varphi_m M_d \left(\coth\xi - \frac{1}{\xi} \right) = \varphi_m M_d L(\xi), \tag{2}$$

with

$$\xi = \frac{\pi \mu_0 M_d D_m^3 H}{6 k_B T}, \tag{3}$$

representing the Langevin parameter. Herein $M_d = 480$ kA/m is the monodomenial magnetization of magnetite, $\mu_0 = 4\pi \times 10^{-7}$ H/m is the magnetic permeability of vacuum, D_m is the magnetic diameter of the dispersed magnetite particles, H is the applied magnetic field, $k_B = 1.38 \times 10^{-23}$ J/K is the Boltzmann constant and T is the absolute temperature.

In low magnetic fields (< 1 mT), with $\xi \to 0$, the Langevin function becomes $L(\xi) \to \frac{\xi}{3}$, that is a linear variation in sample magnetization with the applied field. Knowing that the initial magnetic susceptibility is $\chi_{iL} = M/H$, one can obtain:

$$\chi_{iL} = \frac{\pi \mu_0 \varphi_m M_d^2 D_m^3}{18 k_B T} \tag{4}$$

On the other hand, in the region of intense magnetic fields ($\xi >> 1$), the Langevin function is given by $L(\xi) \to 1 - \frac{1}{\xi}$ and static magnetization of the magnetic nanofluid is approximated by the following relationship [33],

$$M \cong \varphi_m M_d \left(1 - \frac{6k_B T}{\pi \mu_0 M_d D_m^3 H}\right), \tag{5}$$

where $\varphi_m = M_s/M_d$ is the magnetic volume fraction, with M_s representing the saturation magnetization of the magnetic nanofluid sample. The above relationship shows that the magnetization reaches saturation for very high values of the magnetic field $(H \to \infty)$.

In fact, in real ferrofluids, the dimensional polydispersity of the magnetic particles is a characteristic that cannot be neglected and, in the absence of inter-particle interactions, an accurate expression of magnetization is obtained [34],

$$M = M_s \int_0^{\infty} L(\xi) f(x) dx, \tag{6}$$

where f(x) is the log-normal distribution function (**Figure 7**)

$$f(x) = \frac{1}{xS\sqrt{2\pi}} \exp\left(-\frac{\ln^2 \frac{x}{D_0}}{2S^2}\right), \tag{7}$$

with x is the magnetic diameter of the magnetite particles; f(x)dx representing the probability that the magnetic diameter of the magnetic particles to be in the range of $(x;\ x + dx)$; D_0 is the dimensional distribution parameter, defined by the relationship $\ln(D_0) = \langle \ln(x) \rangle$; S is also a dimensional distribution parameter, representing the deviation of ln(x) value from $\ln(D_0)$ [35].

Another important aspect regarding the magnetization evaluation that should be considered is the dependence of the dispersed nanoparticles magnetic moments with their magnetic diameters [36, 37]. Here, $M_s(x) = nm(x)$ and the ferrofluid magnetization become

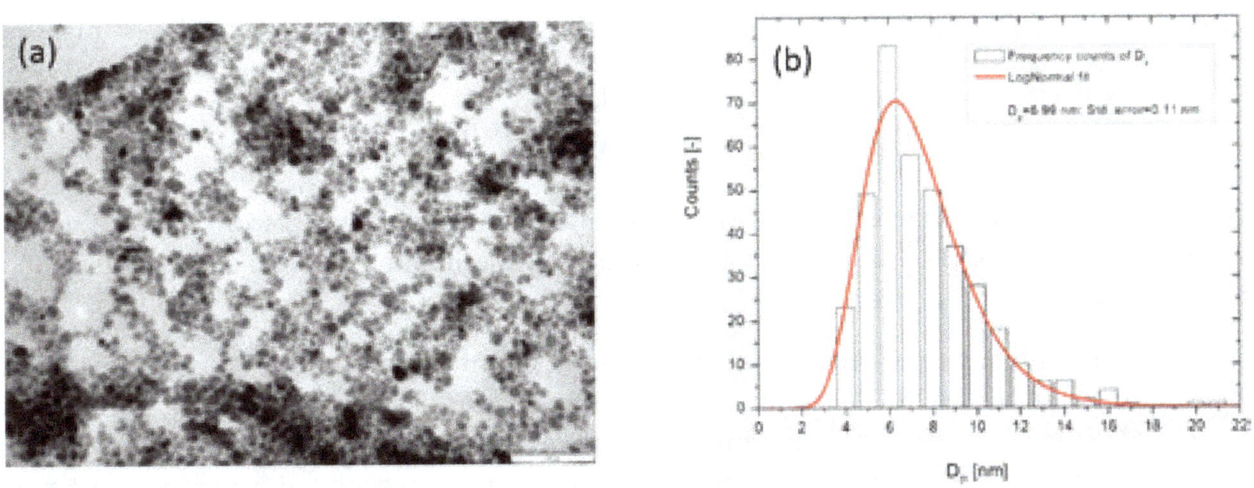

Figure 7. (a) TEM image of mono-layer covered magnetite nanoparticles with oleic acid and stably dispersed in hexane; (b) dimensional distribution of the magnetic particles physical diameters is well approximated by the log-normal distribution function.

$$M = n\int_0^\infty m(x)L(\xi)f(x)dx, \tag{8}$$

with n representing the density of the dispersed magnetite particles in magnetic fluid and m is the dipolar magnetic moment.

The linear dependence of the initial magnetic susceptibility versus magnetic particle concentration in low fields, rel. (4), and the asymptotic variation in magnetization in intense magnetic fields, rel. (5), form the basic instruments of magnetogranulometric analysis, in order to determine the mean magnetic diameter of the dispersed magnetic particles,

$$\langle D_m \rangle = D_0 \exp\left(\frac{S^2}{2}\right), \tag{9}$$

and standard deviation

$$\sigma = \sqrt{\langle D_m^2 \rangle - \langle D_m \rangle^2},$$

$$\sigma = D_0 \exp\left(\frac{S^2}{2}\right)\left[\exp S^2 - 1\right]^{1/2}, \tag{10}$$

where the dimensional distribution parameters are evaluated first,

$$S = \frac{1}{3}\sqrt{\ln\frac{3\chi_{iL}H_0}{M_s}}, \tag{11}$$

$$D_0^3 = \frac{6k_B T}{\pi\mu_0 M_d H_0}\sqrt{\frac{M_s}{3\chi_{iL}H_0}}, \tag{12}$$

and

$$n = \frac{\mu_0 M_s H_0}{k_B T}, \tag{13}$$

The expression of the dimensional distribution parameters was obtained from rel. (8), considering the above presented limit cases and evaluating the initial magnetic susceptibility and saturation magnetization, respectively.

In the linear region of small magnetic fields ($H < 1$ kA/m) from static magnetization curve, measured for the MNF/UTR 40 sample, the initial magnetic susceptibility has been evaluated. M_s and H_0 magnetic field were determined in the quasi-saturation region ($H > 700$ kA/m), where the contribution to the ferrofluid magnetization is given by the magnetic particle interactions with the applied magnetic field, inter-particle interactions being neglected. As a result, in the saturation region can be considered that the magnetic nanofluid magnetization varies linearly with $1/H$ according to Langevin's law, even for concentrated samples. In

Figure 8, which represents the final part (quasi-saturation) of the magnetization curve in $M = f(1/H)$ representation, M_s is obtained as the intersection with the ordinate axis. Also, the ratio of the magnetization curve slope and the corresponding absolute value of M_s is the value of H_0 field, where R^2 represents the measure of accuracy of the fit in linear regression (**Table 3**).

Magnetogranulometric analysis of the MNF/UTR 40 sample has revealed a mean magnetic diameter of the magnetite particles of $\langle D_m \rangle = 6.46$ nm and a standard deviation of $\sigma = 2.18$ nm. The log-normal distribution parameters (rel. (11)–(13)), evaluated directly from the magnetization curve, were obtained through non-linear regression, using rel. (8). Considering the thickness of non-magnetic layer at the surface of magnetic nanoparticles of $\delta_m = 0.83$ nm [33],

$$\langle D_p \rangle = \langle D_m \rangle + 2 \cdot \delta_m, \tag{14}$$

a value of 8.12 nm was determined for the mean physical diameter. Furthermore, a thickness of 1.9 nm of the oleic acid monolayer cover of magnetite particles δ_s [38] leads to a hydrodynamic diameter of the particles of $\langle D_h \rangle = 11.92$ nm,

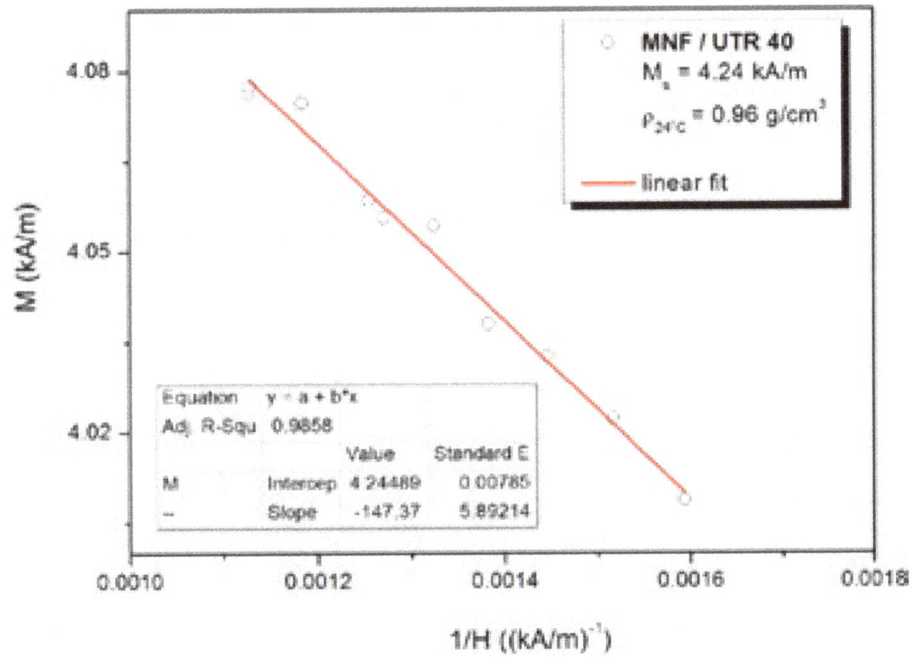

Figure 8. Obtaining M_s and H_0 field in the quasi-saturation region of the magnetization curve, in $M = f(1/H)$ representation, where the ferrofluid magnetization varies linearly with 1/H, according to Langevin's law.

Sample	χ_{iL} (-)	R^2 (χ_{iL}) (-)	M_s (kA/m)	M_s (Gs)	H_0 (kA/m)	R^2 (M_s) (-)
MNF/UTR 40	0.11	0.99467	4.24	53.23	34.72	0.98580

Table 3. Initial magnetic susceptibility and saturation magnetization of the transformer oil-based magnetic fluid sample.

$$\langle D_h \rangle = \langle D_p \rangle + 2 \cdot \delta_s \tag{15}$$

Main results obtained through magnetogranulometric analysis of the MNF/UTR 40 sample are summarized in **Table 4**.

2.2.2.2. *Rheological properties*

Rheological investigations, carried out with an Anton Paar Physica MCR 300 rheometer using a double-gap concentric cylinder geometry, consisted of measuring the dynamic viscosity curves of the samples in the absence of the magnetic field. The shear rate, $\dot{\gamma}$, varied from 1 s^{-1} to 1000 s^{-1}, at different values of working temperature t = (20; 40; 60; 80)°C. The viscosity curves (**Figure 9**) measured for both the transformer oil-based magnetic nanofluid (MNF/UTR 40) and the carrier liquid-transformer oil (UTR 40) showed that adding a small volume fraction of magnetic particles in the carrier ($\phi_{Fe_3O_4} \cong 1.67$ %) leads to a very mild increase in the dynamic viscosity and the Newtonian behavior of the samples is preserved throughout the investigated temperature range.

This behavior indicates the absence of the magnetic particle interactions, that is, a very good stability of the sample, mainly due to the efficient steric stabilization. An Arrhenius-type relationship describes the sample viscosity behavior with temperature,

Sample	D_0 (nm)	S (-)	n (× 10^{22} partic./m^3)	$\langle D_m \rangle$ (nm)	σ (nm)	$\langle D_p \rangle$ (nm)	$\langle D_h \rangle$ (nm)
MNF/UTR 40	6.12	0.33	4.50	6.46	2.18	8.12	11.92

Table 4. Main properties of MNF/UTR 40 sample obtained through magnetogranulometric analysis, including the log-normal distribution parameters values.

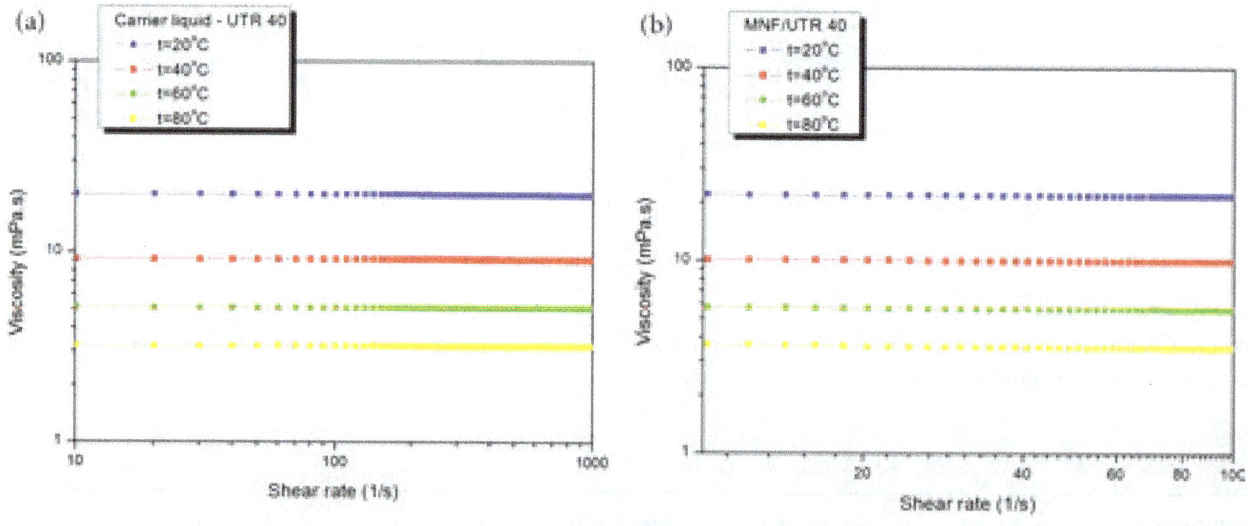

Figure 9. Viscosity curves showed that the Newtonian behavior of the carrier (UTR 40) (a) is preserved throughout the temperature range, also for the magnetic nanofluid (MNF/UTR 40), (b) that contains a small amount of the dispersed magnetic particles, indicating an efficient steric stabilization.

$$\eta = \eta_{ref}\exp\left[-\frac{Ea}{R}\left(\frac{1}{T}-\frac{1}{T_{ref}}\right)\right], \tag{16}$$

where $T_{ref} = 20°C$ is considered the reference temperature; $\eta_{ref}(Pa \cdot s)$ is the dynamic viscosity corresponding to the reference absolute temperature; E_a $(J \cdot mol^{-1})$ means the activation energy and $R = 8.31447\ J \cdot mol^{-1}K^{-1}$ is the ideal gas constant.

Considering a shear rate of 100 s^{-1}, rel. (16) was used to fit the dependence $\eta = \eta(T)$, with activation energy as fit parameter (**Figure 10**).

In **Table 5**, it can be observed that the value of the viscous flow activation energy does not change after dispersing a small amount of surfacted magnetite particles in the carrier liquid. It can be concluded that the adding of small volume fractions of surfacted magnetite particles in a carrier liquid($\varphi_{Fe_3O_4} < 5$), as transformer oil, has no significant influence on the rheological properties of the samples.

2.2.2.3. *Thermal properties*

The addition of metallic nanoparticles (magnetic or non-magnetic) or non-metallic (e.g., diamond nanoparticles) in transformer oils in order to improve their cooling performances is a solution that was demonstrated by several patents and associated research works (e.g. [39–45]). This paragraph analyzes the thermal properties of the magnetic nanofluid, which was tested for use as cooling and insulating medium in power transformers.

Determination of the effective thermal properties that characterize the cooling fluids in our study as well as their modeling using analytical formulae is representing a main problem of the topic in discussion. Irrespective of the considered approach, either theoretical or experimental, a representative element of the studied medium has to be chosen. It has to be

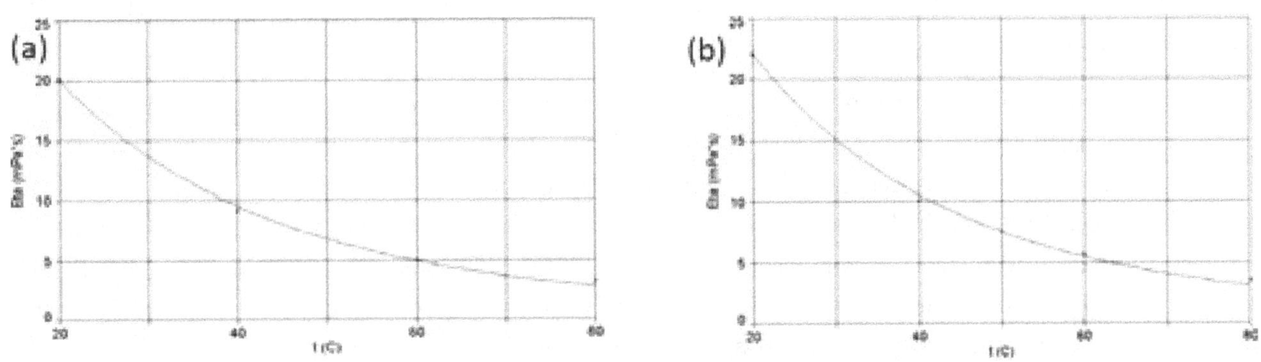

Figure 10. Arrhenius type dependence of the samples dynamic viscosity with temperature: (a) carrier liquid UTR 40 and (b) transformer oil-based magnetic nanofluid MNF/UTR 40.

Sample	$\phi_{Fe_3O_4}$ (%)	$E_a(J \cdot mol^{-1})$
MNF/UTR 40	1.67	28.27 × 10^3
UTR 40	–	28.38 × 10^3

Table 5. Viscous flow activation energy of the carrier liquid (UTR 40) and the magnetic nanofluid (MNF/UTR 40).

underlined that the determination of the properties of heterogeneous materials has to be made by complying with the request of representativeness of the studied volume, that is, the considered volume of the heterogeneous material must be sufficiently large in order to be statistically representative irrespective of the type of the carried out experiment [46, 47].

The main properties that are influencing the thermal behaviour of a material are the heat capacity, thermal conductivity, density, thermal expansion coefficient and thermal diffusivity (a property that depends on the first three). Various experimental studies determined that the effective properties of nanofluids (magnetic or non-magnetic) are dependent on the following characteristics of their components [25, 48–50]: the thermo-physical properties of the carrier fluid, nanoparticles and surfactant; nanoparticles volume fraction, size distribution, mean diameter and shape; temperature; magnetic field (in the case of magnetic nanoparticles).

A review of the reference literature regarding the main properties of the transformer oil-based fluids and their dependence with the temperature outlined the followings: specific heat is increasing as linear function with temperature; thermal conductivity is decreasing as a quasi-linear function with temperature especially for transformer oils; dynamic viscosity is decreasing with the increasing temperature; the dielectric constant has relatively low values and decreases with the increasing temperature [51–53].

2.2.2.3.1. *Thermal conductivity*

The thermal conductivity of magnetic nanofluids can be described as a function of several parameters, among the most important are the thermal conductivities of the carrier liquid and magnetic nanoparticles and their dependence on temperature and pressure, the volume fraction, the shape and the size distribution of the nanoparticles. The interfacial thermal resistance between the nanoparticles and the surrounding liquid is also considered. It has been proved numerically and experimentally that an applied magnetic field can affect the thermal conductivity of a magnetic nanofluid due to the consequent ordering of magnetic dipoles of the nanoparticles along the field lines [25, 54].

There are many Maxwell-type models developed for the thermal conductivity of mixtures (also named effective thermal conductivity—ETC), either solid matrix—solid filler or liquid carrier and dispersed nanoparticles that are based on the Maxwell model, which is recommended for low volume fraction of the filler/nanoparticles and considers that the nanoparticles are identical, spherical and non-interacting. The Holotescu-Stoian model, developed initially for the solid matrix—solid filler mixture and presented in Refs. [55, 56], introduced for the first time the filler particle size distribution in the Maxwell model. The expression for the effective thermal conductivity of the Holotescu-Stoian model, rel. (17), in the case of a magnetic nanofluid is [57],

$$\mathrm{k_e} = \mathrm{k_f}\,\frac{\mathrm{k_p} + 2\,\mathrm{k_f} + 2\,\varphi_\mathrm{e}\,(\mathrm{k_p} - \mathrm{k_f})}{\mathrm{k_p} + 2\,\mathrm{k_f} - \varphi_\mathrm{e}\,(\mathrm{k_p} - \mathrm{k_f})}, \tag{17}$$

where $\mathrm{k_e}$ is the effective thermal conductivity of the magnetic nanofluid, $\mathrm{k_p}$ is the thermal conductivity of the magnetic nanoparticles, $\mathrm{k_f}$ is the thermal conductivity of the carrier fluid, φ_e is the equivalent volume fraction, defined by

$$\varphi_e = \varphi_{Fe_3O_4} \frac{\langle (D_m + \delta)^3 \rangle^2}{\langle (D_m + \delta)^2 \rangle^3}, \tag{18}$$

with $\varphi_{Fe_3O_4}$ is the solid nanoparticles volume fraction (magnetite in this case),D_m is the magnetic diameter, $\delta = 2\delta_m$ is the double thickness of the non-magnetic layer, and

$$\langle u \rangle = \int_0^\infty u f(x) dx \tag{19}$$

with u being a magnetic diameter dependent function and f(x) is the log-normal distribution function.

The relationship between physical (geometrical) diameter D_p, magnetic diameter D_m and δ is given by

$$D_p = D_m + \delta. \tag{20}$$

This model, confirmed by the experimental data [57], was applied to determine the thermal conductivity of the analyzed magnetic nanofluid sample. The results (at room temperature) are given in **Table 6**, and we observe that the addition of magnetite nanoparticles alone is increasing the thermal conductivity of the magnetic nanofluid. The transformer oil thermal conductivity is decreasing with the increasing temperature, thus the cooling performance can be diminished at normal operating conditions in power transformers (and other electrical equipments). We can conclude that the addition of the magnetite nanoparticles can counteract this disadvantage. Moreover, during the operation of a power transformer, for instance, the magnetic field is acting on the magnetic nanofluid, influencing its physical properties, and generating the magneto-convection that can enhance the heat transfer, as shown in the next section.

2.2.2.3.2. *Specific heat*

The specific heat of the magnetic nanofluid, at constant pressure, was determined using the following mixture formula [58]

$$\rho_{MNF}(T) \cdot c_{p,MNF}(T) = (1-\varphi) \cdot \rho_{UTR}(T) \cdot c_{p,UTR}(T) + \varphi_m \cdot \rho_{NP}(T) \cdot c_{p,NP}(T), \tag{21}$$

where the transformer oil specific heat was determined by using

$$c_{p,UTR}(T) = 5.025 \cdot T + 1789.50 \tag{22}$$

and the specific heat of the magnetite by using [59],

Property/sample	UTR	NP	MNF_UTR	$\Delta X/X_{UTR}$
k_e (W/m K)	0.127	1.39	0.135	+6.29%
c_p (J/kg K)	1910.1	0.892	1867.8	−2.2%
β (1/K)	7.15×10^{-4}	1.2×10^{-4}	6.44×10^{-4}	−9.94%

Table 6. Thermal expansion coefficients.

$$c_{p,\,NP}(T) = 0.6334 + 0.871 \cdot 10^{-3}T, \tag{23}$$

where T (K) is the absolute temperature of the solid and ρ is the mass density of the material.

The numerical results obtained for the magnetic nanofluid specific heat indicate a slightly decrease compared to that of the carrier liquid. In what it concerns the effect of an applied magnetic field on the specific heat capacity of a magnetic nanofluid, for a certain range of temperature, the reference literature indicates the influence of the nanoparticles volume fraction and nanofluid composition (carrier liquid and nanoparticles). Also, the magnitude and the applied field orientation relative to the gravitational field (as the experiments were conducted in gravitational field) should be considered. Korolev et al. [60] analyzed the influence of an applied magnetic field (oriented perpendicularly on gravity) on a transformer oil-based magnetic nanofluid with magnetite nanoparticles, having a solid volume fraction of 7.4%, in the temperature range from 15 to 80°C. The experiment showed that, for a certain temperature, the specific heat capacity has a maximum in the investigated range of the applied magnetic field, which, according to the authors, indicates the presence of a magneto-caloric effect.

2.2.2.3.3. *Thermal expansion coefficient*

Similarly, to determine the thermal expansion coefficient of the magnetic nanofluid, a corresponding mixing formula was used [61],

$$\beta_{MNF} = \beta_{UTR}\left[\frac{1}{1+\frac{(1-\varphi_{Fe_3O_4})\rho_{UTR}}{\varphi_{Fe_3O_4}\rho_{NP}}}\frac{\beta_{NP}}{\beta_{UTR}} + \frac{1}{1+\frac{\varphi_{Fe_3O_4}}{1-\varphi_{Fe_3O_4}}\cdot\frac{\rho_{NP}}{\rho_{UTR}}}\right], \tag{24}$$

where β_{MNF} [1/K] is the thermal expansion coefficient of MNF, β_{UTR} is the thermal expansion coefficient of the transformer oil [62], β_{NP} is the thermal expansion coefficient of the magnetite nanoparticles [63], $\rho_{UTR} = 0.867\ g/cm^3$ (at 20°C) and $\rho_{MNF} = 0.96\ g/cm^3$ (at 24°C) are the measured densities of the transformer oil and the magnetic nanofluid, respectively.

The results of the calculations, summarized in **Table 6**, give the thermal properties at room temperature. We observe that the analyzed thermal properties have a diverging behavior. While the thermal expansion coefficient and specific heat are decreasing, the thermal conductivity is increasing. The last column is indicating the relative variation compared to the corresponding property of the carrier liquid (transformer oil).

2.2.2.3.4. *Evaluation of the heat transfer potential of the magnetic nanofluid*

To determine the potential performance of the magnetic nanofluid for heat transfer, we considered the figure-of-merit (FOM), as defined for natural convection [64],

$$FOM_{NC} = \left[\beta\,\rho^2\,c_p\,k^{\frac{1}{n-1}}/\eta\right]^n, \tag{25}$$

where $n = 0.25$ for laminar flow and $n = 0.33$ for turbulent flow.

Property/sample	UTR	MNF_UTR	$\Delta X/X_{UTR}$
FOM, n = 0.25	18.61	19.57	+5.15%
FOM, n = 0.33	91.85	96.25	+4.8%

Table 7. FOM results, obtained for the carrier liquid and for the magnetic nanofluid.

Property/sample	UTR	NP	MNF_UTR	$\Delta\varepsilon$ / ε_{UTR}
ε (F/m)	$2.2 \times \varepsilon_0$	$81 \times \varepsilon_0$	$2.26 \times \varepsilon_0$	2.85%

Table 8. Effective electric permittivity of the Fe_3O_4 transformer oil-based magnetic nanofluid.

The results obtained for the FOM of the carrier liquid and the magnetic nanofluid, using the properties determined above, are presented in **Table 7**. The comparison of the relative increase of FOM indicates that the addition of nanoparticles is advantageous for both laminar and turbulent flow.

2.2.2.4. *Electric permittivity*

As underlined above, the use of a magnetic nanofluid in electrical engineering applications imposes restrictions regarding its insulating properties. Transformer oils are known to be electrical insulators so they are an appropriate carrier liquid for a magnetic nanofluid used in such applications. If the magnetic nanoparticles volume fraction is kept in certain limits, the magnetic nanofluid preserves its insulating properties within the required limits, too [49, 50].

We estimated the effective electric permittivity of the magnetic nanofluid ε_{MNF}, using the Maxwell-Garnett equation for mixtures:

$$\varepsilon_{MNF} = \varepsilon_{UTR} + 3\varphi_{Fe_3O_4}\varepsilon_{UTR}\frac{\varepsilon_{NP}-\varepsilon_{UTR}}{\varepsilon_{NP} + 2\varepsilon_{UTR}-\varphi_{Fe_3O_4}(\varepsilon_{NP}-\varepsilon_{UTR})}, \quad (26)$$

with ε_{UTR} is the electric permittivity of the transformer oil and ε_{NP} is the electric permittivity of the magnetite nanoparticles, $\varphi_{Fe_3O_4}$ being the volume fraction of the magnetite nanoparticles.

The results are presented in **Table 8**, along with the relative difference between the values corresponding to the transformer oil and magnetic nanofluid, ε_0 being the free space permittivity, approximate equal to 8.85×10^{-12} F/m.

We observed that for the current volume fraction of magnetic nanoparticles, the insulating properties of the magnetic nanofluid remain very close to those of the carrier liquid (UTR 40). In what concerns the effect of working temperatures in the power transformer, experimental studies showed that electrical permittivity decreases with increasing temperature in the case of transformer oils [51].

3. Heat transfer and electromagnetic field by numerical simulation for the electrical transformer cooled by a specific nanofluid

A colloidal Fe_3O_4 specific nanofluid dispersed in oil transformer UTR 40, named MNF/UTR 40, is utilized in a number of technologically relevant applications where external magnetic fields are used to adjust their flow. A mathematical model and numerical simulation results are useful for investigating the heat transfer properties of the magnetic nanofluid. Based on this study, it was built and tested the experimental model: low power, medium voltage, single-phased transformer type TMOf-24-5″ (**Figure 18**) and low power, medium voltage, single-phased transformer type TMOf2-36kV-40 kVA (**Figure 19**). First of all, it was used for the transformer the transformer oil UTR 40 as cooling and insulating liquid. After that, this oil was drained and the experimental model was filled with magnetic nanofluid based on transformer oil MNF/UTR 40. The MNF/UTR 40 specific nanofluid has been shown to provide both thermal and dielectric benefits to transformers, and can be utilized to improve cooling by enhancing fluid circulation within transformer windings, to increase transformer capacity to withstand lightning impulses, while also minimizing the effect of moisture on typical insulating fluids. Magnetic nanofluid flow may be influenced by external magnetic fields, and the retention force of a magnetic nanofluid can be adjusted by changing either the magnetization of the fluid or the magnetic field in the region. Opposite to usual magnetic fluids, the magnetizable nanofluids destined to heat transfer should have a low concentration of magnetic nanoparticles in order to make them competitive with the non-magnetic fluids.

3.1. Mathematical model

Several simplifying assumptions aimed and keeping the physical system within approachable software and hardware limits are requested and 2D models are best candidates, providing numerical simulation relevant results, of satisfactory accuracy. Following this path, we consider a 2D, Cartesian cross-sectional model, as shown in **Figure 11**.

The heat transfer and transport processes under the influence of the magnetic field for two prototypes electric transformer: low power mono-phased transformer (24 kVA), at medium voltage (20/√3//0,4/√3kV), TMOf-24-5 and low power mono-phased transformer (40 kVA), at medium voltage (30/√3//0,4/√3kV), prototype TMOf 2-36kV-40 kVA, are described by the following set of coupled partial differential equations [44]:

- electromagnetic field—quasi-steady, harmonic diffusion,

$$(j\omega\sigma - \omega^2\varepsilon_0\varepsilon_r)\mathbf{A} + \nabla\times(\mu_0^{-1}\mu_r^{-1}\nabla\times\mathbf{A}) - \sigma u\times(\nabla\times\mathbf{A}) = J^e, \tag{27}$$

- momentum balance (Navier-Stokes),

$$\rho\left[\frac{\partial\mathbf{u}}{\partial t} + (\mathbf{u}\cdot\nabla)\mathbf{u}\right] = -\nabla p + \underbrace{\mu_0(\mathbf{M}\cdot\nabla)\mathbf{H}}_{f_{mg}} + \mu_f\nabla^2\mathbf{u} + \mathbf{f}_T, \tag{28}$$

- mass conservation (incompressible flow),

$$\nabla \mathbf{u} = 0, \tag{29}$$

- heat transfer (energy equation),

$$\rho c_p \left[\frac{\partial T}{\partial t} + (\mathbf{u} \cdot \nabla) T\right] = k\nabla^2 T + \mathbf{E} \cdot \mathbf{J}^e_\varphi. \tag{30}$$

Here: u is the velocity; p is the pressure; **T** is the absolute temperature; **A** is the magnetic vector field; **M** is the magnetization (in the magnetic nanofluid); **H** is the magnetic field strength; E is the electric field strength; $\mathbf{J}^e_\varphi$ is the angular component of the (external) current density (in the coil); $\mathbf{f}_T$ is the buoyancy body force term; $\mathbf{f}_{mg}$ is the magnetic body force term; σ is the electrical conductivity; k is the thermal conductivity; μ_0 is the magnetic permeability of vacuum ($\mu_{air} = \mu_{r\ air} \cdot \mu_0 \cong \mu_0$; $\mu_{r\ air} \cong 1$); μ_f is the kinematic viscosity; ρ is the mass density; c_p is the specific heat. Heat transfer occurs by conduction in the solid regions of the system and by convection and diffusion in the fluid region. The temperature variation in the fluid region is responsible for a gravitational flow (Boussinesq approximation), whose structure depends on the cell geometric aspect ratio and thermal conditions. All subdomains have linear physical

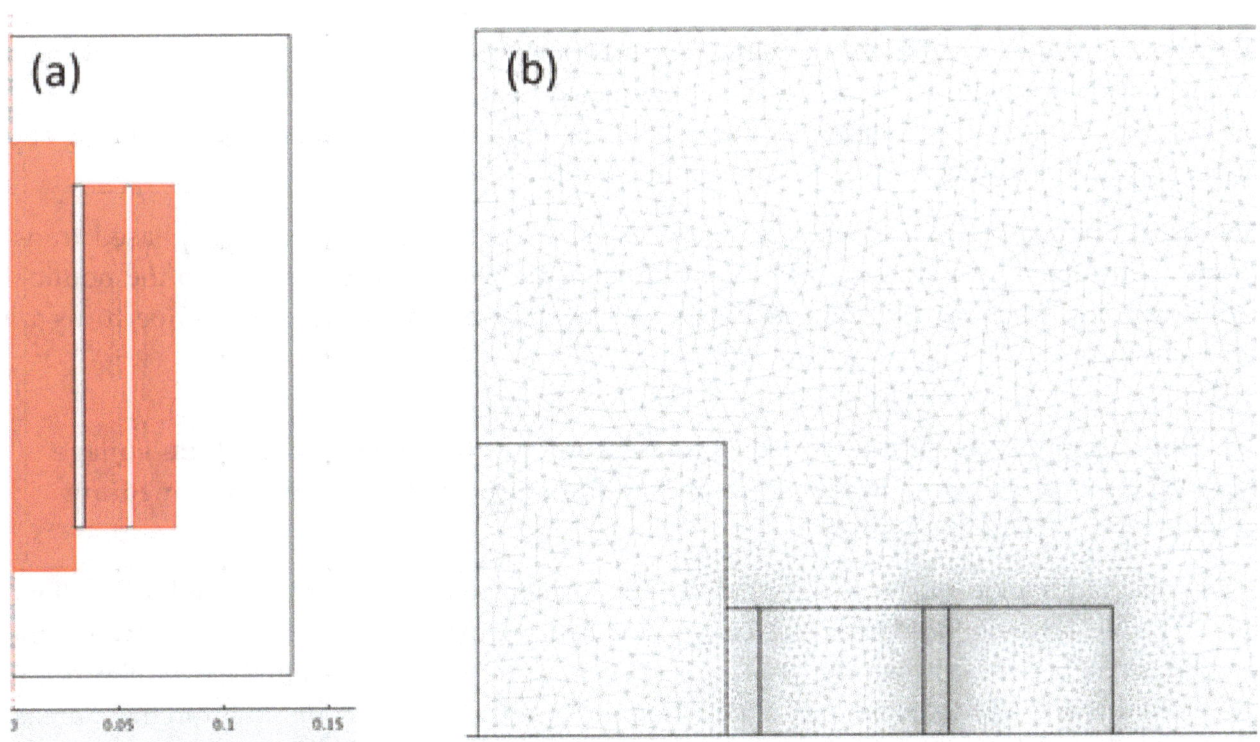

Figure 11. The 2D simplified model and the FEM mesh made of triangular elements: (a) computational domain and (b) detailed view—windings, iron core, case.

properties, except for the magnetic nanofluid (based on transformer oil), whose magnetization characteristic is nonlinear.

The magnetic field magnetizes the magnetic nanofluid, and the corresponding body force term adds to the gravitational, thermal flow of the magnetic nanofluid coolant. The flow structure is the result of the two competing forces: thermal force and magnetic force. In this study, we deal with a super-paramagnetic magnetic nanofluid where the influence of the coercive magnetic field intensity, H_c or the remnant induction, B_r is discarded. The constitutive law for the magnetic field of the magnetic fluid is then

$$\mathbf{B} = \mu_0(\mathbf{H} + \mathbf{M}). \tag{31}$$

The magnetic field produced by the electrical current in the coil magnetizes the fluid and is responsible for the magnetic body forces that influence the thermally induced flow. The magnetization of the magnetic fluid is approximated here by the analytic formula

$$\mathrm{M}_{\mathrm{x,y}} = \mathrm{a} \cdot \arctan(\mathrm{b} \cdot \mathrm{H}_{\mathrm{x,y}}), \tag{32}$$

with $a = 10^4 A/m$ and $b = 3\times10^{-5} m/A$ are empiric constants. The magnetic body forces are then obtained out of the magnetic energy, by taking its derivatives with respect to the coordinates

$$\mathrm{f}_{\mathrm{mg}} = \mu_0(\mathbf{M} \cdot \nabla)\mathbf{H}. \tag{33}$$

The strategy that we used in the numerical simulation consists of solving for the magnetic field first, and then using the active power thus obtained as heat source in the heat transfer and flow parts of the problem. The obtained solutions are steady state [65] for heat transfer and flow and quasi-steady (harmonic) for the electromagnetic field.

The main dimensions (windings, iron core, case sizes) are those of the single-phased transformer considered in our study. The amperturns of the windings correspond to the nominal working point, when the iron core exhibits lower levels of magnetization—the amperturns are compensated.

3.2. Simulation results for mono-phased transformer of low power and medium voltage type TMOf-24-5 compared to the mono-phased transformer of low power and medium voltage type TMOf2-36kV-40 kVA

Numerical simulation of the mono-phased transformer of low power and medium voltage type TMOf-24-5 evidenced that the convective heat transfer in the channels between the windings and between the windings and core (3–5 mm) is less important in the overall process, therefore it was discarded, and only conduction heat transfer was accounted for in these areas. **Figure 12** shows simulation results for a non-magnetic, regular cooling fluid—the temperature field (surface color map, **Figure 12a–d**), the thermal flow (streamlines and velocity vectors, **Figure 12a–c**), magnetic flux density (**Figure 12d**, surface color map, iso-lines of magnetic

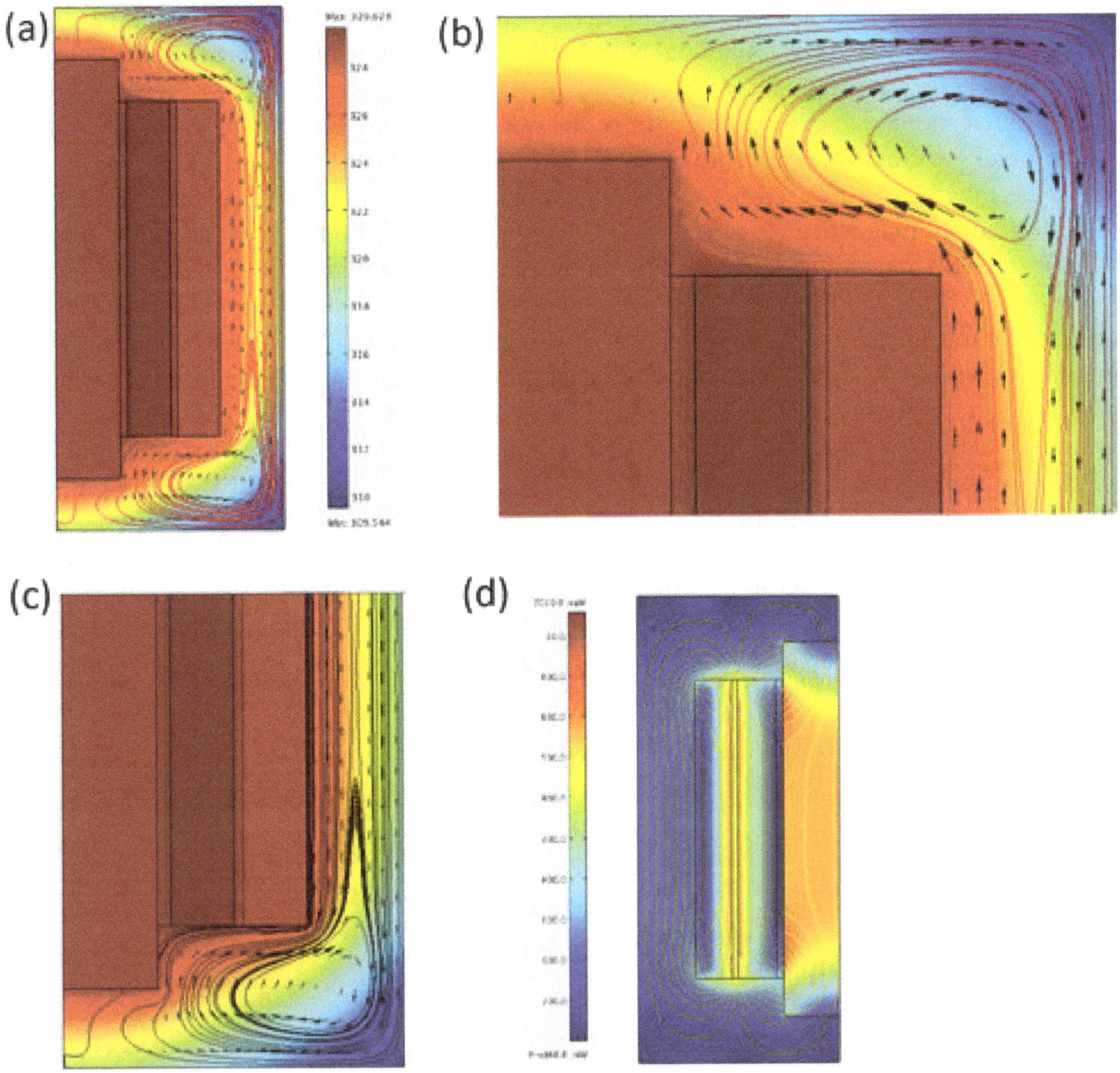

Figure 12. Magnetic field, temperature and flow fields for mono-phased transformer of low power and medium voltage, type TMOf-24-5, (a) Heat transfer and thermal flow—upper part, (b) detail—temperature, buoyancy flow, (c) heat transfer and thermal flow—bottom part and (d) magnetic field.

vector field). Apparently, the thermal flow exhibits two minor recirculation areas—in the upper and lower part of the windings, by the top and bottom covers—trapped within a larger recirculation cell that develops by the lateral wall. For the heat transfer part of the problem, we assumed a convection (Robin) type boundary condition (the ambient temperature was assumed to be $T_{amb} = 300$ K, with a heat transfer coefficient of $h = 2$ W/m^{2}K, that is, moderate natural convection.

The iron core, although less magnetized in this particular regime (compensated primary and secondary amperturns), plays a crucial role in the heat transfer problem.

When a colloidal Fe_3O_4 specific nanofluid MNF/UTR 40 is utilized as a coolant, magnetic body forces add to the thermal, gravitational body forces. **Figure 13** displays the magnetic field and forces (magnetic and thermal). Apparently, the magnetic forces contribute differently to the overall convection flow: in the upper part of the cell they add to the buoyancy forces, whereas in the lower part they are opposite. However not unexpected, Eq. rel. (33), another important finding is the effect that the en-parts of the windings and the iron core have: these are regions of high gradient magnetic field strength, and it is here that the body magnetization forces are significant. The orientation of the magnetic forces versus the thermal forces is an important factor in providing an optimal design. We observe that the thermal gravitationally driven forces and the magnetic forces act concurrently in this plane, their combined effect being greater at the left and right end regions. Similarly, the heat transfer direction is from the hotter regions (core and windings) to the case, but with enhanced convection and increased heat removal efficiency. Comparing the two cooling options, that is, specific nanofluid MNF/UTR 40 (**Figure 14b**) versus regular coolant UTR 40 (**Figure 14a**) apparently the MNF/UTR 40 may do better in cooling the transformer. This essentially means a lower hot spot temperature by approximately 10° in this model and a more uniform temperature distribution. These results suggest that the vertical design of the low-power mono-phased transformer (40 kVA), at medium voltage (30/√3//0,4/√3kV) prototype TMOf 2-36kV-40kVA (**Figure 15**) may be advisable [66, 67]. This prototype TMOf 2-36kV-40kVA realized in compliance with the constructive

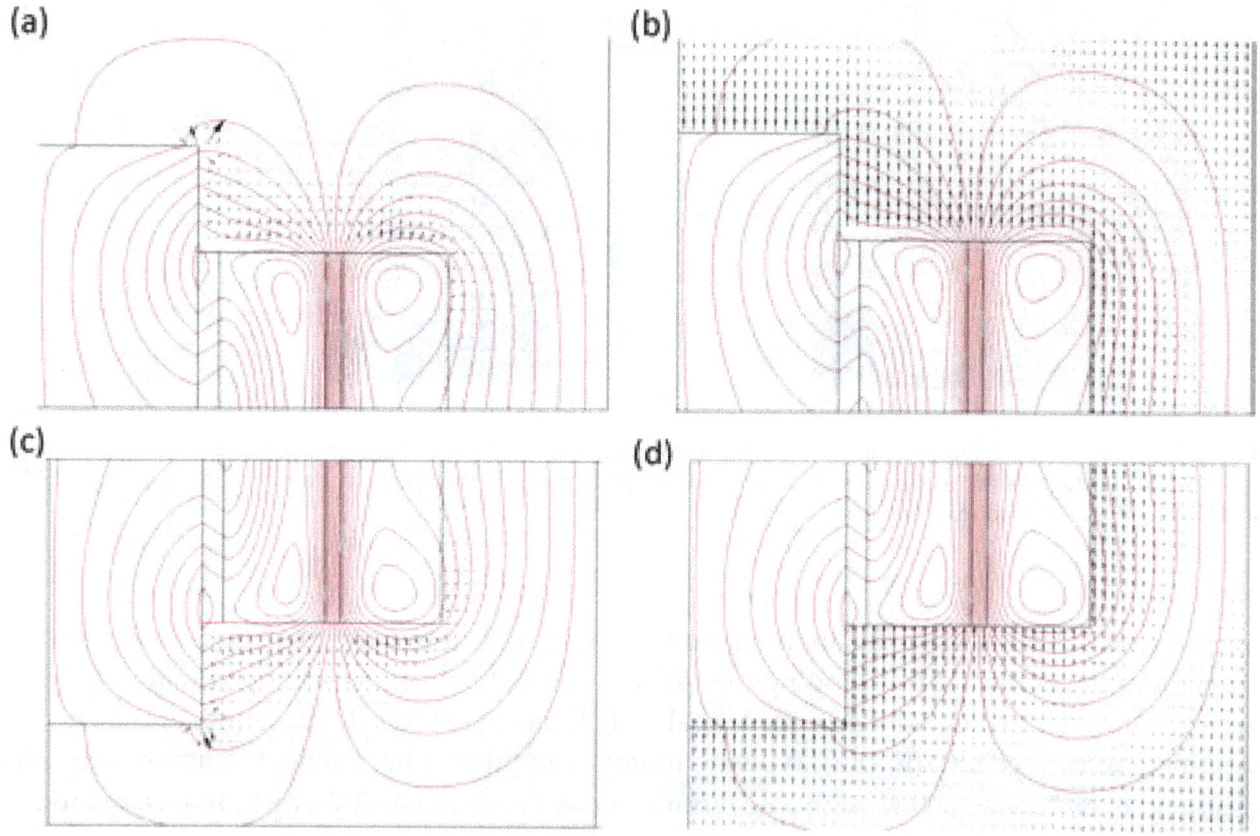

Figure 13. Magnetic field and body forces when the coolant is a magnetic nanofluid, MNF/UTR 40: (a) detail (top)—magnetic body forces, (b) detail (top)—buoyancy forces, (c) detail (bottom)—magnetic body forces, (d) detail (bottom)—buoyancy forces.

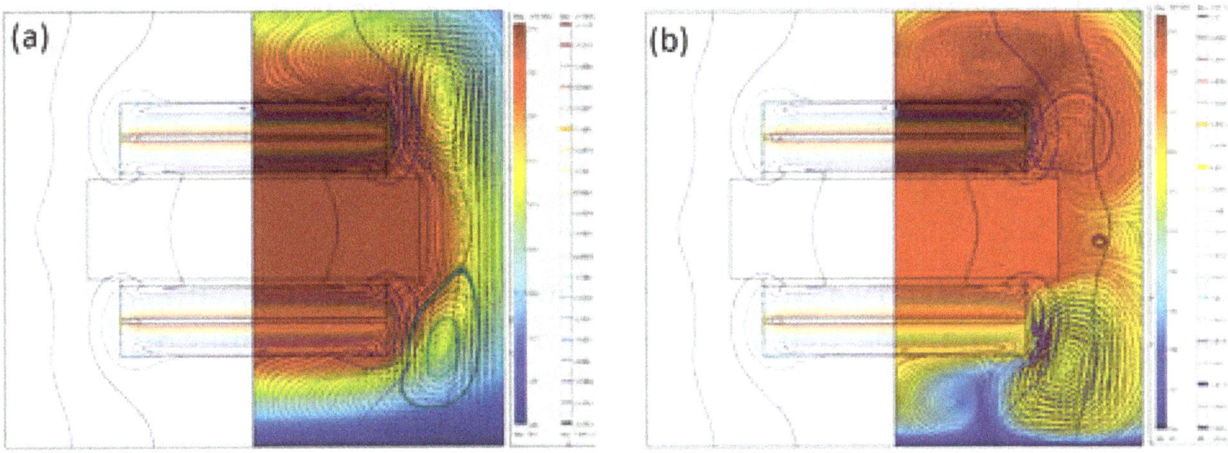

Figure 14. 2D, axial model - magnetic flux density field, temperature distribution, and flow field for the horizontal design TMOf-24-5 type transformer: (a) the coolant is regular UTR 40 transformer oil and (b) the coolant is a magnetic nanofluid, MNF/UTR 40.

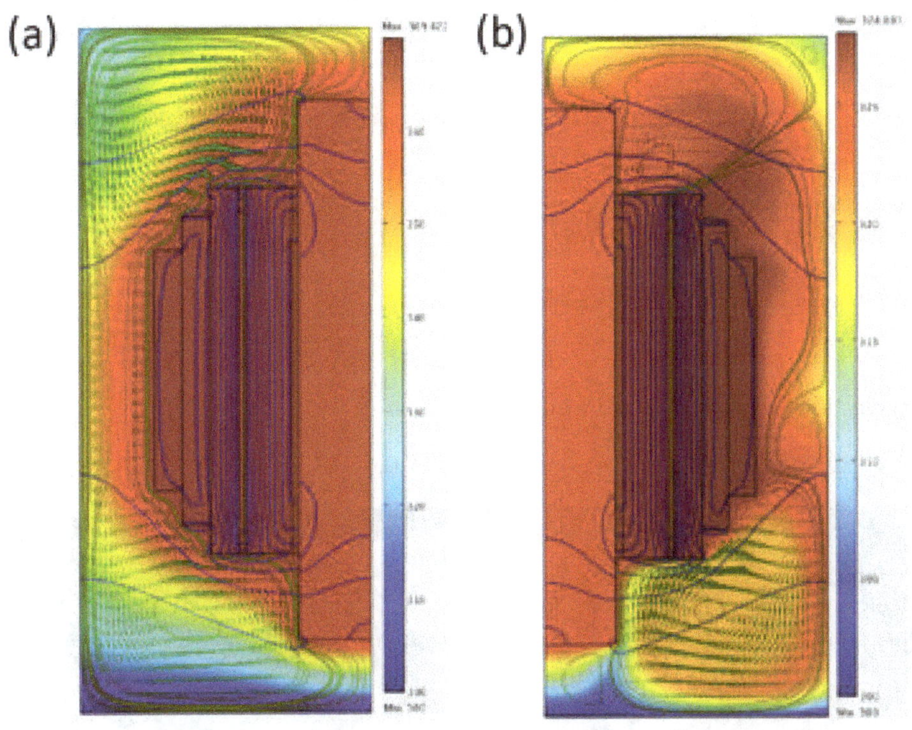

Figure 15. 2D, axial model - magnetic flux density field, temperature distribution, and flow field for the vertical design TMOf 2-36kV-40kVA type transformer: (a) the coolant is regular UTR 40 transformer oil and (b) the coolant is a magnetic nanofluid, MNF/UTR 40.

solutions with characteristics specific for the aimed purpose presents the following advantages: reduced weight and dimensions in comparison with the current power transformers that have the same rated voltage and rated power, following the intensification of the cooling effect in the presence of the specific nanofluid MNF/UTR 40 (**Figure 15b**). The number of convection zones is greater when the coolant is magnetic nanofluid MNF/UTR 40

(**Figure 15b**) as compared to the regular coolant, that is, the UTR 40 transformer oil (**Figure 15a**). Also, because of the execution form of the magnetic circuit and of the metallic construction (tank-bottom-lid), the construction of the power transformer in the aggregate is realized with a smaller consumption of the main materials: copper, magnetic steel sheet and the specific nanofluid MNF/UTR 40 are included.

4. Designing the electrical transformer cooled by nanofluid with colloidal magnetic Fe_3O_4 nanoparticles dispersed in UTR 40 transformer oil

The low-power mono-phased transformer (40 kVA), at medium voltage (30/√3//0,4/√3kV), prototype vertical design TMOf 2-36kV-40kVA type transformer has the active part (**Figures 16** and **17**), magnetic iron core with the high voltage (HV) and low voltage (LV) windings fixed in

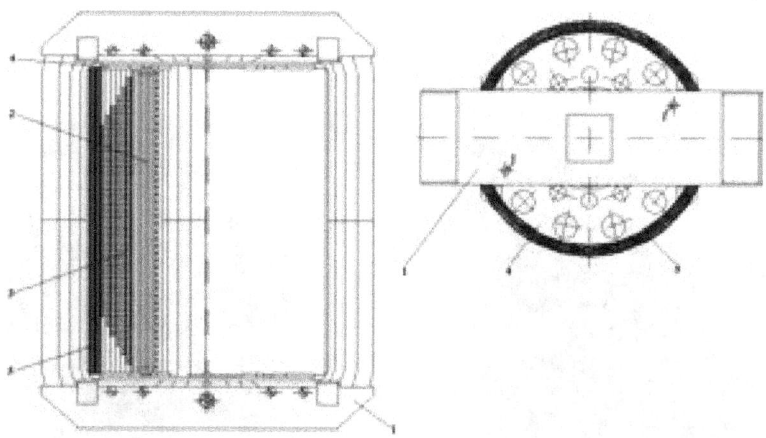

Figure 16. The aggregate active parts, core and windings [44].

Figure 17. Magnetic cores and the core rolling device [44].

a finned metallic tank constituted from two parts air-proof assembled through a soldering that is soft and capable of elastic deformation for the taking-over of the variation with temperature of the cooling liquid volume (**Figure 19**). The magnetic circuit is coating type, constituted of two identical cores of rectangular shape (flat-core), back to back disposal (**Figure 17**). The aggregate active parts (**Figure 16**) are the magnetic core with the high voltage and low voltage windings, fixed on a metallic lid with their axes in a vertical position, the most convenient situation for the heat transfer enhancement by the nanoparticles in the presence of the electromagnetic field. With a view to performing comparative tests related to the use of magnetic nanofluid MNF/UTR 40 as cooling and insulating fluid transformers and regular UTR 40 transformer oil cooling, the mono-phased transformer of low power and medium voltage, the horizontal design TMOf-24-5 type transformer (**Figure 18**) and mono-phased transformer of low power and medium voltage the vertical design TMOf 2-36kV-40kVA type transformer (**Figure 19**) has been achieved [44]. The numerical simulation results show that the direction of the magnetizing force in comparison with the gravitational thermal force is an important

Figure 18. Mono-phased transformer of low power and medium voltage, the horizontal design TMOf-24-5 type transformer.

Figure 19. Mono-phased transformer of low power and medium voltage, the vertical design TMOf 2-36kV-40kVA type transformer [44].

element in assuring of an optimal heat transfer. Both numerical simulations as well as laboratory measurements [65–67] confirm the following aspects: about the usage of a magnetic nanofluid MNF/UTR 40 as cooling and insulating fluid for transformers, this provides for magnetization body forces that add to the thermal, gravitational forces. In the vertical layout of the transformer, these forces act concurrently with the thermal flow, and the overall effect is the enhancement of the heat transferred from the aggregate active parts (core and windings) to the ambient.

In both cases, first of all, the regular UTR 40 transformer oil as cooling and insulating fluid was used for the transformers. After that, this oil was drained and the transformers were filled with magnetic nanofluid MNF/UTR 40 as cooling and insulating fluid.

Figures 20 and **21** show the temperature on the surface of the ribbed tank when magnetic nanofluid MNF/UTR 40 is used for the vertical design TMOf 2-36kV-40kVA type transformer after 1 h of operation. Monitoring of the temperature was achieved with the thermographic camera, FLUKE Ti 20. The temperature does not exceed the value of 54°C. Magnetic nanofluid

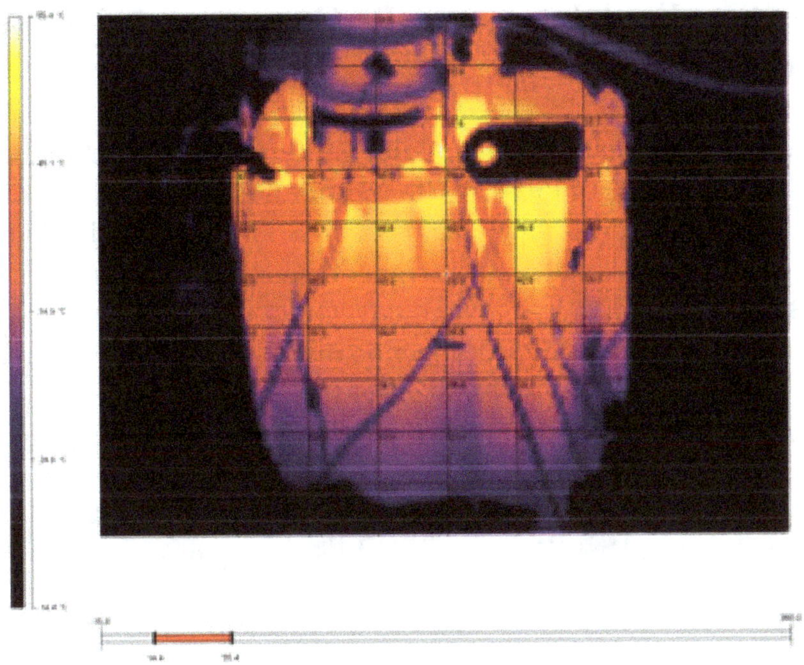

Figure 20. Temperature distribution by thermographic imaging-the tank- for mono-phased transformer of low power and medium voltage, type TMOf2-36kV-40 kVA, after 1 h of operation.

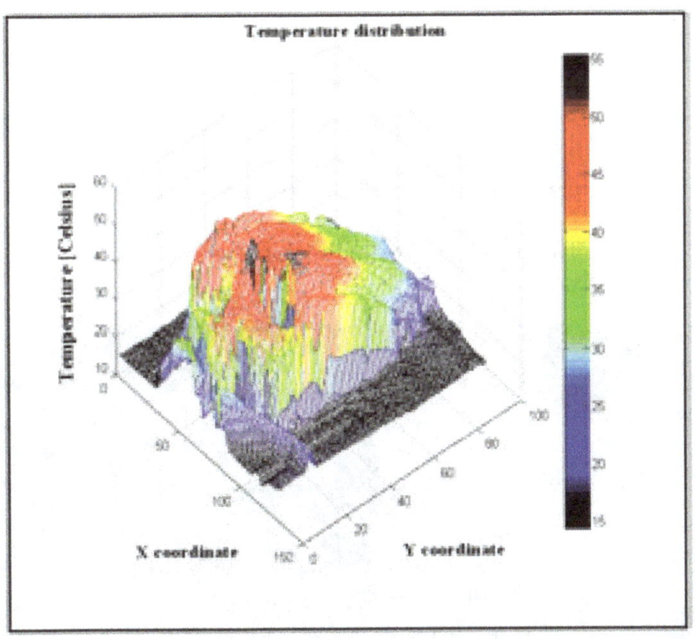

Figure 21. 3D characteristic of the temperature depending on the *X* and *Y* coordinates, associated with the thermographic image in **Figure 20**.

MNF/UTR 40 provides also the increase of the transformer's capacity to sustain over-voltages and withstand better to degradation in time due to humidity, as compared to the regular UTR 40 transformer oil coolant. Thus, transformers with reduced dimensions and higher efficiency with loading capacity and extended life duration may be designed.

5. Nanofluid with colloidal magnetic Fe_3O_4 nanoparticles used in microactuation process

Based on the afore described magnetic nanofluid, we can make a microactuator whose operation complies with the principle of Pulse Width Modulation (PWM) [23, 68]. The output PWM rectangular pulse form for a pulse duty factor of 14% is presented in **Figure 22**. The PWM generator discharges on the microactuator magnetic nanofluid impedances, two windings L_1 and L_2 (**Figure 23**). The electromagnetic force developed by the microactuator and implicitly the movement of the magnetic nanofluid depends mainly on the windings excitation voltage pulse duty factor *Ku* %. Passing an electric current by the microactuator windings results a magnetic field. The net effect of this magnetic field is a mass transfer of the magnetic nanofluid.

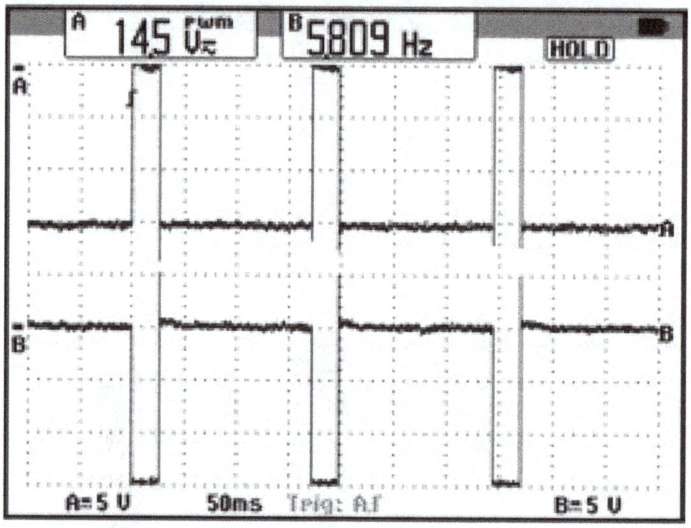

Figure 22. The output PWM rectangular pulse form, for a pulse duty factor of 14%.

Figure 23. The microactuator with magnetic nanofluid during testing [68].

Also, the maximum amplitude of the excitation voltage is constant, U_{max} = 15 V. The RMS value of the current that goes through the coils of the actuator, for a fixed frequency of the PWM voltage, depends mainly on the pulse duty factor. Two windings, L_1 and L_2, are excited with a rectangular waveform, counter phase, in compliance with **Figure 22**.

Acknowledgements

The authors express special thanks to Prof. Alexandru Mihail Morega, corresponding Member of the Romanian Academy, for valuable results concerning the numerical simulations. The research was performed with the support of UEFISCDI, PNCDI II Programme—Joint Applied Research Projects, Romania, Contract 63/2014, Environment energy harvesting hybrid system by photovoltaic and piezoelectric conversion, DC/DC transformation with MEMS integration and adaptive storage. Also, the numerical simulations were conducted in the Laboratory for Multiphysics Modeling, at UPB, with the support of the PNCDI-II "Parteneriate" Contract 21-043/2008.

Nomenclature

D (nm)	crystallites medium size
λ (nm)	wavelength of the Cu-Kα radiation (0.154059 nm)
B^* (rad)	full width at half maximum (FWHM)
ϑ (°)	half diffraction angle of crystal orientation peak (Bragg angle)
M (A/m)	magnetization
M_r (A/m)	remnant magnetization
H_c (A/m)	coercive magnetic field
ρ (kg/m^3)	mass density
$\varphi_{Fe_3O_4}$ (-)	solid volume fraction of the dispersed magnetite
M_d (A/m)	monodomenial magnetization of magnetite (480 kA/m)
ξ (-)	Langevin parameter
$L(\xi)=\coth\xi-1/\xi$ (-)	Langevin function
μ_0 (H/m)	magnetic permeability of vacuum ($4\pi \times 10^{-7}$ H/m)
μ (H/m)	magnetic permeability of the medium
μ_r (-)	relative magnetic permeability (μ/μ_0)
D_m (nm)	magnetic diameter
H (A/m)	magnetic field strength
k_B (J/K)	Boltzmann constant (1.38×10^{-23} J/K)
T (K)	absolute temperature
χ_{iL} (-)	initial magnetic susceptibility
M_S (A/m)	saturation magnetization
$\varphi_m = M_S/M_d$ (-)	magnetic volume fraction
f(x) (-)	log-normal distribution function

x (nm)	magnetic diameter of the magnetite particles – argument of the log-normal distribution function
f(x)dx (-)	probability that the magnetic diameter of the magnetic particles to be in the range of (x; x + dx)
D_0 (nm)	dimensional distribution parameter, defined by $\ln(D_0) = \langle\ln(x)\rangle$
S (-)	dimensional distribution parameter, representing the deviation of ln(x) value from $\ln(D_0)$
n (partic./m^3)	density of the dispersed magnetite particles
m (A m^2)	dipolar magnetic moment
σ (nm)	standard deviation
H_0 (A/m)	magnetic field strength, representing the ration of the magnetization curve slope and saturation magnetization, in M = f(1/H) representation
R^2 (-)	measure of accuracy of the fit in linear regression
δ_m (nm)	thickness of non-magnetic layer
D_p (nm)	physical (geometrical) diameter
δ_s (nm)	surfactant cover thickness
D_h (nm)	hydrodynamic diameter
$\dot{\gamma}$ (s^{-1})	shear rate
η (Pa s)	dynamic viscosity
η_{ref} (Pa s)	dynamic viscosity corresponding to the reference temperature, T_{ref}
T_{ref} (K)	reference temperature
E_a (J/mol)	activation energy
R (J/mol K)	ideal gas constant (8.31447 J/mol K)
k (W/m K)	thermal conductivity
k_e (W/m K)	effective thermal conductivity of the magnetic nanofluid
k_p (W/m K)	thermal conductivity of the magnetite nanoparticles
k_f (W/m K)	thermal conductivity of the carrier fluid
φ_e (-)	equivalent volume fraction
$\delta = 2\delta_m$ (nm)	double thickness of non-magnetic layer
u (-)	magnetic diameter dependent function
c_p (J/kg K)	specific heat at constant pressure
$c_{p,MNF}$ (J/kg K)	specific heat of the magnetic nanofluid
$c_{p,UTR}$ (J/kg K)	specific heat of the transformer oil
$c_{p,NP}$ (J/kg K)	specific heat of the magnetite nanoparticles
ρ_{MNF} (kg/m^3)	mass density of the magnetic fluid (0.960 g/cm^3, at 24°C)
ρ_{UTR} (kg/m^3)	mass density of the transformer oil (0.867 g/cm^3, at 20°C)
ρ_{NP} (kg/m^3)	mass density of the magnetite nanoparticles
β (1/K)	thermal expansion coefficient
β_{MNF} (1/K)	thermal expansion coefficient of the magnetic nanofluid
β_{UTR} (1/K)	thermal expansion coefficient of the transformer oil
β_{NP} (1/K)	thermal expansion coefficient of the magnetite nanoparticles
FOM_{NC} (-)	figure-of-merit determined for natural convection

n (-)	constant whose value depends on flow type; $n = 0.25$ for laminar flow and $n = 0.33$ for turbulent flow
ε (F/m)	electric permittivity
ε_0 (F/m)	electric permittivity of free space ($8.8541878176 \times 10^{-12}$ F/m)
ε_r (-)	relative electric permittivity ($\varepsilon_r = \varepsilon/\varepsilon_0$)
ε_{MNF} (F/m)	effective electric permittivity of the magnetic fluid
ε_{UTR} (F/m)	electric permittivity of the transformer oil
ε_{NP} (F/m)	electric permittivity of the magnetite nanoparticles
u (m/s)	velocity
ω (rad/s)	angular speed
p (N/m^2)	pressure
A (-)	magnetic vector field
E (V/m)	electric field strength
J_{φ}^{e} (A/m^2)	angular component of the current density
f_T (N)	buoyancy body force term
f_{mg} (N)	magnetic body force term
σ (S/m)	electrical conductivity
μ_f (m^2/s)	kinematics viscosity
B (T)	magnetic field induction
B_r (T)	remnant induction
a (A/m)	empiric constant (10^4 A/m)
b (m/A)	empiric constant (3×10^{-5} m/A)
T_{amb} (K)	ambient temperature (300 K)
h (W/m^2 K)	heat transfer coefficient
U_{max} (V)	excitation voltage at maximum amplitude

Author details

Lucian Pîslaru-Dănescu[1*], Gabriela Telipan[1], Floriana D. Stoian[2], Sorin Holotescu[2] and Oana Maria Marinică[2]

*Address all correspondence to: lucian.pislaru@icpe-ca.ro

1 National Institute for Electrical Engineering ICPE-CA, Bucharest, Romania

2 Politehnica University of Timisoara, Timisoara, Romania

References

[1] Vékás L, Bica D, Avdeev M. Magnetic nanoparticles and concentrated magnetic nanofluids: Synthesis, properties and some applications. China Particuology. 2007;**5**(1–2):43–49.

[2] Ervithayasuporn V, Kawakawi Y. Synthesis and characterization of core-shell type Fe_3O_4 nanoparticles in poly (organosilsesquixane). Journal of Colloid and Interface Science. 2009;**332**:389–393. DOI: 10.1016/j.jcis.2008.12.061

[3] Gubin S P. Magnetic Nanoparticles. Wiley - VCH Verlag Gmb H&Co. Weinheim, Germany; 2009. 466 p. DOI: 10.1002/9783527627561

[4] Vékás L, Bica D, Marinica O. Magnetic nanofluid stabilized with various chain length surfactants. Romanian Reports in Physics. 2006;**58**(3):257–267.

[5] Arulmurugan R, Vaidyanathan G, Sendhilnathan S, Jeyadevan B. Mn-Zn ferrite nanoparticles for ferrofluid preparation: Study on thermal-magnetic properties. Journal of Magnetism and Magnetic Materials. 2006;**298**(2):83–94.

[6] Hayashi K, Sakamoto W, Yogo T. Magnetic and rheological properties of monodisperse Fe_3O_4 nanoparticles/organic hybrid. Journal of Magnetism and Magnetic Materials. 2009;**321**:450–457. DOI: 10.1016/j.jmmm.2008.10.004

[7] Gupta A K, Gupta M. Synthesis and surface engineering of iron oxide nanoparticles. Biomaterials. 2005;**26**:3995–4021. DOI: 10.1016/j.biomaterials.2004.10.012

[8] Ozkaya T, Toprak M S, Baykal A, Kavas H, Koseoglu Y, Aktas B. Synthesis of Fe_3O_4 nanoparticles at 100°C and its magnetic characterization. Journal of Alloys and Compounds. 2009;**472**:18–23. DOI: 10.1016/j.jallcom.2008.04.101

[9] Paddar P, Fried T, Markovich G, Sharoni A, Katz D, Wizansky T, Mills O. Manifestation of the Verwey transition in the tunneling spectra of magnetite nanocrystals. Europhysics Letters. 2003;**64**(1):98–103. DOI: 10.1209/epl/i2003-00141-0

[10] Jordan K, Cazacu A, Manai G, Ceballos S F, Murphy S, Shvets I V. Scanning tunneling spectroscopy study of the electronic structure of Fe_3O_4 surfaces. Physical Review. 2006;**74**:085416. DOI: 10.1103/PhysRevB.74.085416

[11] Chastellain M, Petri A, Hofmann H. Particles size investigations of a multistep synthesis of PVA coated superparamagnetic nanoparticles. Journal of Colloid and Interface Science. 2004;**278**:353–360. DOI: 10.1016/j.jcis.2004.06.025

[12] Marques R F C, Garcia C, Lecante P, Ribeiro S J L, Noe L, Silva N J O, Amaral V S, Millan A, Verelst M. Electro-precipitation of Fe_3O_4 nanoparticles in ethanol. Journal of Magnetism and Magnetic Materials. 2008;**320**:2311–2315. DOI: 10.1016/j.jmmm.2008.04.165

[13] Abu Mukh-Qasem R, Gedanken A. Sonochemical synthesis of stable hydrosol of Fe_3O_4 nanoparticles. Journal of Colloid and Interface Science. 2005;**284**:489–494. DOI: 10.1016/j.jcis.2004.10.073

[14] Cai Y, Shen Y, Xie A, Li S, Wang X. Green synthesis of soya bean sprouts-mediated superparamagnetic Fe_3O_4 nanoparticles. Journal of Magnetism and Magnetic Materials. 2010;**322**:2938–2943. DOI: 10.1016/j.jmmm.2010.05.009

[15] Cabrera L, Gutierrez S, Menendez N, Morales M P, Herrasti P. Magnetite nanoparticles: Electrochemical synthesis and characterization. Electrochimica Acta. 2008;**53**:3436–3441. DOI: 10.1016/j.electacta.2007.12.006

[16] Neuberger T, Schöpf B, Hofmann H, Hofmann M, von Rechenberg B. Superparamagnetic nanoparticles for biomedical applications: Possibilities and limitations of a new drug delivery system. Journal of Magnetism and Magnetic Materials. 2005;**293**:483–496. DOI: 10.1016/j.jmmm.2005.01.064

[17] Maity D, Agrawal D C. Synthesis of iron oxide nanoparticles under oxidizing environment and their stabilization in aqueous and non-aqueous media. Journal of Magnetism and Magnetic Materials. 2007;**308**:46–55. DOI: 10.1016/j.jmmm.2006.05.001

[18] Park G S, Park S H. Design of magnetic fluid linear pump. IEEE Transactions on Magnetics. 1999;**35**:4058–4060. DOI: 10.1109/20.800754

[19] Park G S, Seo K. New design of the magnetic fluid linear pump to reduce the discontinuities of the pumping forces. IEEE Transactions on Magnetics. 2004;**40**(2):916–919. DOI: 10.1109/TMAG.2004.824718

[20] Nethe A, Scholz T, Stahlmann H-D, Filtz M. Ferrofluids in electric motors: A numerical process model. IEEE Transactions on Magnetics. 2002;**38**(2):1177–1180. DOI: 10.1109/20.996301

[21] Choi I H, Hong S P, Chung W E, Kim Y J, Lee M H, Kim J Y. Concentrated anisotropic magnetization for high sensitivity of optical pickup actuator. IEEE Transactions on Magnetics. 1999;**35**(3):1861–1864. DOI: 10.1109/20.767396

[22] Ahn J J, Oh J G, Choi B. A study of the novel type of ferrofluid magnetic pipette. In Nanotech – Technical Proceedings of the 2003 Nanotechnology Conference and Trade Show; 23–27 February 2003; San Francisco. Cambridge, MA: Nano Science and Technology Institute; 2003. p. 226–229.

[23] Pîslaru-Dănescu L, Morega A M, Telipan G, Stoica V. Nanoparticles of ferrofluid Fe_3O_4 synthetised by coprecipitation method used in microactuation process. Optoelectronics and Advanced Materials—Rapid Communications. 2010;**4**(8):1182–1186. WOS:000281734800030

[24] Turcu R, Craciunescu I, Nan A. Magnetic Microgels: Synthesis and Characterization. In Nirschl H, Keller K, Eds., Upscaling of Bio-Nano-Processes: Selective Bioseparation by Magnetic Particles. Springer-Verlag, Berlin Heidelberg, Germany. 2014; p. 57–76.

[25] Nkurikiyimfura I, Wang Y, Pan Z. Heat transfer enhancement by magnetic nanofluids: A review. Renewable and Sustainable Energy Reviews. 2013;**21**:548–561. DOI: 10.1016/j.rser.2012.12.039

[26] Sheikholeslami M, Ganji D D. Ferrohydrodynamic and magnetohydrodynamic effects on ferrofluid flow and convective heat transfer. Energy. 2014;**75**:400–410. DOI: 10.1016/j.energy.2014.07.089

[27] Sheikholeslami M, Rashidi M M, Ganji D D. Numerical investigation of magnetic nanofluid forced convective heat transfer in existence of variable magnetic field using two phase model. Journal of Molecular Liquids. 2015;**212**:117–126. DOI: 10.1016/j.molliq.2015.07.077

[28] Patel J, Parekh K, Upadhyay R V. Prevention of hot spot temperature in a distribution transformer using magnetic fluid as coolant. International Journal of Thermal Sciences. 2016;**103**:35–40.

[29] Sheikholeslami M, Hayat T, Alsaedi A. MHD free convection of Al2O3–water nanofluid considering thermal radiation: A numerical study. International Journal of Heat and Mass Transfer. 2016;**96**:513–524. DOI: 10.1016/j.ijheatmasstransfer.2016.01.059

[30] Tsai T-H, Kuo L-S, Chen P H, Lee D-S, Yang C-T. Applications of ferro-nanofluid on a micro-transformer. Sensors. 2010;**10**:8161–8172.

[31] Dumitru J B, Morega A M, Pîslaru-Dănescu L, Morega M. A parametric study of lumped circuit parameters of a miniature planar spiral transformer. In Proceedings of the Advanced Topics in Electrical Engineering (ATEE), 8th International Symposium; 23–25 May 2013; Bucharest, Romania. IEEEXplore Digital Library, US; 2013. DOI: 10.1109/ATEE.2013.6563477

[32] Shliomis M I. Magnetic fluids. Soviet Physics Uspekhi. 1974;**17**(3):153–169. DOI: 10.1070/PU1974v017n02ABEH004332

[33] Rosensweig R E. Ferrohydrodynamics. Journal of Applied Mathematics and Mechanics. 1986;**67**(6):279. DOI: 10.1002/zamm.19870670626

[34] Chantrell R W, Popplewell J, Charles S W. Measurements particle size distribution parameters in ferrofluids. IEEE Transactions on Magnetics. 1978;**14**:975–977. DOI: 10.1109/TMAG.1978.1059918

[35] Cogoni G, Grosso M, Baratti R, Romagnoli J A. Time evolution of the PSD in crystallization operations: An analytical solution based on Ornstein-Uhlenbeck process. AIChE Journal. 2012;**58**(12):3731–3739. DOI: 10.1002/aic.13760

[36] Susan-Resiga D, Socoliuc V, Boros T, Borbáth T, Marinica O, Han A, Vekas L. The influence of particle clustering on the rheological properties of highly concentrated magnetic nanofluids. Journal of Colloid and Interface Science. 2012;**373**:110–115. DOI: 10.1016/j.jcis.2011.10.060

[37] Ivanov A O. Magnetostatic properties of moderately concentrated ferrocolloids. Magnitnaia Gidrodinamika. 1992;**28**(4):39–46.

[38] Pshenichnikov A F, Mekhonoshin W V, Lebedev A V. Magneto-granulometric analysis of concentrated ferrocolloids. Journal of Magnetism and Magnetic Materials. 1996;**161**:94–102. DOI: 10.1016/S0304-8853(96)00067-4

[39] Raj K, Moskowitz R. Ferrofluid-cooled electromagnetic device and improved cooling method. US Patent 5462685. 1995.

[40] Kader T, Bernstein S, Crowe C. Magnetic fluid cooler transformer. US Patent 5898353 A. 1997.

[41] Pîslaru-Dănescu L, Macamete E, Telipan G, Pintea J, Nouras F, Paduraru N, Vekas L, Stoian F, Borbath I, Borbath T, Morega A, Morega M. Transformer with magnetic nanofluid as cooling agent. RO Patent 126613 B1. 2011.

[42] Segal V, Rabinovich A, Nattrass D, Raj K, Nunes A. Experimental study of magnetic colloidal fluids behavior in power transformers. Journal of Magnetism and Magnetic Materials. 2000;**215–216**:513–515. DOI: 10.1016/S0304-8853(00)00205-5

[43] Morega A M, Morega M, Pîslaru-Dănescu L, Stoica V, Nouraş F, Stoian F D. A novel, ferrofluid-cooled transformer. Electromagnetic field and heat transfer by numerical simulation. In Proceedings of the 12th International Conference on Optimization of Electrical and Electronic Equipment (OPTIM 2010); 20–22 May 2010; Brasov, Romania. IEEEXplore Digital Library, US; 2010. p. 140–146. DOI: 10.1109/OPTIM.2010.5510425

[44] Pîslaru-Dănescu L, Morega A M, Morega M, Stoica V, Marinica O, Nouraş F, Paduraru N, Borbath I, Borbath T. Prototyping a ferrofluid-cooled transformer. IEEE Transactions on Industry Applications. 2012;**49**(3):1289–1298. DOI: 10.1109/TIA.2013.2252872

[45] Stoian F D, Holotescu S, Taculescu A, Marinica O, Susan-Resiga D, Timko M, Kopcansky P, Rajnak M. Characteristic properties of a magnetic nanofluid used as cooling and insulating medium in a power transformer. In Proceedings of the 8th International Symposium on Advanced Topics in Electrical Engineering (ATEE); 23–25 May 2013; Bucharest, Romania. IEEEXplore Digital Library, US; 2013. p. 1–4. DOI: 10.1109/ATEE.2013.6563463

[46] Ostoja-Starzewski M. Microstructural randomness versus representative volume element in thermomechanics. ASME Journal of Applied Mechanics. 2002;**69**:25–35. DOI: 10.1115/1.1410366

[47] Torquato S. Random Heterogeneous Materials. Springer-Verlag: New York; 2002. 703 p. DOI: 10.1007/978-1-4757-6355-3

[48] Yu W, France D M, Routbort J L, Choi S U S. Review and comparison of nanofluid thermal conductivity and heat transfer enhancements. Heat Transfer Engineering. 2008;**29**(5):432–460. DOI: 10.1080/01457630701850851

[49] Timko M, Kopcansky P, Molcan M, Tomco L, Marton K, Molokac S, Rybar P, Stoian F D, Holotescu S, Taculescu A. Magnetodielectric properties of transformer oil based magnetic fluids. Acta Physica Polonica A. 2012;**121**(5–6):1253–1256. DOI: 10.12693/APhysPolA.121.1253

[50] Timko M, Marton K, Tomco L, Kiraly J, Molcan M, Rajnak M, Kopcansky P, Cimbala R, Stoian F D, Holotescu S, Taculescu A. Magneto-dielectric properties of transformer oil based magnetic fluids in the frequency range up to 2 MHz. Magnetohydrodynamics. 2012;**48**(2):427–434.

[51] Perrier C, Beroual A, Bessede J-L. Improvement of power transformers by using mixtures of mineral oil with synthetic esters. IEEE Transaction on Dielectrics and Electrical Insulation. 2006;**13**(5):556–564. DOI: 10.1109/TDEI.2006.1657968

[52] MIDEL. Specific heat – MIDEL 7131 – Specific Heat vs. Temperature [Internet]. Available from: http://www.midel.com/productsmidel/midel-7131/thermal-properties/specific-heat [Accessed: 2016-07-13].

[53] Wang H J, Ma S J, Yu H M, Zhang Q, Guo C M, Wang P. Thermal conductivity of transformer oil from 253 K to 363 K. Petroleum Science and Technology. 2014;**32**(17):2143–2150. DOI: 10.1080/10916466.2012.757235

[54] Philip J, Shima P D, Raj B. Enhancement of thermal conductivity in magnetite based nanofluid due to chainlike structures. Applied Physics Letters. 2007;**91**(20):203108. DOI: 10.1063/1.28126699

[55] Holotescu S, Stoian F D. A study of the influence of filler dimensional distribution on the polymer composite effective thermal conductivity. In Proceedings of the 29th Japan Symposium of Thermophysical Properties; 8–10 October 2008; Tokyo. Japan Society of Thermophysical Properties, Tokyo, Japan; 2008. p. 318–320

[56] Holotescu S, Stoian F D. Evaluation of the effective thermal conductivity for composite polymers by considering the filler size distribution law. Journal of Zhejiang University SCIENCE A. 2009;**10**:704–709. DOI: 10.1631/jzus.A0820733

[57] Holotescu S, Stoian F D, Marinica O, Kubicar L, Kopcansky P, Timko M. Utilization of the magnetogranulometric analysis to estimate the thermal conductivity of magnetic fluids. Journal of Magnetism and Magnetic Materials. 2011;**323**(10):1343–1347. DOI: 10.1016/j.jmmm.2010.11.043

[58] Khanafer K, Vafai K. A critical synthesis of thermophysical characteristics of nanofluids. International Journal of Heat and Mass Transfer. 2011;**54**:4410–4428. DOI: 10.1016/j.ijheatmasstransfer.2011.04.048

[59] Hopstock D M, Zanko L M. Minnesota taconite as a microwave-absorbing road aggregate material for deicing and pothole patching applications. Final Report, CTS Report No. 2004019. University of Minnesota, Duluth, US; 2005. 26 p.

[60] Korolev V V, Aref'ev I M, Ramazanova A G. Magnetocaloric effect and the heat capacity of ferrimagnetic nanosystems in magnetic fluids. Russian Journal of Physical Chemistry A. 2007;**81**(10):1677–1680. DOI: 10.1134/S003602440710024X

[61] Ravnik J, Skerget L, Hribersek M. Analysis of three-dimensional convection of nanofluids by BEM. Engineering Analysis with Boundary Elements. 2010;**34**:1018–1030. DOI: 10.1016/j.enganabound.2010.06.019

[62] Susa D, Lehtonen M, Nordman H. Dynamic modelling of power transformers. IEEE Transactions on Power Delivery. 2005;**20**(1):197–204. DOI: 10.1109/TPWRD.2004.835255

[63] Nikolaev V A, Shipilin A M. On the thermal expansion of nanoparticles. Physics of the Solid State. 2000;**42**(1):109–110. DOI: 10.1134/1.1131176

[64] Bergles A E, Bar-Cohen A. Immersion Cooling of Digital Computers, in Cooling of Electronic Systems. Publisher Springer Netherlands; NATO Advanced Study Institute Series; 1994. p. 539–621. Print ISBN 978-94-010-4476-9; Online ISBN 978-94-011-1090-7; Series ISSN 0168-132X; DOI: 10.1007/978-94-011-1090-7_23

[65] COMSOL MultiPhysics, Comsol A.B. Sweden, v. 3.5a (2010), v. 4.2a (2012), v 4.3 (2013). www.comsol.com

[66] Stoian F D, Holotescu S, Stoica V, Bica D, Vekas L. Comparative study of convective heat transfer in water and water based magnetizable nanofluid for thermal applications. Journal of Optoelectronics and Advanced Materials. 2008;**10**(4):773–776.

[67] Pîslaru-Dănescu L, Morega A M, Telipan G, Morega M, Dumitru J B, Marinescu V. Magnetic nanofluid applications in electrical engineering. IEEE Transactions on Magnetics. 2013;49(11):5489–5497. DOI: 10.1109/TMAG.2013.2271607

[68] Morega A M, Pîslaru-Dănescu L, Morega M. A novel microactuator device based on magnetic nanofluid. In: Proceedings of the 12th International Conference on Optimization of Electrical and Electronic Equipment (OPTIM 2012); 24–26 May 2012; Basov, Romania. IEEEXplore Digital Library, US; 2012. p. 1100–1106. DOI: 10.1109/OPTIM.2012.6231963

4

Heat Transfer of Ferrofluids

Seval Genc

Abstract

Magnetic field-responsive materials are an important group of smart materials. They can adaptively change their physical properties due to external magnetic field. Magnetic liquids or ferrofluids are colloidal systems of ferro or ferrimagnetic single domain nanoparticles that are dispersed either in aqueous or in organic liquids. Currently, the research field is undergoing a transition taking into account the bulk forces in fluids, which are magnetically nonuniform. These researches enable scientists to develop promising new designs. Today, because of the advancement in technology and limited energy sources, engineering innovations are focused on development of alternative resources instead of current systems. In this chapter, it is aimed to give a brief review of the heat transfer of magnetic fluids based on different types of magnetic nanoparticles as well as some of the research and results of the heat transfer of magnetite-based ferrofluids. The heat transfer of these materials was investigated under stationary conditions, and the heat transfer coefficient was calculated.

Keywords: ferrofluid, thermal conductivity, magnetic nanoparticles

1. Introduction

Traditional heat transfer fluids such as water, oil, and ethylene glycol cause problems in the performance of engineering equipment such as heat exchangers and electronic devices due to their low thermal conductivity. To improve the performance of these devices, fluids with higher thermal conductivity have to substitute these fluids. An inventive way to increase the thermal conductivity of these fluids can be achieved by the use of the nanofluids [1]. A nanofluid is a new class of heat transfer fluids containing nanoparticles with the size range under 100 nm that are uniformly and stably suspended in a liquid. Compared to the thermal conductivity of the base fluids, nanofluids showed dramatic increase in the heat transfer due to the higher thermal conductivity of these nanoparticles [2]. Intensive investigations on nanofluid containing metallic or nonmetallic nanoparticles such as TiO_2, Al_2O_3, Cu, CuO, Ag, and carbon nanotubes are being conducted to enhance their potential applications in heat transfer [3, 4].

Among different kinds of researches on nanofluids containing metallic or nonmetallic nanoparticles, some of the studies have been focused on the nanofluids prepared by dispersing magnetic nanoparticles in a carrier liquid. These are called *ferrofluids*. They are colloidal suspensions of ultrafine single domain superparamagnetic nanoparticles of metallic materials (ferromagnetic materials) such as iron, cobalt, and nickel as well as their oxides (ferrimagnetic materials) such as magnetite (Fe_3O_4) and ferrites (MnZn, Co ferrites) in either polar or nonpolar liquid carriers [5–7]. These magnetic fluids are a specific subset of smart materials that can adaptively change their physical properties under an externally applied magnetic field [6].

The research and development on the preparation, characterization, and application of the ferrofluids have been studied since mid-1960s which involve multidisciplinary sciences of chemistry, fluid mechanics, and magnetism. The most important advantage of these fluids is their ability to achieve a wide range of viscosity in a fraction of millisecond. The viscosity in the absence of magnetic field is called the "off-state" viscosity. The off-state viscosity of ferrofluids can go up to 2–500 mPa s depending on the concentration of the solid particles and the carrier liquid. Although they can respond to the action of external magnetic fields, stable ferrofluids show a relatively modest magneto-rheological effect such as an increase in yield strength. Since the particle size of the magnetic phase is very small, under ordinary field strengths, thermal agitation gives rise to Brownian forces that can overcome the alignment of the dipoles. Therefore, ferrofluids exhibit field dependent viscosity, but they exhibit no yield stress ($\tau_y = 0$) under magnetic field. Some properties of the ferrofluid are given in **Table 1**. The field dependent viscosity is given by Eq. 1 [8].

$$\frac{\Delta\eta}{\eta} = \frac{3}{2}\varphi \frac{\frac{1}{2}\alpha L(\alpha)}{1+\frac{1}{2}\alpha L(\alpha)} sin^2\beta \tag{1}$$

where $\alpha = \mu_0 M_d HV/kT$.

Magnetic control of the properties and behavior of these fluids are promising fields for advanced applications and a challenge for basic research and what makes these materials interesting. They are widely used in dynamic loudspeakers, computer hardware, dynamic sealing, electronic packaging, aerospace, and bioengineering [9]. Another important technological application of magnetic fluids, which depends on the heat transfer, is its use as a voice coil coolant for modern loudspeakers, high power electric transformers, and in advanced energy conversion systems like solar collectors and magnetically controlled thermosyphons [10–12]. In some of the home appliances, such as refrigerators and ovens, heat transfer techniques are used in order to provide heating or cooling. Controlling the heat transfer in these appliances with these fluids may decrease the energy consumption.

	Ferrofluid
Particle material	Ceramics, ferrites, iron, cobalt, etc.
Particle size	5–10 nm
Suspending fluid	Oils, water
Density (gr/cm^3)	1–2
Off state viscosity (mPa.s)	2–500
Required field	~1 kOe
Field induced fluids	$\Delta\eta\ (B)/\eta\ (0)\sim 2$
Device excitation	Permanent magnet

Table 1. Some of the properties of ferrofluids [6].

In the conventional nanofluids, the origin of the enhancement of thermal conductivity was thought to be due to the higher thermal conductivity of the nanoparticles (TiO_2, Al_2O_3, Cu, etc.) than the carrier fluid. Since the thermal conductivity of common magnetic materials (Fe_3O_4) is relatively low, the investigations did not gain much attention until it was understood that the thermal conductivity of the solid material did not have much effect in the enhancement of the thermal conductivity of the dispersion [13]. By understanding the control of the thermal conductivity of the ferrofluid by magnetic field increased the intensity of the research. Using ferrofluids under an applied magnetic field for the heat transfer enhancement is more advantageous compared with the conventional nanofluids (nonmagnetic nanofluids). The advantages of the ferrofluids over conventional nanofluids can be summarized as thermomagnetic convection is more intense than the gravitational one, and the thermal conductivity and viscosity are tunable under magnetic field.

2. Preparation of ferrofluids

Although pure metals (Fe, Co, Ni) possess the highest saturation magnetization, they are extremely sensitive to oxidation, hence the magnetic particles such as ferrites like magnetite

3 4 2 3 2 3

divalent ion, M = Mn, Zn, Ni, Co, Fe) are commonly used in ferrofluids. And among these, nano-sized iron oxide is the most widely used magnetic phase in ferrofluids. Various approaches have been explored for synthesis and characterization of high quality magnetic iron oxide nanoparticles. For example, sol-gel pyrolysis method was performed by Laokul et al. [14]. Synthesis of nanoparticles by thermal reductive decomposition method was performed by various scientists [15, 16]. Waje et al. performed mechanical alloying technique [17]. Hydrothermal technique was also used by various scientists to synthesize ferrite nanoparticles [18, 19]. However, the chemical method of coprecipitation of ferrous and ferric ions from solutions by addition of an alkali is a method which is very often used to prepare nanoparticles due to its low cost and simplicity [20]. Size reduction could be another method where magnetic powder of micron size is mixed with a solvent and a dispersant in a ball mill in order to grind for a period of several weeks [21].

The function of the carrier liquid is to provide a medium in which the magnetic powder is suspended. Ferrofluids used in different research and technology fields have been synthesized in carrier liquids such as water, silicone oil, synthetic or semi-synthetic oil, mineral oil, lubricating oil, kerosene, and combinations of these and many other polar liquids [22–24]. Boiling temperature, vapor pressure at elevated temperature, and freezing point are important parameters to be considered when choosing the carrier liquid. The carrier liquid should be non-reactive with the magnetic phase and also with the material used in the device. In terms of the heat transfer applications, the choice of the carrier fluid for the ferrofluid needs some additional requirements such as high conductivity, high heat capacity, and high thermal expansion coefficient. Water, oils, and ethylene glycol are considered as conventional heat transfer fluids and these can be good candidates for the carrier liquids. In recent years, studies on ferrofluids using ionic liquids have been reported, which seems to be a promising field of study [23, 24].

Colloidal stability of the ferrofluids is important in the technological applications. The stability is obtained by minimizing the agglomeration, which is maintained by the addition of the surfactants. The additives must be chosen to match the dielectric properties of the carrier liquid. Various surfactants such as silica, chitosan, polyvinyl alcohol (PVA), and ethylene glycol are usually used to coat the nanoparticles and to enhance dispersibility in aqueous medium [25–28]. Oleic acid (OA) is a commonly used surfactant to stabilize magnetic nanoparticles synthesized by traditional coprecipitation method [22]. Antioxidation additives may also be added to prevent oxidation. In water-based MR fluids, pH control additives are also used.

Magnetic nanoparticles tend to aggregate due to strong magnetic dipole-dipole attraction between particles. Stability of the magnetic colloid depends on the thermal contribution and the balance between attractive (van der Waals and dipole-dipole) and repulsive (steric and electrostatic) interactions. Under the magnetic field, the magnetic energy derives the particles to higher intensity regions; on the other hand, thermal energy forces the particles to wander around in the whole liquid. The stability against segregation is favored by the high ratio of the thermal energy to the magnetic energy. Stability against settling due to gravitational field is given by the ratio between gravitational energy and magnetic energy [8]. The two basic attractive interactions between the magnetic particles are dipole-dipole and van der Waals-London interactions. The ratio of thermal energy (kT) to dipole-dipole contact energy ($E_{\text{dipole}} = (\mu_0 M^2/12)V$) must be greater than unity. The particle diameter is given by $D \leq (72\, kT/\pi\mu_0 M^2)^{1/3}$, and the particle size is calculated as $D \leq 7.8$ nm. The normal ferrofluids with the particle size of 10 nm are in the limits of agglomeration. Van der Waals forces arise due to the fluctuating electric dipole-dipole forces. Preventing the contact of the particles is another necessity if a stable colloid is to be obtained [8].

The Brownian motion, electrostatic repulsion, and steric repulsion are the main mechanisms supporting the ferrofluid colloidal stability. Electrostatic interaction is the dominant mechanism in ionic ferrofluids, whereas steric repulsion is the dominant mechanism supporting the colloidal stability in organic-based ferrofluids [23]. The agglomeration of particles suspended in a liquid can be prevented by creating mutually repelling charged double layers or by physically preventing the close approach of particles by steric hindrance which is provided by the surfactant molecules adsorbed onto the particle surface [8, 23]. As the thickness of the adsorbed polymer is increased, the stability of the dispersion increases [8, 23]. Wang and Huang showed that by retaining excess oleic acid in their ferrofluid, stable magnetic colloid was achieved by steric repulsion [8, 23].

3. Thermal conductivity of ferrofluids

3.1. Experimental investigations

Thermal conductivity of ferrofluids has gained much attention in the last decade due to the significant enhancement compared to the nonmagnetic nanofluids. The increase in the thermal conductivity can occur both with and without the applied magnetic field. Experimental studies show that the change in the off-state (when there is no magnetic field) thermal conductivity of ferrofluids could be due to volume fraction of magnetic phase, particle size distribution, temperature, surfactant, etc. On the other hand, when the magnetic field is applied, besides the factors mentioned above, the magnitude and direction of the applied magnetic field affect the thermal conductivity of the ferrofluids.

In the experimental studies of the thermal conductivity of ferrofluids, it has been observed that both the on-state and off-state thermal conductivities increase with the increase in the volume fraction of the magnetic phase. When the literature was reviewed, it was seen that most of the ferrofluids synthesized with magnetite (Fe_3O_4) which was produced by coprecitation method has been studied. Abareshi and coworkers synthesized ferrofluids by dispersing Fe_3O_4 nanoparticles in water [29]. They reported an increase of 11.5% in the off-state thermal conductivity as the particle loading increased to 3 vol% at 40°C. This increase was observed when the magnetic field was applied parallel to the heat flux. The study of Li et al. was also performed with water-based ferrofluids and an increase in the on-state (magnetic field, H = 19 kA/m) thermal conductivity of 11% for 1 vol% and 25% for 5 vol% magnetic nanoparticles was reported [30]. Philip et al. and Shima et al. [30, 31] investigated the thermal conductivity of kerosene-based ferrofluids synthesized with Fe_3O_4 nanoparticles. When the magnetic field was applied parallel to the heat flux, they discovered a dramatic increase in the thermal conductivity. For a volume fraction of 6.3%, the increase was 300% at H = 7 kA/m field strength. No increase was observed when the magnetic field was applied perpendicular to the heat flux. The reason may be that the different directions of external magnetic field lead to quite different morphologies of the magnetic fluids that exerted quite different effects on the energy transport process inside the magnetic fluid [31]. They further explained that the chains formed by the particles provided more effective bridges for energy transport inside the ferrofluid along the direction of temperature gradient and as a result, the thermal process in the ferrofluid was enhanced. The anisotropic property of thermal conductivity was addressed in the theoretical study by Fu et al., and Blums et al [10, 32]. Blums et al. predict anisotropy of thermal conductivity in ferrofluids in the presence of a magnetic field [10]. In the research conducted by Gavili et al. the thermal conductivity of ferrofluids containing Fe_3O_4 nanoparticles suspended in deionized water under magnetic field was experimentally investigated [33]. According to their results, a ferrofluid with 5.0% volume fraction of nanoparticles with an average diameter of 10 nm enhanced the thermal conductivity more than 200% at 1000 Gauss magnetic field [33].

As it is seen from all the research and reports, it is evident the experimental results have been heterogeneous. The difference in the experimental outcomes may be due to magnetization, size distribution of the particles, the type of the carrier liquid, etc. The effect of the carrier liquid showed that the thermal conductivity ratio is higher for carrier liquid with a low thermal conductivity like common hydrocarbons. However, the absolute thermal conductivity of ferrofluid is higher for a carrier liquid with a high thermal conductivity.

In recent years, studies have been carried out to understand the thermal conductivity of ferrofluids synthesized by magnetic phase other than Fe_3O_4, especially with carbon nanotubes (CNT) has also been investigated by many scientists. Hong et al. [34] and Wensel et al. [35] experimentally measured the thermal conductivity of single wall carbon nanotubes coated by Fe_2O_3 nanoparticles suspended in water and they observed an approximately 10% increase in the thermal conductivity with 0.02% particle loading. Wright et al. reported thermal conductivity enhancement of single wall carbon nanotubes coated by Ni nanoparticles suspended in water [36]. Sundar et al. measured the thermal conductivity enhancement of the hybrid ferrofluid which was composed of carbon nanotube (CNT)—Fe_3O_4 and water [37]. They observed a thermal conductivity enhancement of 13.88–28.46% at 0.3% volume concentration in the temperature range of 25–60°C. Shahsavar et al. analyzed the thermal conductivity behavior of Fe_3O_4 and CNT hybrid ferrofluids and observed that the highest enhancement in the thermal conductivity was about 151% for 0.9% ferrofluid and 1.35% CNT [38].

The review of the literature on the experimental studies of the thermal conductivity of ferrofluids revealed that the thermal conductivity is enhanced by the volume fraction of the magnetic phase and the applied magnetic field. Next chapter will discuss the reasons for the abnormal enhancement in the thermal conductivity of the ferrofluid under the influence of applied magnetic field.

3.2. Mechanisms of heat transfer enhancement

In the thermal conductivity of conventional nanofluids and ferrofluids, the most discussed mechanisms have been Brownian motion and formation of particle chain/cluster structure [39]. The Brownian motion indicates the random movement of particles dispersed in liquid or gas, and the motion is due to collision with base fluid molecules, which makes particles undergo a random walk motion [40]. The Brownian motion could contribute to the thermal conductivity enhancement in two ways, namely, the direct contribution due to motion of nanoparticles that transports heat (diffusion of nanoparticles) and the indirect contribution due to the so called micro-convection of fluid surrounding individual nanoparticles [40]. The diffusion of magnetic nanoparticles plays an important role at a low volume fraction ($\varphi < 2\%$), which could be explained by the effective medium (Maxwell) theory rather than the effects associated with the Brownian motion-induced hydrodynamics. The effective medium or mean-field theory of Maxwell, which describes the effective macroscopic properties of the composite material as a function of the particle fraction and the material properties of the components, is most often used to analyze the thermal conductivity results of nanofluid experiments. For a nanofluid with non-interacting spherical nanoparticles with low volume fraction, the theory predicts (Eq. 2)

$$\frac{\kappa}{\kappa_f} = \frac{1+2\beta\Phi}{1-\beta\Phi} \tag{2}$$

where φ is the nanoparticle volume fraction, κ_p and κ_f are thermal conductivity of the particle and the fluid, respectively. $\beta = (\kappa_p - \kappa_f)/(\kappa_p + 2\kappa_f)$, and $(\kappa_p - \kappa_f)$ is the difference between the thermal conductivities of the nanoparticle and the base fluid. However, in the study done by Vadasz et al. [41] and Keblinski et al. [42], the results of the thermal conductivity measurements showed divergence from the effective medium theory. One possible discrepancy between effective medium theory and the experimental results is the interparticle interactions, which can result in the formation of chain and cluster-like formations. Philip and coworkers showed that the micro-convection of the fluid medium around randomly moving nanoparticles did not affect the thermal conductivity of a nanofluid and the microconvection model overestimated the thermal conductivity values [43]. According to them, the conductivity enhancement in the ferrofluid at high volume fraction ($\varphi < 2\%$) was due to the presence of dimmers or trimmers in the fluid. These results were in a reasonable agreement with the Maxwell-Gannet model, especially at higher volume fractions. Clusters or chains of the particles may form heat bridges [31]. The form and magnitude of these structures vary and depend not only on the material of the carrier medium and the particles but also on the shape and size of the particles [31].

In ferrofluids, the interparticle interactions have even more important impact on the properties of the fluid, due to the chain-like formation of the particles caused by the magnetic dipole interactions. The effect of this interaction can be seen, for example, in changes in the viscosity of ferrofluids, which depends on interparticle interaction [7, 44].

Magnetically induced structure formation only arises if the magnetic energy of the particles is larger than their thermal energy. The mechanism of thermal conductivity enhancement can be explained as follows: The magnetic particles in the ferrofluid are single domain and super-paramagnetic with magnetic moment $\boldsymbol{m}$ as mentioned above [28]. The interparticle dipole-dipole interaction, which is also called dipolar coupling, refers to the interaction between magnetic dipoles. The potential energy of the interaction U_d is given by Eq. (3),

$$U_d(ij) = \left[\frac{3(m_i \cdot r_{ij})(m_j \cdot r_{ij})}{r_{ij}^5} - \frac{(m_i \cdot m_j)}{r_{ij}^3} \right] \tag{3}$$

Suppose m_i and m_j are two magnetic moments in space and r_{ij} $(= r_i - r_j)$ is the distance between the ith and the jth particles. The magnetic moments are oriented in random directions in the absence of magnetic field and the nanoparticles are influenced by the Brownian motion as the thermal energy exceeds the magnetic dipole attraction ($U_d(ij) < k_B T$). In the presence of a magnetic field, the magnetic dipolar interaction becomes strong enough to dominate the thermal energy so that the magnetic particles start aligning in the direction of the magnetic field [31]. **Figure 1** gives the schematic drawing of the clustering/chain-like formation under magnetic field.

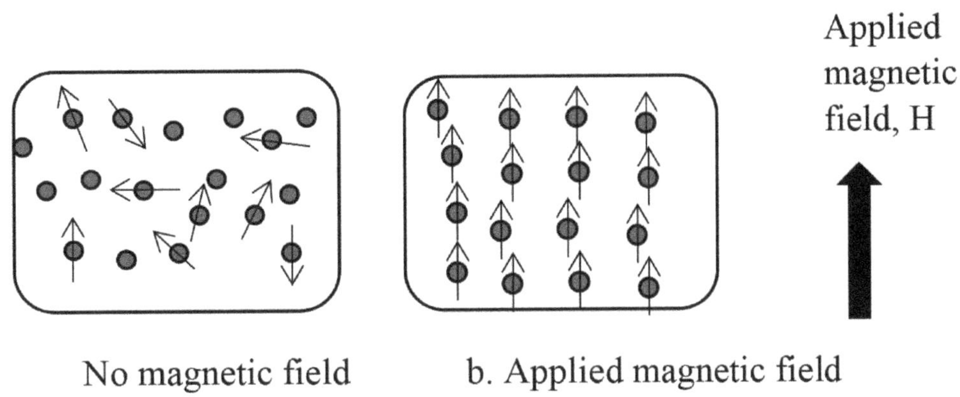

Figure 1. Schematic drawing of the chain-like formation of the magnetic particles in the fluid. (a) No magnetic field. (b) Applied magnetic field.

The lengths of the chains depend on the magnitude of the magnetic field. As the magnetic field increases, the particles start forming short chains along the direction of the magnetic field and the chains get longer as the magnetic field increases. Based on Philip's theory due to linear chain-like structures of the magnetic nanoparticles, the percolation theory could support the abnormal enhancement of the ferrofluid. They stated that the maximum enhancement was observed when the chain-like aggregates were well dispersed without clumping [31].

Although the thermal conductivity of ferrofluids enhances with increasing magnetic field, there are reports regarding a decrease in the thermal conductivity of these fluids. Shima et al. observed decrease in the thermal conductivity above 82 Gauss magnetic field [31]. They attributed this decrease to the "zippering" of the chains. The linear and thick aggregates with the aspect ratio due to zippering can collapse to the bottom of the cell, and hence the thermal conductivity cannot be measured. Gavili and coworkers observed that the thermal conductivity dramatically decreased in the presence of magnetic field with increasing temperature. When the temperature of the ferrofluid increases, the chain-like structure is broken due to the increase in the thermal velocity and consequently the thermal conductivity decreases [33].

Theoretical and experimental studies related with the thermal radiation and convection of the heat transfer of nanofluids have started to gain more attention in the recent years especially, in the field of engineering applications such as solar collectors and in space applications [45–47].Thermal convection in magnetic fluids heated from below subjected to an external magnetic field causes a convection-driving mechanism. The temperature difference causes a gradient in the magnetic field and as a result a magnetic force appears. Beyond a certain threshold a thermomagnetic convection is generated.

4. Experimental study

4.1. Apparatus and data analysis

An experimental setup is built according to the ISO 8301 numbered "Thermal insulation-determination of steady state thermal resistance and related properties—Heat flow meter apparatus" standard has been used in the heat transfer experiments. Experimental setup has been established as single-specimen asymmetrical configuration according to the standard shown in **Figure 2**.

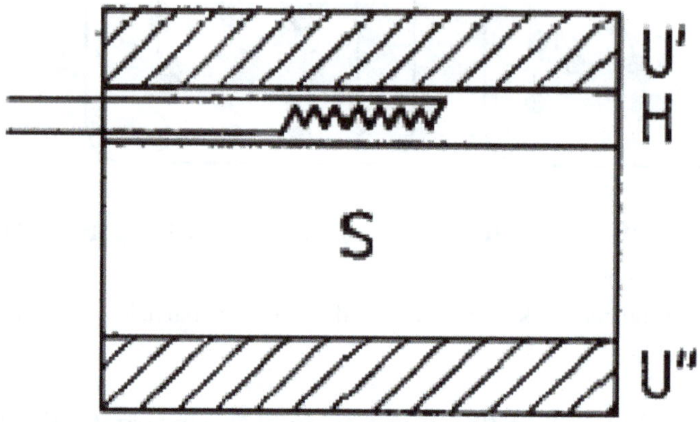

Figure 2. Single-specimen asymmetrical configuration (U', U" are the cooling and heating units, respectively, and H = heat flux meter).

Schematic drawing containing the requirements of experimental setup such as testing unit, two water baths, data acquisition system, and computer is given in **Figure 3**. The testing unit was heated and cooled by water bath.

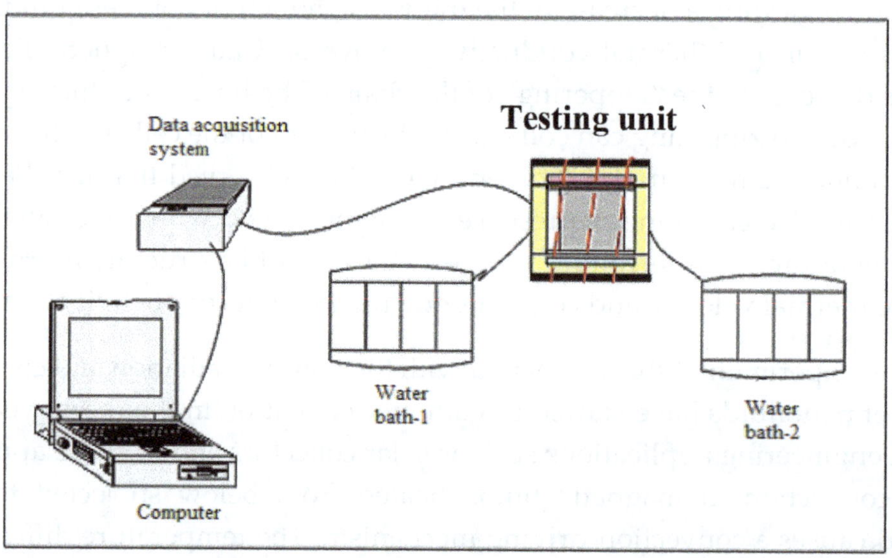

Figure 3. The illustration of experimental setup.

A more detailed representation of the testing unit given in **Figure 4**, which consisted of a space for sample, heat flux sensors, thermocouples, heating-cooling sources, and polyurethane insulating material from outside to inside. The magnetic field was applied by neodymium permanent magnets parallel to the temperature gradient. The maximum magnetic field obtained was approximately 140 Gauss. One of the uncertainties of the setup could be the non-uniformity of magnetic field. The magnetic field was calculated as the average of the field from three different points on the radial direction. Another uncertainty could be heat loss. Although the experimental setup was isolated, there could still be some heat loss. For reproducibility of the data, each measurement was performed for five ferrofluid samples.

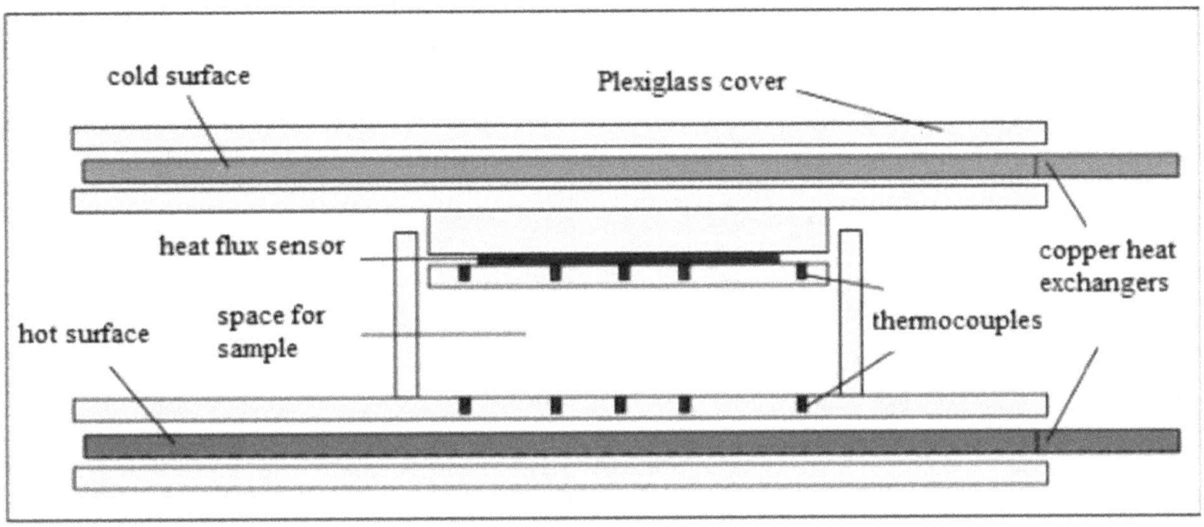

Figure 4. Testing unit.

In order to determine the temperature range at which the thermal conductivity of ferrofluids is more effective for different applications, the experiments were done in two different temperature intervals; from -20 to 0°C, and from 0 to 50°C and two different temperature differences. The temperatures difference between the hot and cold surfaces of the setup are given in **Table 2**.

Between 0 and -20°C interval	Between 0 and 50°C interval
Temperature difference, 20 K	Temperature difference, 35 K
Temperature difference, 15 K	Temperature difference, 20 K

Table 2. Temperature intervals and temperature differences at which the experiments were performed.

Besides the effect of temperature, following variables were considered during investigation of heat transfer of ferrofluids, different carrier liquid and concentration of the magnetic phase.

The thermal conductivity measurements have been carried out when the water baths came a steady state temperature. The heat transfer coefficient has been calculated by using Eq. 4,

$$\kappa = \frac{A}{Q}\frac{\Delta T}{\Delta x} \qquad (4)$$

where A is the cross-sectional area, Q is heat flux, ΔT is the temperature difference, and Δx is the height of the measuring cell.

4.2. Ferrofluid preparation

Synthesis of stable and well dispersible MR fluid is extremely important for mechanical and heat transfer applications. Synthesis of stable ferrofluid depends on concentration and viscosity of carrier fluid, concentration of magnetic phase, particle size, and surfactants. The ferrofluids in this study were synthesized in water and silicone oil (viscosity 350 cSt) with as received magnetite Fe_3O_4 nanopowder from Aldrich. The particle size was approximately 50 nm. Samples were synthesized as volumetric percentages of 5 and 20%. Surfactant was also added to prevent sedimentation. The names and description of the ferrofluids are given in **Table 3**.

Name	Description
5 Fe_3O_4-S	5 vol% Fe_3O_4 + 350 cSt silicone oil + surfactant
5 Fe_3O_4-W	5 vol% Fe_3O_4 + water
20 Fe_3O_4-W	20 vol% Fe_3O_4 + water
20 Fe_3O_4-S	20 vol% Fe_3O_4 + 350 cSt silicone oil + surfactant

Table 3. Description of the MR fluids used in this research.

In order to have a stable dispersion ball milling which could break up the agglomerated, was applied. Ball milling procedure was conducted with yttria stabilized zirconia grinding media with 0.5 mm diameter.

5. Results and discussion

5.1. Analysis of thermal conductivity in the temperature interval between 0 and -50°C.

In this part of the study, base liquid, volume fraction (5 and 20 vol% Fe_3O_4), temperature (ΔT = 20 and 35 K), and dependence of the thermal conductivity of the ferrofluids were investigated.

5.2. Carrier liquid dependence of the 5 vol% Fe_3O_4-based ferrofluids

Heat transfer coefficients of ferrofluids were analyzed for two different temperature intervals, and in each temperature interval, the thermal conductivity of the ferrofluids was investigated for two different temperature differences. The temperature intervals were chosen as 0 to 50°C and -20 to 0°C, and the temperature differences are $\Delta T = 35$ K and $\Delta T = 20$ K for the first interval and $\Delta T = 15$ K and $\Delta T = 20$ K for the second interval mentioned above.

In the first part of the study, the heat transfer was investigated for $5Fe_3O_4$-S and $5Fe_3O_4$-W at a temperature difference of 20 K. The increase in heat transfer coefficient in the presence of magnetic field of 134 Gauss was 7 and 18% for 5 Fe_3O_4-S and 5 Fe_3O_4-W-type ferrofluids, respectively (**Figure 5**).

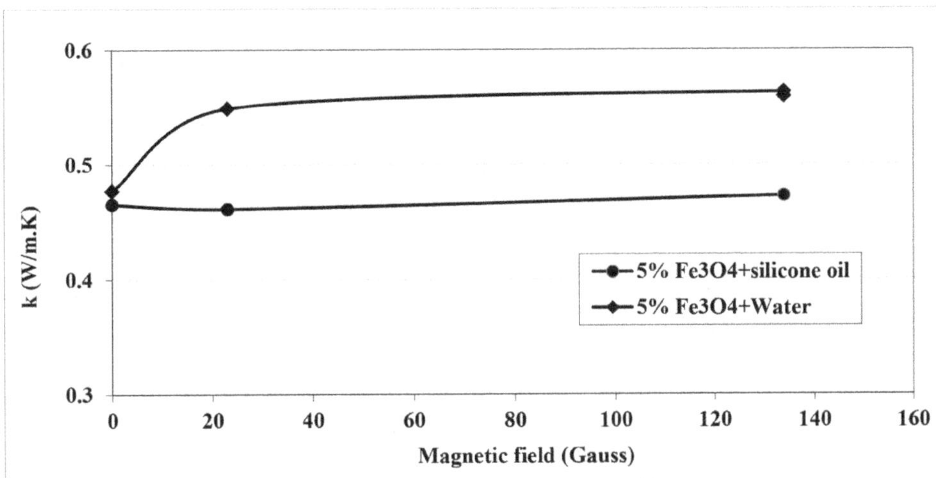

Figure 5. Thermal conductivity at 0–50°C interval and the temperature difference $\Delta T = 20$ K.

As discussed above, the heat transfer increased as we increased the magnetic field. Although this increase could be attributed to the chain formation of the iron particles in the fluid, the effect of magnetic field is still not very clear. In **Figure 6**, the percent change in the thermal conductivity is given at temperature difference of $\Delta T = 35$ K in temperature interval of 0–50°C

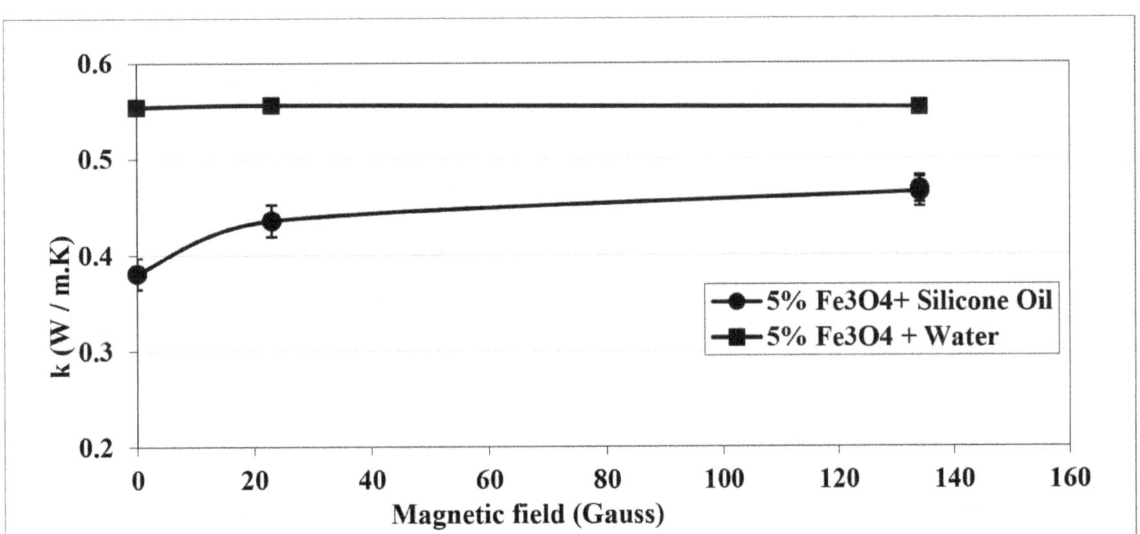

Figure 6. Thermal conductivity at 0–50°C interval and the temperature difference $\Delta T = 35$ K.

As given in **Figure 6**, heat transfer coefficients for 5 Fe_3O_4-S and 5 Fe_3O_4-W at the highest magnetic field were measured as 0.47 and 0.56 W/m K, respectively. In addition, the increase in the thermal conductivity for 5 Fe_3O_4-S was about 23% and water based was 5%. In either temperature differences (ΔT = 35 and 20 K), the thermal conductivity coefficient of silicone-based ferrofluid is less than that of the water-based ferrofluid. Since the volume concentration of the magnetic phase is small, the thermal conductivity of the base liquid could be a factor in the increase of the thermal conductivity of ferrofluid. The thermal conductivity of water is more than that of the silicone oil.

5.3. Volume percent and temperature dependence of thermal conductivity

In this part of the study, the volume dependence of the thermal conductivity was investigated. This investigation was performed at ΔT = 35 and 20 K as well. In **Figure 7**, it is observed that as the volume fraction is increased, the thermal conductivity coefficients for 5 Fe_3O_4-S and 20 Fe_3O_4-S ferrofluid also increased for ΔT = 35 K. At the highest magnetic field, the thermal conductivity coefficient, **k**, is measured as 0.51 and 0.47 W/mK for 20 and 5 vol%, respectively. The percent increase in both of the fluids was the same, which was approximately 24%.

When the temperature difference is kept smaller (ΔT = 20 K) in the same temperature interval, the thermal conductivity coefficient, **k**, is 0.51 and 0.47 W/m K (**Figure 10**) at 134 Gauss which were the same as the coefficients in ΔT = 35 K and the percent change of the thermal conductivity for these two intervals was almost the same. When we compare **Figures 7** and **8** at a magnetic field of 134 Gauss we saw that the **k** values were the same for 20 and 35 K temperature differences. However, at zero magnetic field, they were different which made a difference in the percent increase. The percent increase in the 35 K difference is more than that of the 20 K difference.

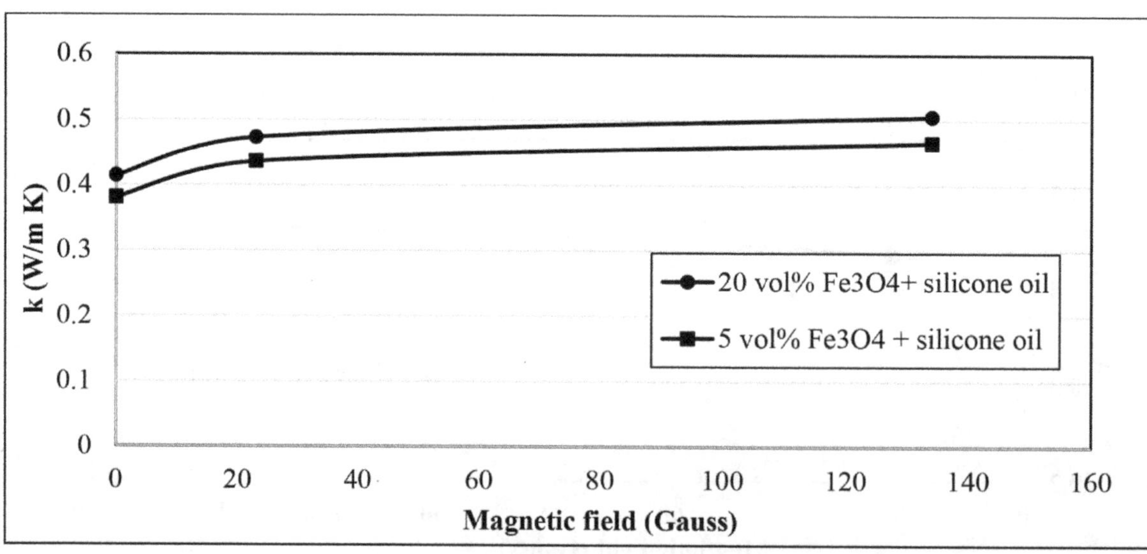

Figure 7. Volume dependence of thermal conductivity between 0 and 50°C temperature interval and 35 K temperature difference.

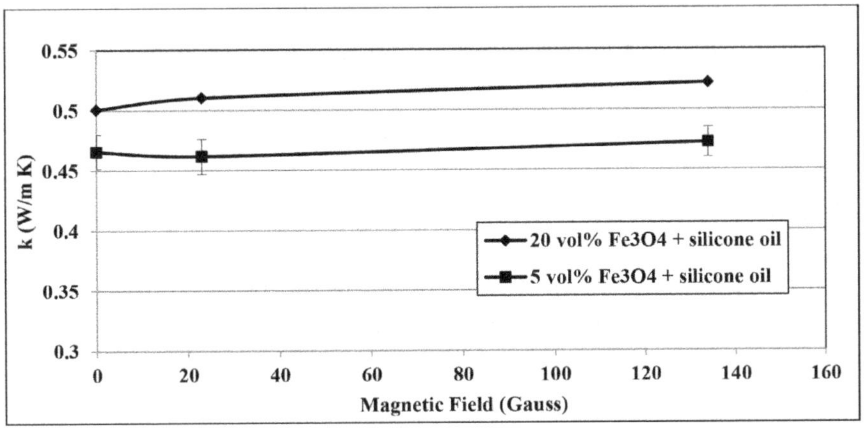

Figure 8. Volume dependence of thermal conductivity between 0 and 50°C temperature interval and 20 K temperature difference.

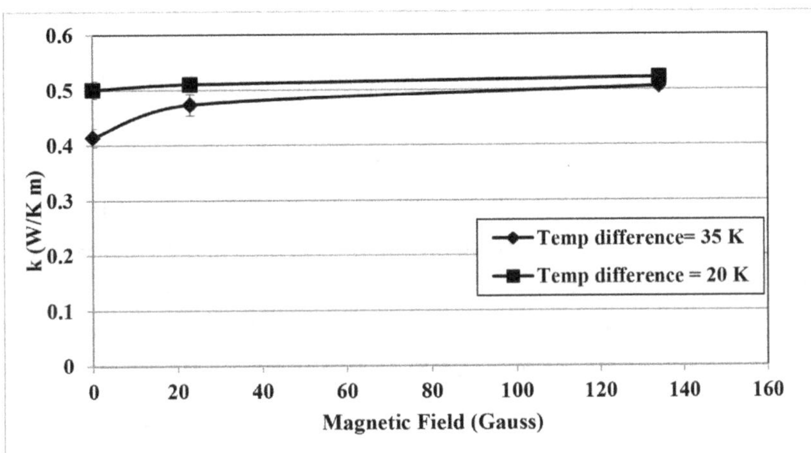

Figure 9. Temperature difference dependence of 20 vol% Fe_3O_4 and silicone oil-based ferrofluid.

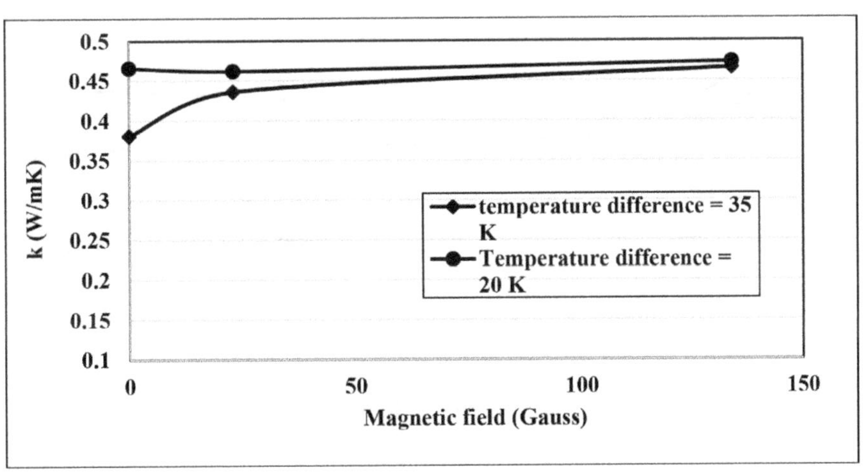

Figure 10. Temperature difference dependence of 5 vol% Fe_3O_4 and silicone oil-based ferrofluid.

Figures 9 and **10** give a clearer view of the temperature difference dependence of the thermal conductivity for 20 Fe_3O_4-S and 5 Fe_3O_4-S-type ferrofluid, respectively. In both graphs, the thermal conductivity enhancement is more for 35 K temperature difference. As the magnetic field increased, the k values reached the same value irrespective of the temperature difference.

5.4. Analysis of thermal conductivity in the temperature interval between -20 and 0°C.

In this part of the study, the thermal conductivity of the for 5 Fe_3O_4-S and 20 Fe_3O_4-S ferrofluids were investigated in the temperature interval of -20 to 0°C in which the temperature differences were kept as 15 and 20 K.

5.4.1. Volume percent and temperature dependence of thermal conductivity

The next set of measurements involved a lower temperature interval, such as from -20 to 0°C. In this interval, the temperature differences were taken as 15 and 20 K. **Figure 11** shows the change in the thermal conductivity with respect to the magnetic field when the temperature gradient was 20 K. The percentage decrease in the thermal conductivity of ferrofluids for 5 and 20 vol% ferrofluids at 134 Gauss was 33 and 34%, respectively. The volume dependence of the thermal conductivity was also observed in these measurements. The 20 vol% ferrofluid had a higher thermal conductivity. **Figure 12** shows the change in the thermal conductivity with respect to the magnetic field under a 15 K temperature difference. Unlike the results obtained in the higher temperature intervals, there was a very small increase in the conductivity, followed by a slight decrease as the magnetic field increased. All the fluids showed a similar trend. The same amount of decrease was observed in this range for two different magnetic phase concentrations. The percentage decrease in the thermal conductivity for 5 Fe_3O_4-S and 20 Fe_3O_4-S type ferrofluids at 134 Gauss was 2 and 3%, respectively. The volume dependence

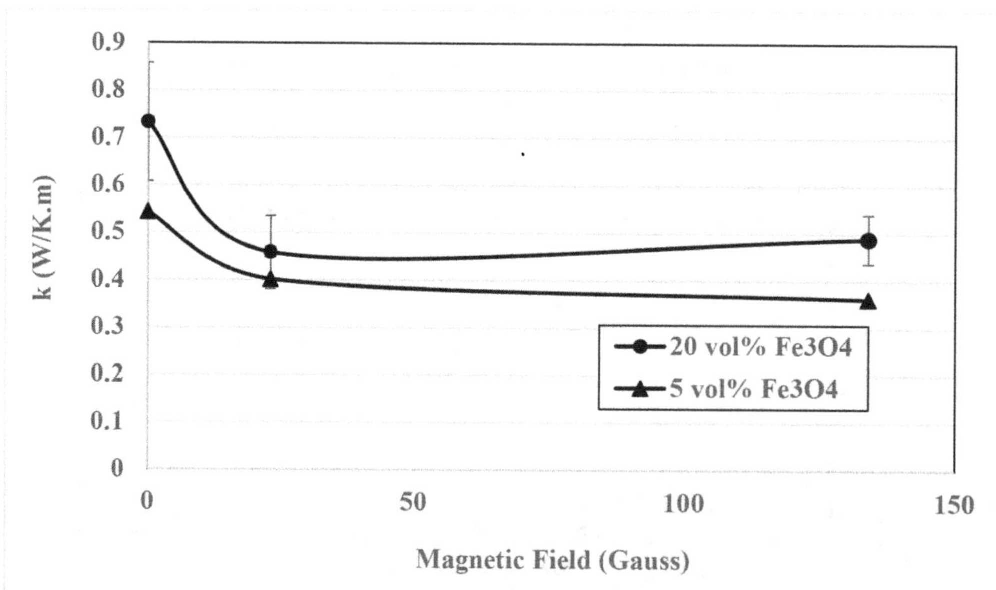

Figure 11. Volume dependence of thermal conductivity between -20 and 0°C temperature interval and 20 K temperature difference.

of the thermal conductivity was also observed in these measurements. The for 20 Fe_3O_4-S type ferrofluid had a thermal conductivity of 42 W/K.m at 134 Gauss whereas for 5 Fe_3O_4-S type ferrofluid had 37 W/K.m.

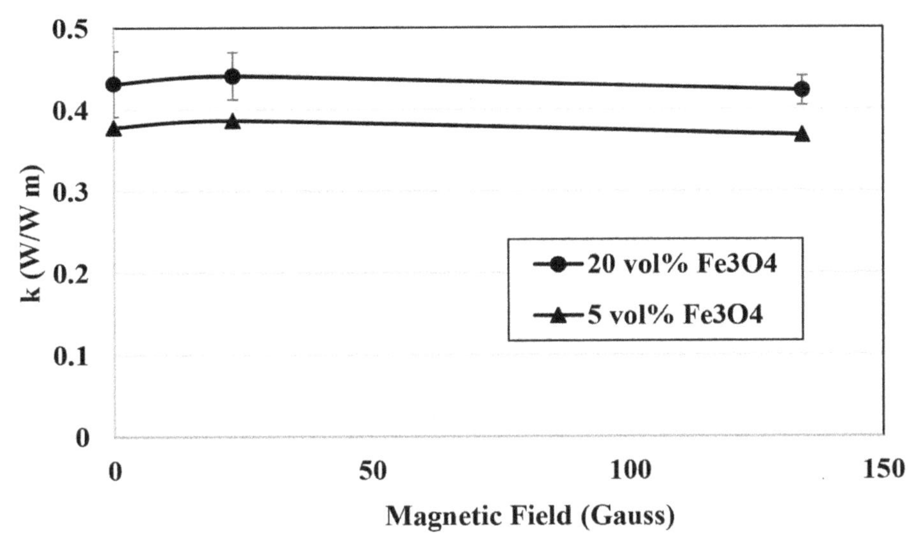

Figure 12. Volume dependence of thermal conductivity between -20 and 0°C temperature interval and temperature difference of 15 K.

The thermal conductivity depended on the temperature difference at very low magnetic fields. As the magnetic field increases the thermal conductivity became irrespective of the temperature. This behavior is observed both for 5 Fe_3O_4-S and 20 Fe_3O_4-S (**Figures 13** and **14**, respectively).

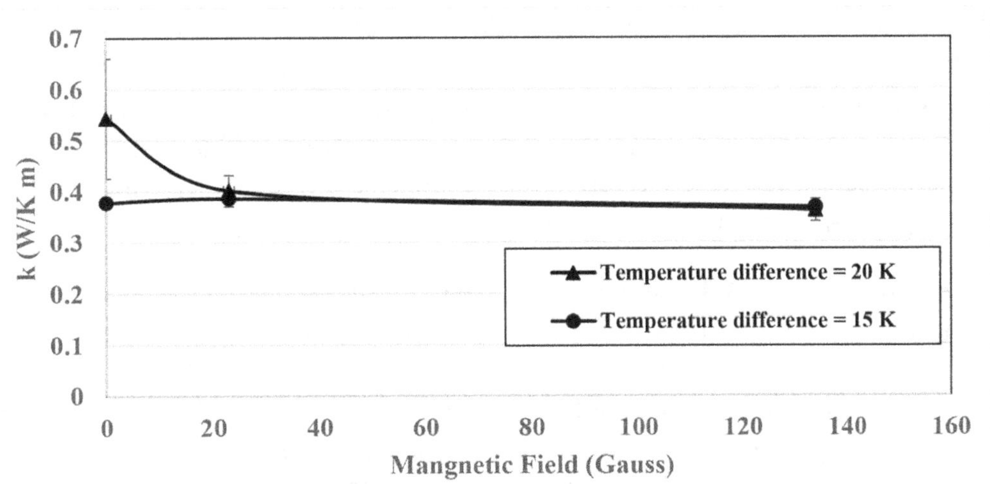

Figure 13. Temperature difference dependence of 5 vol% Fe_3O_4 and silicone oil-based ferrofluid ($5Fe_3O_4$-S).

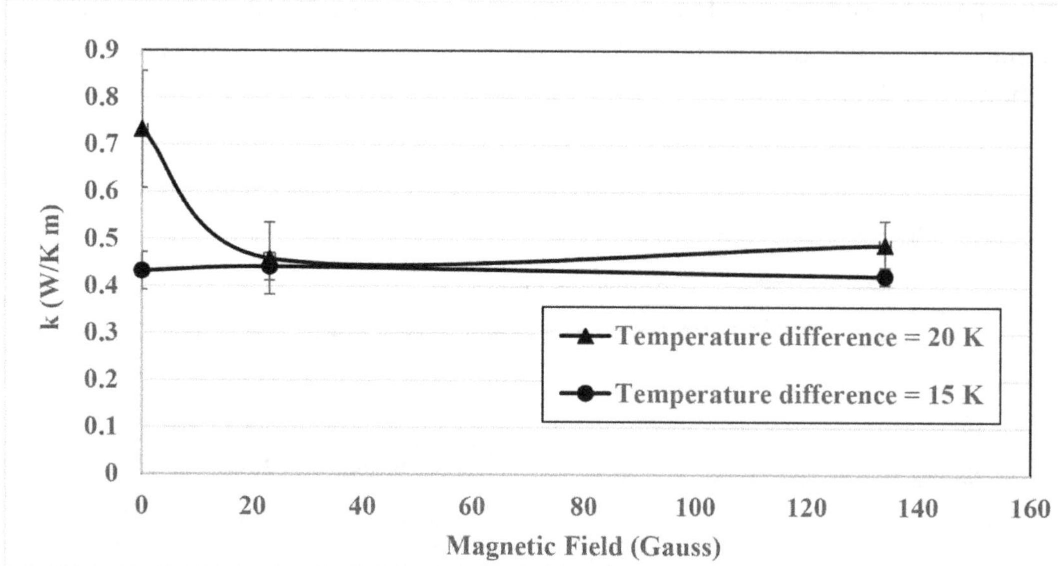

Figure 14. Temperature difference dependence of 20 vol% Fe_3O_4 and silicone oil-based ferrofluid ($20Fe_3O_4$-S).

6. Conclusion

Thermal conductivity mechanism in liquid involves collision of the molecules and transfer of the energy and momentum with one another. Transfer of the kinetic energy occurs in the lower temperature part of the system when a molecule moves from a high temperature region to a region of lower temperature and this molecule gives up this energy via collision with lower energy molecules. In solids, on the other hand, thermal energy may be conducted by lattice vibrations. At low-temperature ranges (-20 to 0°C), the energy of the molecules in the liquid is not enough to cause collisions and lattice vibrations could be insufficient to conduct heat. The decrease of the thermal conductivity observed in the temperature range between 0 and -20°C could be due to the less energetic particles. We consider that the thermal conductivity of the ferrofluid is determined by the factors such as stability of the ferrofluid particle size and the viscosity of the base liquid. Another important point in the heat transfer of the ferrofluids could be the thermo-convective instability of the magnetic fluids. The instability arises due to the stronger magnetization of the colder fluid which is drawn to the higher region and is displaced by the warmer fluid [5, 46, 47]

As the temperature decreases the density and viscosity of the base fluid, silicone oil, increase. The denser and more viscous fluid hinders the motion of the magnetic particles due to the temperature difference. Thus, lower temperatures inhibit the motion of particles in the ferrofluid which will prevent settling of the particles. The stability of the fluid can be affected in a negative way at higher temperatures and at lower magnetic fields because the density and viscosity of the silicone oil decrease at high temperature. The instability of the ferrofluid at low temperature could be the reason for the different thermal conductivities at low magnetic fields. At low temperature, chain formation between the magnetic particles is activated by the increase in magnetic field. Consequently, this increases the thermal conductivity of the magnetic fluid at low temperatures. As the magnetism increases, the effect of the temperature range diminishes and thermal conductivity reaches almost the same value.

The heat transfer characteristics of $5Fe_3O_4$-S and $20Fe_3O_4$-S ferrofluid were investigated in the presence of the magnetic field applied parallel to the temperature gradient. The thermal conductivity behavior of the ferrofluids in different temperature ranges was analyzed, and it was seen that the heat transfer was more effective at higher temperatures. The fluids showed an increase in the thermal conductivity in the temperature intervals from 0 to 50°C, and a decrease from -20 to 0°C.

Acknowledgements

The work is supported by Scientific and Technological Council of Turkey, TUBITAK, under Grant 108M473.

Nomenclature

$\Delta\eta$ (B):	Change in viscosity under magnetic field (Pa s)
η (0):	Viscosity without magnetic field (Pa s)
k:	Boltzmann constant (J/K)
T:	Temperature (K)
E_{dipole}:	Dipole-dipole energy
μ_0:	Vacuum permeability (Vs/A m)
M:	Magnetization (A/m)
M_d:	Domain magnetization
H:	Magnetic field (Gauss)
V:	Volume (m^3)
D:	Particle diameter (m)
κ_p:	Thermal conductivity of the particle (W/m K)
κ_f :	Thermal conductivity of the fluid
β:	Magnetic induction (T)
Φ:	Volume fraction
U_d:	Potential energy of the interaction
m_i, m_j:	Magnetic moments in space
r_{ij} : $(= r_i - r_j)$	distance between the ith and the jth particle

A:	Cross-sectional area
Q:	Heat flux
ΔT:	Temperature gradient
Δx:	Distance between the plates (m)
$L(\alpha)$:	Langevin function

Author details

Seval Genc

Address all correspondence to: sgenc@marmara.edu.tr

Engineering Faculty, Metallurgical and Materials Engineering, Marmara University, Istanbul, Turkey

References

[1] Nkurikiyimfura, I., Wang, Y., and Zhidong, P. Heat transfer enhancement by magnetic nanofluids—a review. Renewable and Sustainable Energy Reviews. 2013; 21: 548–561.

[2] Vadasz, J. J., Govender, S., and Peter, V. Heat transfer enhancement in nano-fluids suspensions: possible mechanisms and explanations. International Journal of Heat and Mass Transfer. 2005; 48.13: 2673–2683.

[3] Chen, H., et al. Predicting thermal conductivity of liquid suspensions of nanoparticles (nanofluids) based on rheology. Particuology. 2009; 7.2: 151–157.

[4] Singh, A. K. Thermal conductivity of nanofluids. Defence Science Journal. 2008; 58.5: 600.

[5] Berkovsky, B. M., Medvedev, V. F., and Krakov, M. S. Magnetic fluids: Engineering Applications. 1st ed. Oxford University Press; 1993. 256 p

[6] Genc, S., and Bora, D. Synthesis and rheology of ferrofluids: a review. Current Opinion in Chemical Engineering. 2014; 3: 118–124.

[7] López-López, M. T., et al. Colloids on the frontier of ferrofluids. Rheological properties. Langmuir. 2012; 28.15: 6232–6245.

[8] Rosensweig, R. E. Ferrohydrodynamics. Courier Corporation. 2013. 344 p

[9] Scherer, C., and Antonio Martins, F. N. Ferrofluids: properties and applications. Brazilian Journal of Physics. 2005; 35.3A: 718–727.

[10] Blums, E. Heat and mass transfer phenomena. Stefan Odenbach, editor. Ferrofluids. Springer: Berlin, Heidelberg, 2002. pp. 124–139.

[11] Trahms, L. Biomedical applications of magnetic nanoparticles. Stefan Odenbach, editor. Colloidal Magnetic Fluids. Springer: Berlin, Heidelberg, 2009. pp. 1–32

[12] Kuzubov, A. O., and Ivanova, O. I. Magnetic liquids for heat exchange. Journal of Physics III France. 1994; 4.1: 1–6.

[13] Hong, T-K., Ho-Soon, Y., and Choi, C. J. Study of the enhanced thermal conductivity of Fe nanofluids. Journal of Applied Physics. 2005; 97.6: 064311.

[14] Laokul, P., et al. Characterization and magnetic properties of nanocrystalline $CuFe_2O_4$, $NiFe_2O_4$, $ZnFe_2O_4$ powders prepared by the aloe vera extract solution. Current Applied Physics. 2011; 11.1: 101–108.

[15] Xu, Z., et al. Oleylamine as both reducing agent and stabilizer in a facile synthesis of magnetite nanoparticles. Chemistry of Materials. 2009; 21.9: 1778–1780.

[16] Bica, I. Nanoparticle production by plasma. Materials Science and Engineering: B. 1999; 68.1: 5–9.

[17] Waje, S. B., et al. X-ray diffraction studies on crystallite size evolution of $CoFe_2O_4$ nanoparticles prepared using mechanical alloying and sintering. Applied Surface Science. 2010; 256.10: 3122–3127.

[18] Cote, L. J., et al. Continuous hydrothermal synthesis and crystallization of magnetic oxide nanoparticles. Journal of Materials Research. 2002; 17.09: 2410–2416.

[19] Cabuil, V., et al. Ionic magnetic fluid based on cobalt ferrite nanoparticles: influence of hydrothermal treatment on the nanoparticle size. Journal of Magnetism and Magnetic Materials. 2011; 323.10: 1238–1241.

[20] Velmurugan, K., Vellaiyappan Sangli, K. V., and Sechassalom, S. Synthesis of nickel zinc iron nanoparticles by coprecipitation technique. Materials Research. 2010; 13.3: 299–303.

[21] Chakka, V. M., et al. Magnetic nanoparticles produced by surfactant-assisted ball milling. Journal of Applied Physics. 2006; 99.8: 08E912.

[22] López-López, M. T., et al. Stability and magnetic characterization of oleate-covered magnetite ferrofluids in different nonpolar carriers. Journal of Colloid and Interface Science. 2005; 291.1: 144–151.

[23] Huang, W., and Xiaolei, W. Study on the properties and stability of ionic liquid-based ferrofluids. Colloid and Polymer Science. 2012; 290.16: 1695–1702.

[24] Rodríguez-Arco, L., et al. Steric repulsion as a way to achieve the required stability for the preparation of ionic liquid-based ferrofluids. Journal of Colloid and Interface Science. 2011; 357.1: 252–254.

[25] Kroell, M., et al. Magnetic and rheological characterization of novel ferrofluids. Journal of Magnetism and Magnetic Materials. 2005; 289: 21–24.

[26] Donadel, K., et al. Synthesis and characterization of the iron oxide magnetic particles coated with chitosan biopolymer. Materials Science and Engineering: C. 2008; 28.4: 509–514.

[27] Khosroshahi, M. E., and Ghazanfari, L. Preparation and rheological studies of uncoated and PVA-coated magnetite nanofluid. Journal of Magnetism and Magnetic Materials. 2012; 324.24: 4143–4146.

[28] Brullot, W., et al. Versatile ferrofluids based on polyethylene glycol coated iron oxide nanoparticles. Journal of Magnetism and Magnetic Materials. 2012; 324.11: 1919–1925.

[29] Abareshi, M., et al. Fabrication, characterization and measurement of thermal conductivity of Fe_3O_4 nanofluids. Journal of Magnetism and Magnetic Materials. 2010; 322.24: 3895–3901.

[30] Philip, J., Shima, P. D., and Baldev, R. Enhancement of thermal conductivity in magnetite based nanofluid due to chainlike structures. Applied Physics Letters. 2007; 91.20: 203108.

[31] Shima, P. D., John, P., and Baldev, R. Magnetically controllable nanofluid with tunable thermal conductivity and viscosity. Applied Physics Letters. 2009; 95.13: 133112.

[32] Fu, H. L., and Gao, L. Theory for anisotropic thermal conductivity of magnetic nanofluids. Physics Letters A. 2011; 375.41: 3588–3592.

[33] Gavili, A., et al. The thermal conductivity of water base ferrofluids under magnetic field. Experimental Thermal and Fluid Science. 2012; 41: 94–98.

[34] Hong, H., et al. Heat transfer nanofluids based on carbon nanotubes. Journal of Thermophysics and Heat Transfer. 2007; 21.1: 234–236.

[35] Wensel, J., et al. Enhanced thermal conductivity by aggregation in heat transfer nanofluids containing metal oxide nanoparticles and carbon nanotubes. Applied Physics Letters. 2008; 92.2: 023110.

[36] Wright, B., et al. Magnetic field enhanced thermal conductivity in heat transfer nanofluids containing Ni coated single wall carbon nanotubes. Applied Physics Letters. 2007; 91.17: 173116.

[37] Sundar, L. S., Manoj, K. S., and Antonio, C. M. S. Enhanced heat transfer and friction factor of MWCNT–Fe_3O_4/water hybrid nanofluids. International Communications in Heat and Mass Transfer. 2014; 52: 73–83.

[38] Shahsavar, A., et al. Effect of magnetic field on thermal conductivity and viscosity of a magnetic nanofluid loaded with carbon nanotubes. Journal of Mechanical Science and Technology. 2016; 30.2: 809–815.

[39] Hong, K. S., Tae-Keun, H., and Ho-Soon, Y. Thermal conductivity of Fe nanofluids depending on the cluster size of nanoparticles. Applied Physics Letters. 2006; 88.3: 031901.

[40] Nkurikiyimfura, I., Yanmin, W., and Zhidong, P. Effect of chain-like magnetite nanoparticle aggregates on thermal conductivity of magnetic nanofluid in magnetic field. Experimental Thermal and Fluid Science. 2013; 44: 607–612.

[41] Vadasz, J. J., Saneshan, G., and Peter, V. Heat transfer enhancement in nano-fluids suspensions: possible mechanisms and explanations. International Journal of Heat and Mass Transfer. 2005; 48.13: 2673–2683.

[42] Keblinski, P., Ravi, P., and Jacob, E. Thermal conductance of nanofluids: is the controversy over? Journal of Nanoparticle Research. 2008; 10.7: 1089–1097.

[43] Philip, J., Shima, P. D., and Baldev, R. Evidence for enhanced thermal conduction through percolating structures in nanofluids. Nanotechnology. 2008; 19.30: 305706.

[44] Odenbach, S. Recent progress in magnetic fluid research. Journal of Physics: Condensed Matter. 2004; 16.32: R1135.

[45] Sheikholeslami, M., and Ali, J. C. Flow and convective heat transfer of a ferro-nanofluid in a double-sided lid-driven cavity with a wavy wall in the presence of a variable magnetic field. Numerical Heat Transfer, Part A: Applications. 2016; 69.10: 1186–1200.

[46] Sheikholeslami, M., et al. Effect of thermal radiation on magnetohydrodynamics nanofluid flow and heat transfer by means of two phase model. Journal of Magnetism and Magnetic Materials. 2015; 374: 36–43.

[47] Sheikholeslami, M., and Davood, D. G. Ferrohydrodynamic and magnetohydrodynamic effects on ferrofluid flow and convective heat transfer. Energy. 2014; 75: 400–410.

5

Numerical Analysis of Three-Dimensional MHD Nanofluid Flow over a Stretching Sheet with Convective Boundary Conditions through a Porous Medium

Stanford Shateyi

Abstract

Numerical analysis has been carried out on the problem of three-dimensional magneto-hydrodynamic boundary layer flow of a nanofluid over a stretching sheet with convictive boundary conditions through a porous medium. Suitable similarity transformations were used to transform the governing partial differential equations into a system of ordinary differential equations. We then solved the resultant ordinary differential equation by using the spectral relaxation method. Effects of the dimensionless parameters on velocity, temperature and concentration profiles together with the friction coefficients, Nusselt and Sherwood numbers were discussed with the assistance of graphs and tables. The velocity was found to decrease with increasing values of the magnetic, stretching and permeability parameters. The local temperature was observed to rise as the Brownian motion, thermophoresis and Biot numbers increased. The concentration profiles diminish with increasing values of the Lewis number and chemical reaction parameter.

Keywords: numerical analysis, MHD nanofluid, stretching sheet, convective boundary conditions, porous medium

1. Introduction

Many researchers have over the past few years paid significant attention to the study of boundary layer flow heat and mass transfer over a stretching sheet due to its industrial and engineering applications. These applications include cooling of papers, glass-fibre production, plastic sheets and polymer extrusion, hot rolling wire drawing, metal spinning, stretching of rubber sheets and crystal growing. The quality and final product formation in these processes are dependent on the rate of stretching and cooling.

Since the pioneering study by Crane [1] who presented an exact analytical solution for the steady two-dimensional flow due to a stretching surface in a quiescent fluid many studies on stretched surfaces have been done [1–5].

Thermal conductivity of nanoparticles has been shown in recent research on nanofluid to change the fluid characteristics. The thermal conductivity of the base liquid with the enhanced conductivity of nanofluid and the turbulence induced by their motion contribute to a remarkable improvement in the convective heat transfer coefficient. This feature of nanofluid makes them attractive to a wide variety of industries, ranging from transportation to energy production and supply to electronics. They can be used in welding equipment, high heat flux and to cool car engines, among other applications. Many researchers [6–10] have studied the boundary layer flow of a nanofluid caused by a stretching surface.

Shateyi and Prakash [11] carried out a numerical analysis on the problem of magneto hydrodynamic boundary layer flow of a nanofluid over a moving surface in the presence of thermal radiation. Kuznetsov and Nield [12] examined the influence of nanoparticles on natural convection boundary layer flow past a vertical plate, using a model in which Brownian motion and thermophoresis are accounted for. Aziz and Khan [13] investigated using a similarity analysis of the transport equations by their numerical computations to the natural convective flow of a nanofluid over a convectively heated vertical plate.

Makinde and Aziz [14] numerically studied the boundary layer flow induced in a nanofluid due to a linearly stretching sheet. Hayat et al. [15] addressed the MHD flow of second grade nanofluid over a nonlinear stretching sheet. Zhao et al. [16] studied the three-dimensional nanofluid bio-convection near a stagnation attachment. Sheikholeslami and Ganji [17] studied two-dimensional laminar-forced convection nanofluids over a stretching surface in a porous medium. The study used different models of nanofluid based on different formulas for thermal conductivity and dynamic viscosity. Nayak et al. [18] did a numerical study on the mixed convection of copper-water nanofluid inside a differentially heated skew enclosure. Recently, Mabood and Das [19] analysed MHD flow and melting heat transfer of a nanofluid over a stretching surface. Naramgari and Sulochana [20] analysed the momentum and heat transfer of MHD nanofluid embedded with conducting dust particles past a stretching surface in the presence of volume fraction of dust particles. Sandeep et al. [21] analysed the unsteady MHD radiative flow and heat transfer characteristics of a dusty nanofluid over an exponentially permeable stretching in the presence of volume fraction of dust and nanoparticles. Sheikholeslami et al. [22] computationally investigated nanofluid flow and heat transfer in a square heated rectangular body. Sheikholeslami and Ganji [23] provided a review of researches on nanofluid flow and heat transfer via semi-analytical and numerical methods. Lastly, Naramgari and Sulochana [20] analysed the three-dimensional MHD Newtonian and non-Newtonian fluid flow over a stretching surface in the presence of thermophoresis and Brownian motion.

The main objective of this chapter is to numerically analyse the influence of convective boundary conditions on the model of three-dimensional magnetohydrodynamic, nanofluid flow over a stretching sheet through a porous medium in the presence of thermophoresis and Brownian motion as well as thermal radiation. The governing partial differential equations use suitable similarity transformations. The transformed governing equations are solved numerically using the spectral relaxation method (SRM). The effects of dimensionless parameters on velocity components, temperature and concentration profiles together with the skin friction coefficients, local Nusselt and Sherwood numbers are discussed with the aid of tables and graphs.

2. Mathematical formulation

We consider a three-dimensional steady incompressible MHD nanofluid flow, heat and mass transfer over a linearly stretching sheet through a porous medium. The sheet is assumed to be stretched along the xy -plane while the fluid is placed along the z-axis. A uniform magnetic field B_0 is applied normally to the stretched sheet and the induced magnetic field is neglected by assuming very small Reynolds number. We assume that the sheet is stretched with linear velocities $u = ax$ and $v = by$ along the xy-plane, respectively, with constants a and b. Under the above assumptions and the boundary approximation, the governing equations for the current study are given by:

$$\frac{\partial u}{\partial x}+\frac{\partial v}{\partial y}+\frac{\partial w}{\partial z}=0, \tag{1}$$

$$u\frac{\partial u}{\partial x}+v\frac{\partial u}{\partial y}+w\frac{\partial u}{\partial z}=v\frac{\partial^2 u}{\partial z^2}-\frac{\sigma B_0^2}{\rho}u-\frac{v}{k_1}u, \tag{2}$$

$$u\frac{\partial v}{\partial x}+v\frac{\partial v}{\partial y}+w\frac{\partial v}{\partial z}=v\frac{\partial^2 v}{\partial z^2}-\frac{\sigma B_0^2}{\rho}v-\frac{v}{k_1}v, \tag{3}$$

$$u\frac{\partial T}{\partial x}+v\frac{\partial T}{\partial y}+w\frac{\partial T}{\partial z}=\alpha\frac{\partial^2 T}{\partial z^2}-\frac{\partial q_r}{\partial z}+\tau\left[D_B\frac{\partial C}{\partial z}\frac{\partial T}{\partial z}+\frac{D_T}{T_\infty}\left(\frac{\partial T}{\partial z}\right)^2\right], \tag{4}$$

$$u\frac{\partial C}{\partial x}+v\frac{\partial C}{\partial y}+w\frac{\partial C}{\partial z}=D_B\frac{\partial^2 C}{\partial z^2}+\frac{D_T}{T_\infty}\frac{\partial^2 T}{\partial z^2}, \tag{5}$$

where u, v and w are the velocity components in the x, y and z- directions, respectively, T is the fluid temperature, C is the fluid concentration, k_1 is the permeability, v is kinematic viscosity, ρ is the fluid density, τ is the ratio of the heat capacitances, D_B and D_T are the Brownian motion and thermopheric diffusion coefficients and c_p is the specific heat capacity.

The corresponding boundary conditions for the flow model are:

$$u = ax, v = by, w = 0, -k\frac{\partial T}{\partial z}=h_f(T_f-T),\ -D_B\frac{\partial C}{\partial z}=h_s(C_s-C) \text{ at } z=0, \tag{6}$$

$$u\to 0, v\to 0, w\to 0, T\to T_\infty, C\to C_\infty, \text{as} z\to\infty. \tag{7}$$

We have h_f as the convective heat transfer coefficient, h_s is the convective mass transfer coefficient and T_f and C_f are the convective fluid temperature and concentration below the moving sheet.

3. Similarity transformation

In order to non-dimensionalise the governing equations, we introduce the following similarity equations [24]. These transformations also transform the partial differential equations into a system of ordinary differential equations which is then solved using the spectral relaxation method:

$$\eta = \sqrt{\frac{a}{v}}z, u = axf'(\eta), v = byg'(\eta), w = -\sqrt{av}[f(\eta) + cg(\eta)], \theta(\eta) = \frac{T - T_\infty}{T_f - T_\infty}, \phi(\eta) = \frac{C - C_\infty}{C_s - C_\infty}. \quad (8)$$

Upon substituting the similarity variables into Eqs. (2)–(5), we obtain the following system of ordinary equations

$$f''' + (f + cg)f'' - f'^2 - (M + K)f' = 0, \quad (9)$$

$$g''' + (f + cg)g'' - g'^2 - (M + K)g' = 0, \quad (10)$$

$$\left(\frac{3 + 4R}{3P_rR}\right)\theta'' + (f + cg)\theta' + Nb\theta'\varnothing' + Nt(\theta')^2 = 0, \quad (11)$$

$$\varnothing'' + Le(f + cg)\varnothing' + \frac{Nt}{Nb}\theta'' = 0. \quad (12)$$

The corresponding boundary conditions are

$$f = 0, f' = 1, g = 0, g' = 1, \theta' = -Bi_t(1 - \theta), \varnothing' = -Bi_c(1 - \varnothing), \ at\ \eta = 0, \quad (13)$$

$$f'(\infty) \to 0, g'(\infty) \to 0, \theta(\infty) \to 0, \varnothing(\infty) \to 0. \quad (14)$$

Primes denote differentiation with respect to η and parameters appearing in Eqs (9)–(14) are defined as: $Pr = v/\alpha$ is the Prandtl number, $Le = v/D_B$ is the Lewis number, $Nb = \tau D_B(C_s - C_\infty)/v$ is the Brownian motion parameter, $Nt = \tau D_T(T_f - T_\infty)/vT_\infty$ is the thermophoresis parameter, $Bi_t = \frac{h_f}{k}\sqrt{v/a}, Bi_c = \frac{h_s}{D_B}\sqrt{v/a}$ are the Biot numbers and $c = b/a$ is the stretching parameter. The quantities of engineering interest are the skin-friction coefficient C_f along the x- and y-direction (C_{fx} and C_{fy}), the Nusselt number and Sherwood number. These quantities are defined as follows:

$$C_{fx} = \frac{\tau_{wx}}{\rho u_w^2}, C_{fy} = \frac{\tau wx}{\rho u_w^2}, Nu = x\frac{q_w}{k(T_f - T_w)}, Sh = x\frac{q_w}{D_B(C_f - C_w)}, \quad (15)$$

where τ_{wx}, τ_{wy} are the wall shear along x- and y-directions, respectively, and q_w and q_m are the heat flux and mass flux at the surface, respectively.Upon using the similarity variables into the above expressions, we obtain the following:

$$Re^2 C_{fx} = f''(0), Re^{\frac{1}{2}}C_{fy} = g''(0), Re^{\frac{-1}{2}}Nu = -\theta'(0), Re^{\frac{-1}{2}}Sh = -\varnothing(0). \quad (16)$$

4. Method

To solve the set of ordinary differential Eqs. (9)–(12) together with the boundary conditions (13) and (14), we employ the Chebyshev pseudo-spectral method known as spectral relaxation method. This is a recently developed method, and the details of the method are found in Motsa et al. [25]. This method transforms sets of non-linear ordinary differential into sets of linear ordinary differential equations. The entire computational procedure is implemented using a program written in MATLAB computer language. The nanofluid velocity, temperature, the local skin-friction coefficient and the local Nusselt and Sherwood numbers are determined from these numerical computations.

To apply the SRM to the non-linear ordinary differential equations, we first set $f'(\eta) = p(\eta)$ and $g'(\eta) = q(\eta)$. We then write the equations as follows:

$$f' = p, \tag{17}$$

$$p'' + (f + cg)p' - p^2 - (M + K)p = 0, \tag{18}$$

$$g' = q, \tag{19}$$

$$q'' + (f + cg)q' - q^2 - (M + K)q = 0, \tag{20}$$

$$\left(\frac{3 + 4R}{3PrR}\right)\theta'' + (f + cg)\theta' + Nb\theta'\varnothing' + Nt\theta'^2 = 0, \tag{21}$$

$$\varnothing'' + Le(f + cg)\varnothing' + \frac{Nt}{Nb}\theta'' = 0. \tag{22}$$

The boundary conditions become

$$f(0) = 0, g(0) = 0, p(0) = 1, q(0) = 1, \tag{23}$$

$$\theta'(0) = -Bit(1 - \theta), \varnothing'(0) = -Bic(1 - \phi), \tag{24}$$

$$p(\infty) = 0, q(\infty) = 0, \varnothing(\infty) = 0, \theta(\infty) = 0, \tag{25}$$

In view of the SRM, we then obtain the following iterative scheme:

$$f'_{r+1} = p_r, f_{r+1}(0) = 0, \tag{26}$$

$$p''_{r+1} + (f_{r+1} + cg_{r+1})p_{r+1} - (M + K)p_{r+1} = p_r^2, p_{r+1}(0) = 1, p_{r+1}(\infty) = 0, \tag{27}$$

$$g'_{r+1} = q_r, g_{r+1}(0) = 0, \tag{28}$$

$$q''_{r+1} + (f_{r+1} + cg_{r+1})q'_{r+1} - (M + K)q_{r+1} = q_r^2, q_{r+1}(0) = 1, q_{r+1}(\infty) = 0, \tag{29}$$

$$\left(\frac{3 + 4R}{3PrR}\right)\theta''_{r+1} + (f_{r+1} + cg_{r+1})\theta'_{r+1} = -Nb\theta'_r\varnothing'_r - Nt\theta'^2_r, \theta'_{r+1}(0) = -Bit(1 - \theta_{r+1}(0)), \varnothing_{r+1}(\infty) = 0, \tag{30}$$

$$\varnothing''_{r+1} + Le(f_{r+1} + cg_{r+1})\varnothing'_{r+1} = -\frac{Nt}{Nb}\varnothing''_{r+1}, \phi'_{r+1}(0) = -Bic(1-\phi_{r+1}(0)), \varnothing_{r+1}(\infty) = 0. \tag{31}$$

The above equations form a system of linear decoupled equations which can be solved iteratively for $r = 1, 2, \ldots$. Starting from initial guesses $\left(p_0(\eta), q_0(\eta), \theta(\eta), \varnothing(\eta)\right)$. Applying the Chebyshev pseudo-spectral method to the above equations, we obtain

$$A_1 f_{r+1} = B_1, f_{r+1}(\tau N) = 0, \tag{32}$$

$$A_2 p_{r+1} = B_2, p_{r+1}(\tau N) = 1, p_{r+1}(\tau_0) = 0, \tag{33}$$

$$A_3 g_{r+1} = B_3, g_{r+1}(\tau N) = 0, \tag{34}$$

$$A_4 q_{r+1} = B_4, q_{r+1}(\tau N) = 1, q_{r+1}(\tau_0) = 0, \tag{35}$$

$$A_5 \theta_{r+1} = B_4, \theta_{r+1}(\tau N) = \frac{Bit}{1 + Bit}, \theta_{r+1}(\tau_0) = 0, \tag{36}$$

$$A_6 \varnothing_{r+1} = B_6, \varnothing_{r+1}(\tau N) = \frac{Bit}{1 + Bit}, \varnothing_{r+1}(\tau_0) = 0. \tag{37}$$

where, $A_1 = D, B_1 = p_r,\ A_2 = D^2 + \text{diag}(f_{r+1} + cg_{r+1})D-(M + K)I,\ B_2 = q^2_{r+1},\ A_3 = D, B_3 = q_r,$ $A_4 = D^2 + \text{diag}(f_{r+1} + cg_{r+1})D-(M + K)I,\ A_5 = \left(\frac{3+4R}{3PrR}\right)D^2 + \text{diag}(f_{r+1} + cg_{r+1})D,\ B_5 = -N_6\theta'_r\varnothing'_r$ $-Nt\theta'^2_r,\ A_6 = D^2 + \text{diag}[Lef_{r+1} + cLeg_{r+1}]D,\ B_6 = -\frac{Nt}{Nb}\varnothing''_{r+1},$

where I is the identity matrix of size $(N + 1)(N + 1)$. The initial guesses are obtained as:

$$p_0(\eta) = e^{-n}, q_0(\eta) = e^{-n}, \theta_0(\eta) = \frac{Bite^{-n}}{1 + Bit}, \varnothing_0(\eta) = \frac{Bice^{-n}}{1 + Bic}. \tag{38}$$

5. Results and discussion

The system of ordinary differential Eqs. (9)–(12) subject to the boundary conditions (13) and (14) is numerically solved by applying the spectral relaxation method. The SRM results presented in this chapter were obtained using $N = 40$ collocation points, and also the convergence was achieved after as few as six iterations. We also use these default values for the parameters $Pr = 0.71, Nt = Nb = 0.3, Le = 2, R = 1, M = 1, K = 0.5, C = 0.1, Bit = 0.2 = Bic$.

Table 1 displays the validation of the present results with those obtained by the bvp4c results. As can be clearly observed from this table, there is an excellent agreement between the results obtained by bvp4c method giving confidence in the findings of this study. **Table 1** also shows the influence of the magnetic, permeability and stretching parameters on the skin friction coefficients. It is noticed that the skin friction coefficient increase with the increasing values of the parameters.

Table 2 depicts the influence of Brownian motion thermophoresis, parameters and the Biot numbers on the Nusselt and Sherwood numbers. Both the rates of heat transfer and mass

			⟨i⟩$-f''(0)$⟨/i⟩		⟨i⟩$-g''(0)$⟨/i⟩	
M	**K**	**C**	**Bvp4c**	**SRM**	**Bvp4c**	**SRM**
0	0.1	0.1	1.22690	1.22690	0.09495	0.09495
1	0.1	0.1	1.52270	1.52270	0.137834	0.197854
3	0.1	0.1	2.12214	2.12214	0.19750	0.197520
5	0.1	0.1	2.55018	2.55018	0.24290	0.24290
1	0.0	0.1	1.415504	1.415504	0.118270	0.118270
1	0.5	0.1	1.582270	1.582270	0.137834	0.137834
1	1.5	0.1	1.871764	1.871764	0.170320	0.170320
1	3.0	0.1	2.236839	2.236839	0.209800	0.209800
1	0.2	0.2	1.585645	1.585645	0.280922	0.28022
1	0.2	0.4	1.599025	1.599025	0.585064	0.585064
1	0.2	0.6	1.610701	1.610701	0.916198	0.916198
1	0.2	0.8	1.647760	1.647760	0.927968	0.927968

Table 1. Variation of the magnetic, permeability and stretching parameters on the skin friction coefficients.

Nb	**Nt**	**Bit**	**Bic**	⟨i⟩$-\theta'(0)$⟨/i⟩	⟨i⟩$\varnothing'(0)$⟨/i⟩
0.2	0.1	0.2	0.2	0.528756	0.272081
0.4	0.1	0.2	0.2	0.540153	0.283370
0.6	0.1	0.2	0.2	0.627812	0.370495
0.8	0.1	0.2	0.2	0.685700	0.428392
0.1	0.2	0.2	0.2	0.484629	0.277618
0.1	0.4	0.2	0.2	0.418686	0.299623
0.1	0.6	0.2	0.2	0.364816	0.321133
0.1	0.8	0.2	0.2	0.320526	0.342009
0.1	0.1	0.2	0.2	0.523097	0.266480
0.1	0.1	0.4	0.2	0.868307	0.633407
0.1	0.1	0.6	0.2	1.114143	0.926920
0.1	0.1	0.2	0.2	0.523097	0.266480
0.1	0.1	0.2	0.4	0.516229	0.115364
0.1	0.1	0.2	0.6	0.511106	0.002054

Table 2. The influence of the Brownian motion and thermophoresis parameters as well as that of the Biot numbers on the Nusselt and Sherwood numbers.

transfer are increasing functions of the Brownian motion parameter. By definition, thermophoresis is the migration of a colloidal particle in a solution in response to a microscopic temperature gradient. The heat transfer is reduced while the mass transfer increases with increasing values of the thermophoresis parameter. Lastly, **Table 2** shows the influence of the Biot numbers on the heat transfer and mass transfer rates and they both increase with increasing values of the thermal Biot number. But we noticed opposite effects when the solutal Biot number increases.

Figures 1 and **2** display the effect of permeability parameter on the velocity profiles. We observe that the tangential velocity profiles decrease as the values of the permeability parameter. Also, the transverse velocity ($f'(\eta)$) is reduced by the increasing values of the permeability parameter as more nanofluid is taken away from the boundary layer. This explains the thinning of the velocity boundary layers as the values of K increases (**Figure 1**). **Figures 3** and **4** depict the effect of the magnetic field parameter on the velocity profiles. As expected, we observe that both velocity components are greatly reduced as the values of the magnetic parameter increase. This is because physically increasing the values of magnetic field strength produces a drag-like force known as the Lorentz force. This force acts against the flow when the magnetic field is applied in the normal direction, as in this chapter. **Figures 5** and **6** display the influence of the stretching parameter on the velocity fields. It is seen from **Figure 5** that the tangential velocity profiles $f'(\eta)$ are reduced by increasing values of the stretching parameter c. The transverse velocity is enhanced with the increasing values of the stretching parameter.

Figure 7 displays the influence of the Biot number *Bi*, on the temperature profiles. It is clearly observed on this figure that the nanofluid temperature field rapidly increases near the boundary with increasing values of the Biot number, *Bit*, . It is also observed that as the Biot number increases the convective heating of the sheet also increases.

Figures 8 and **9** reveal the effect of the stretching ratio parameter c on the temperature and concentration profile. It is observed that the temperature and concentration profiles are

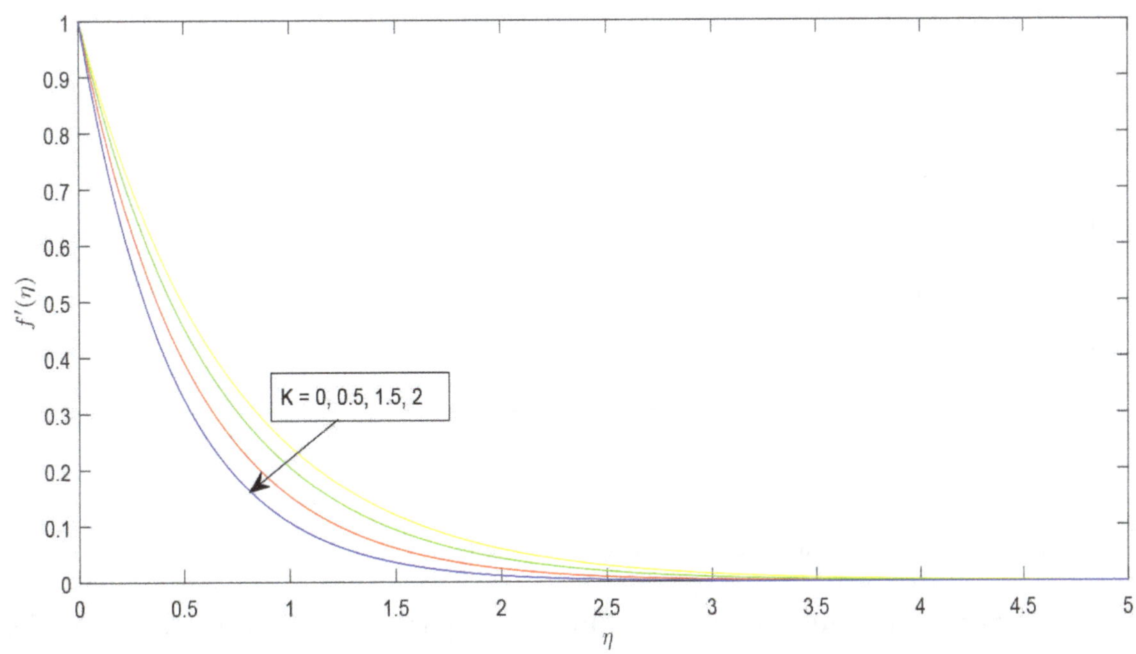

Figure 1. Effect of permeability parameter on the tangential velocity profiles.

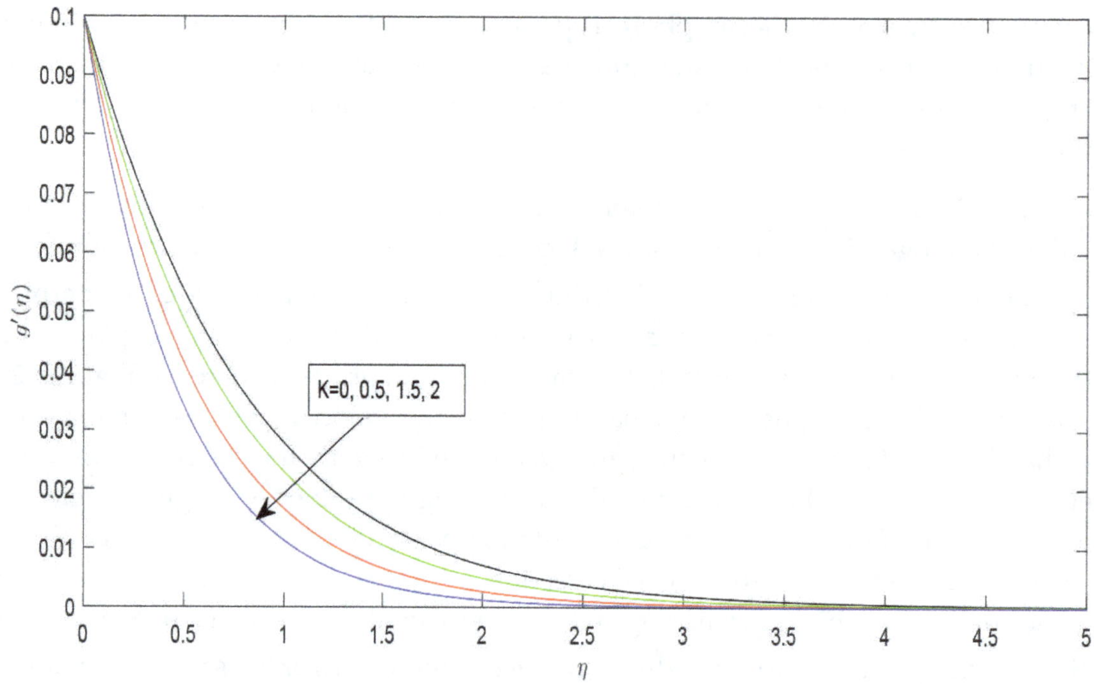

Figure 2. Effect of permeability parameter on the transverse velocity profiles.

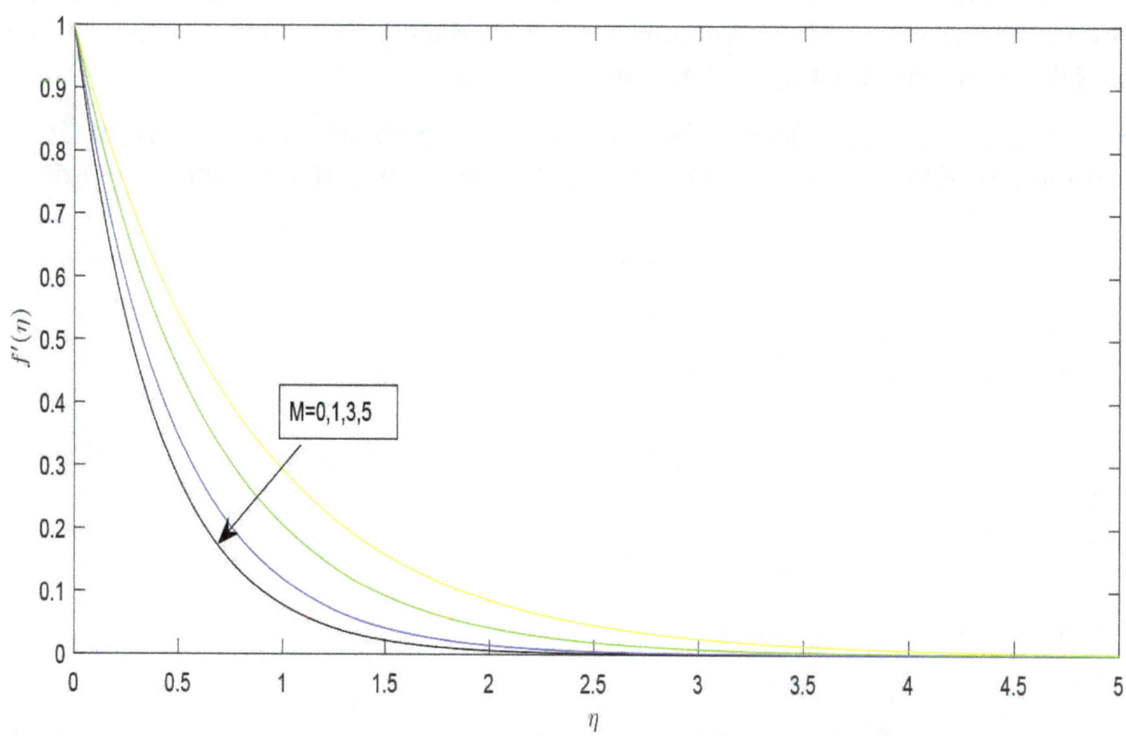

Figure 3. Influence of the magnetic parameter on the tangential velocity.

reduced with increasing values of the stretching ration parameter. **Figures 10** and **11** display the effects of thermophoresis parameter on the dimensionless temperature and concentration profiles. It is observed that the temperature and concentration profiles increase as the values of

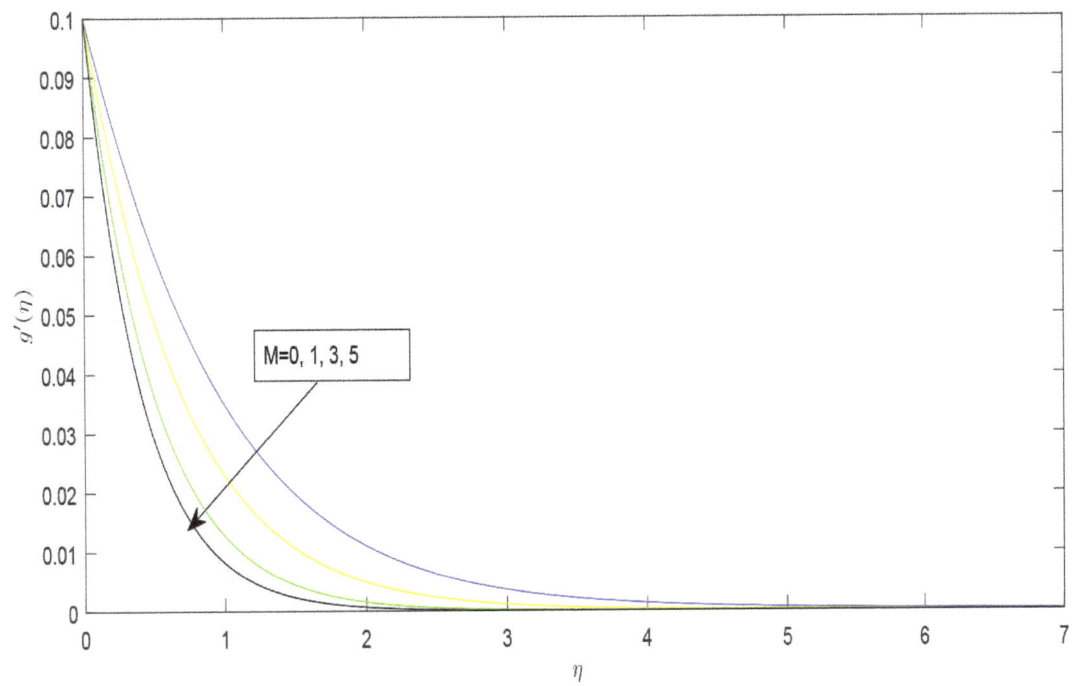

Figure 4. Influence of the magnetic parameter on the transverse velocity.

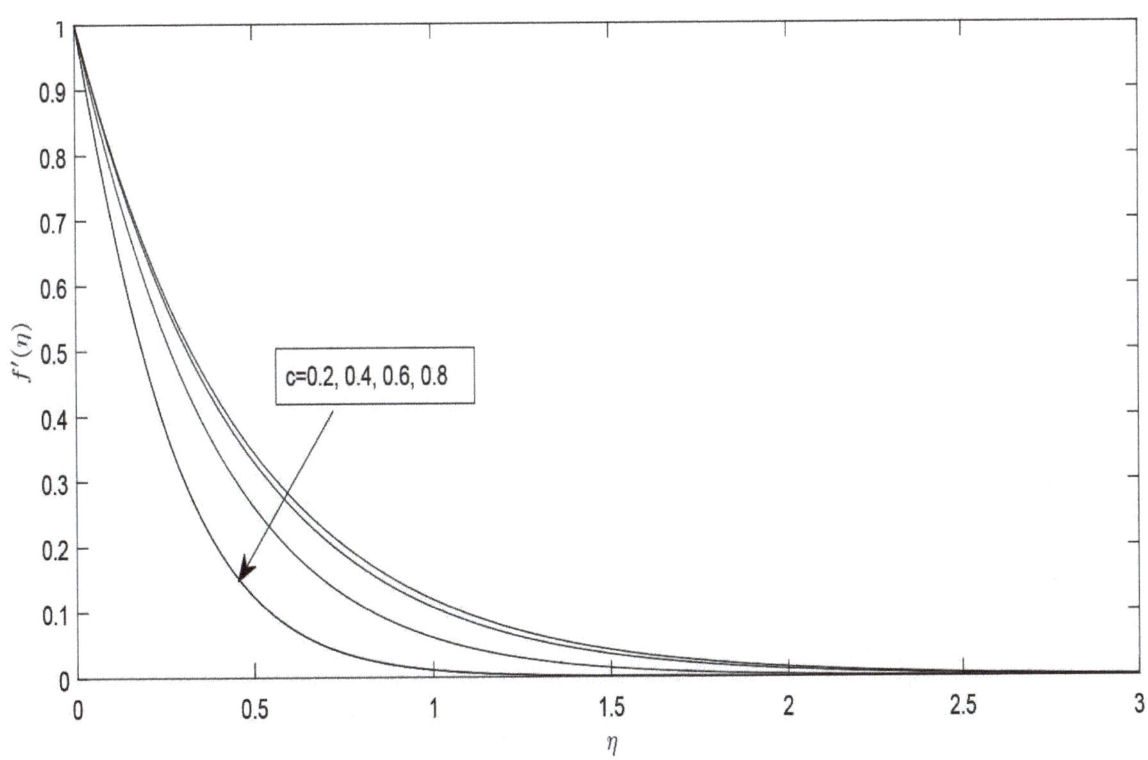

Figure 5. Influence of the stretching parameter on the tangential velocity.

the thermophoresis Nt increase. **Figure 12** depicts the influence of the Brownian motion parameter Nb on the temperature profiles. Increasing the values of the Brownian motion parameter Nb results in thicking of the thermal boundary layer, thus enhancing the

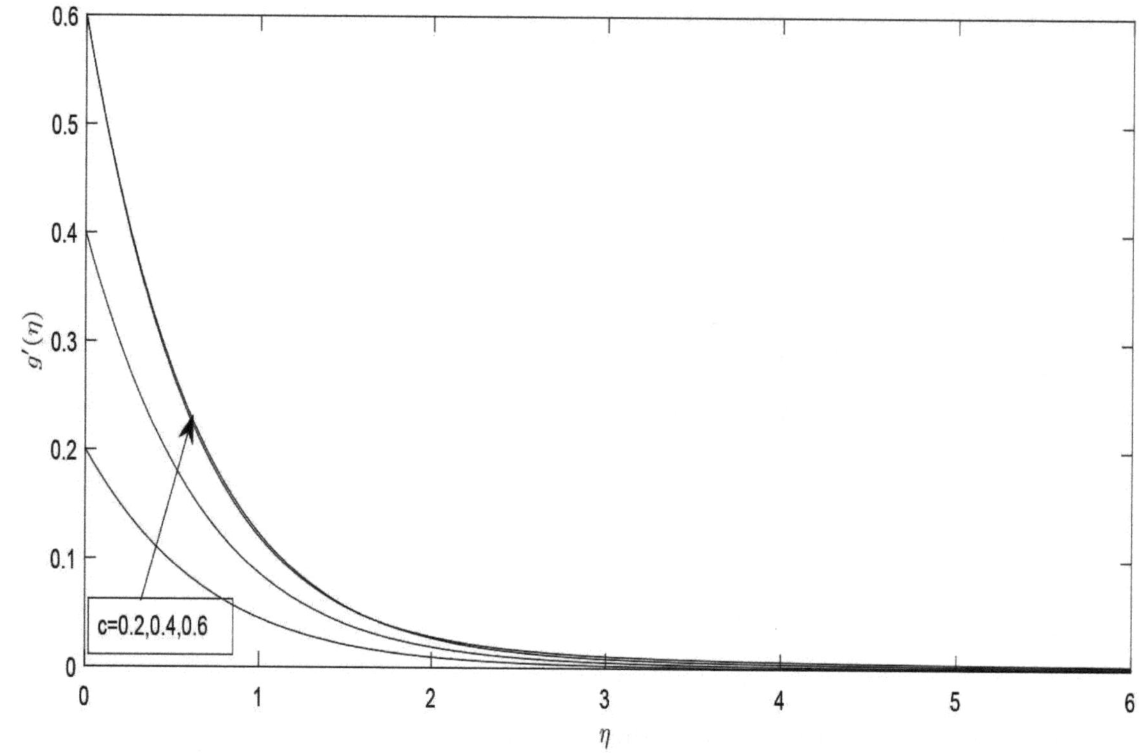

Figure 6. Variation of the stretching parameter on the velocity.

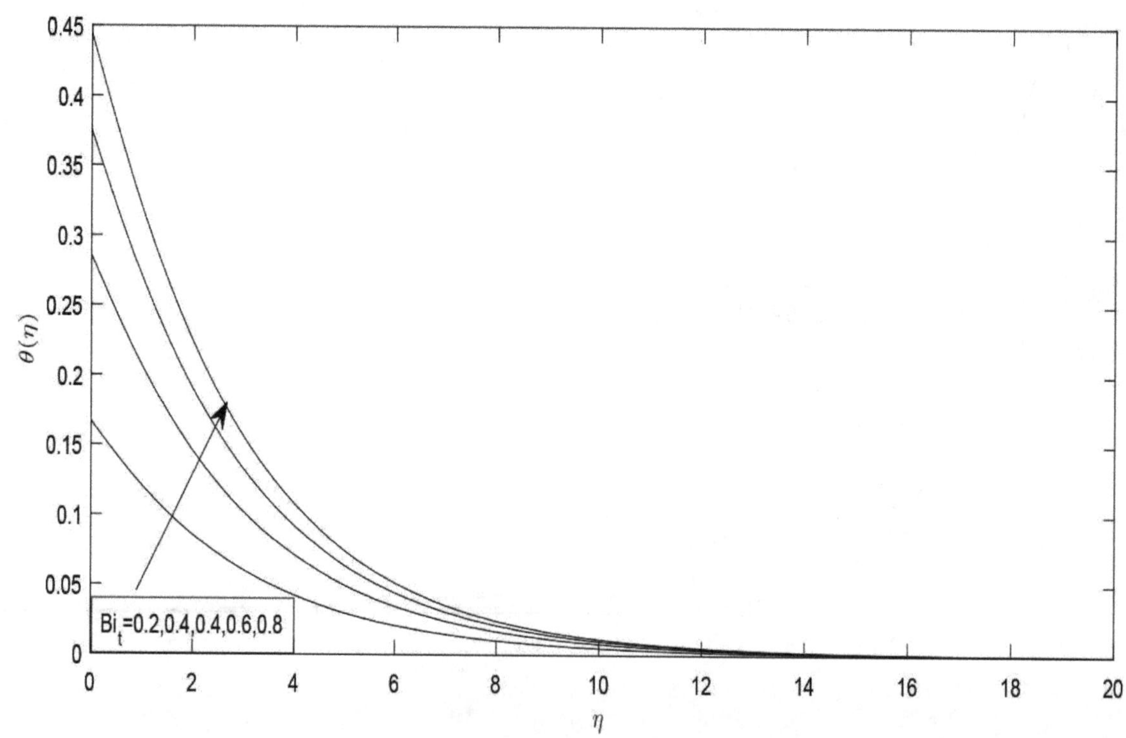

Figure 7. Effect of the Biot number on the temperature.

temperature of the nanofluid. **Figures 13** and **14** are plotted to depict the influence of the permeability K and magnetic M, parameters on the temperature profiles. The temperature of

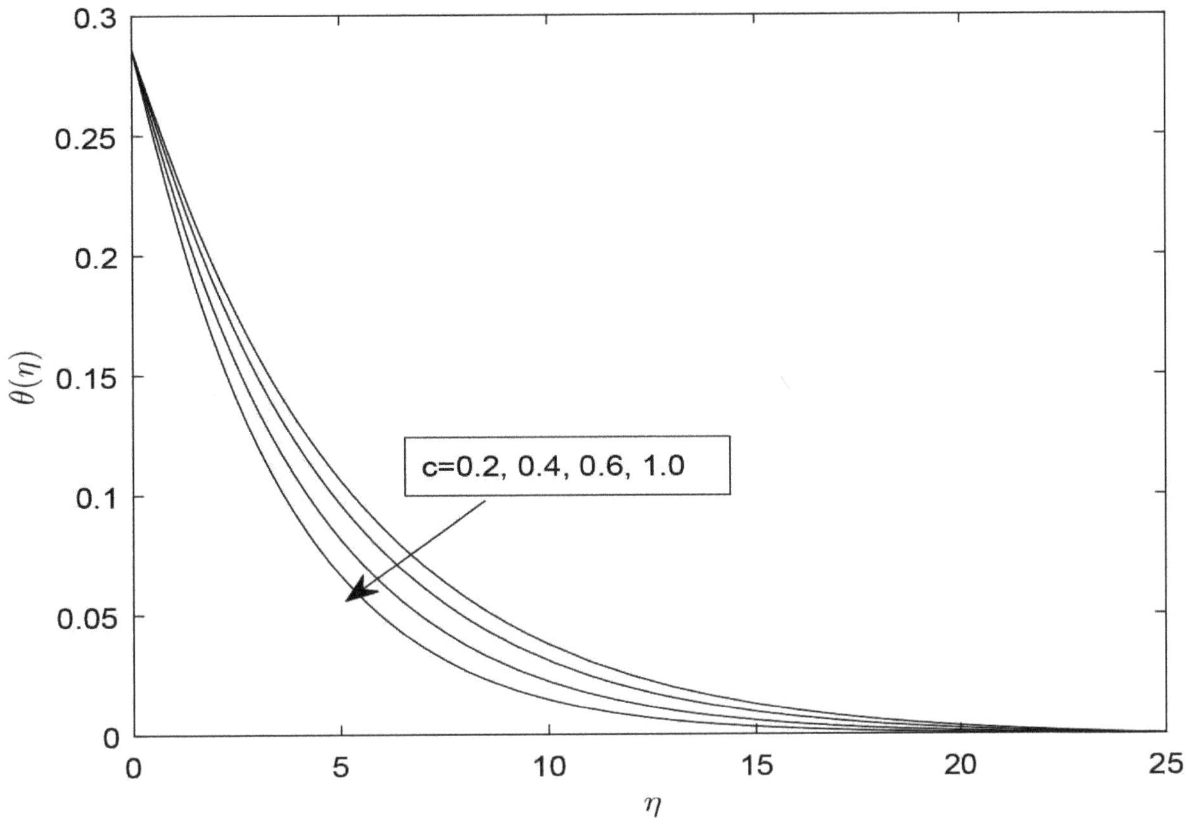

Figure 8. Effect of varying the stretching parameter on the temperature.

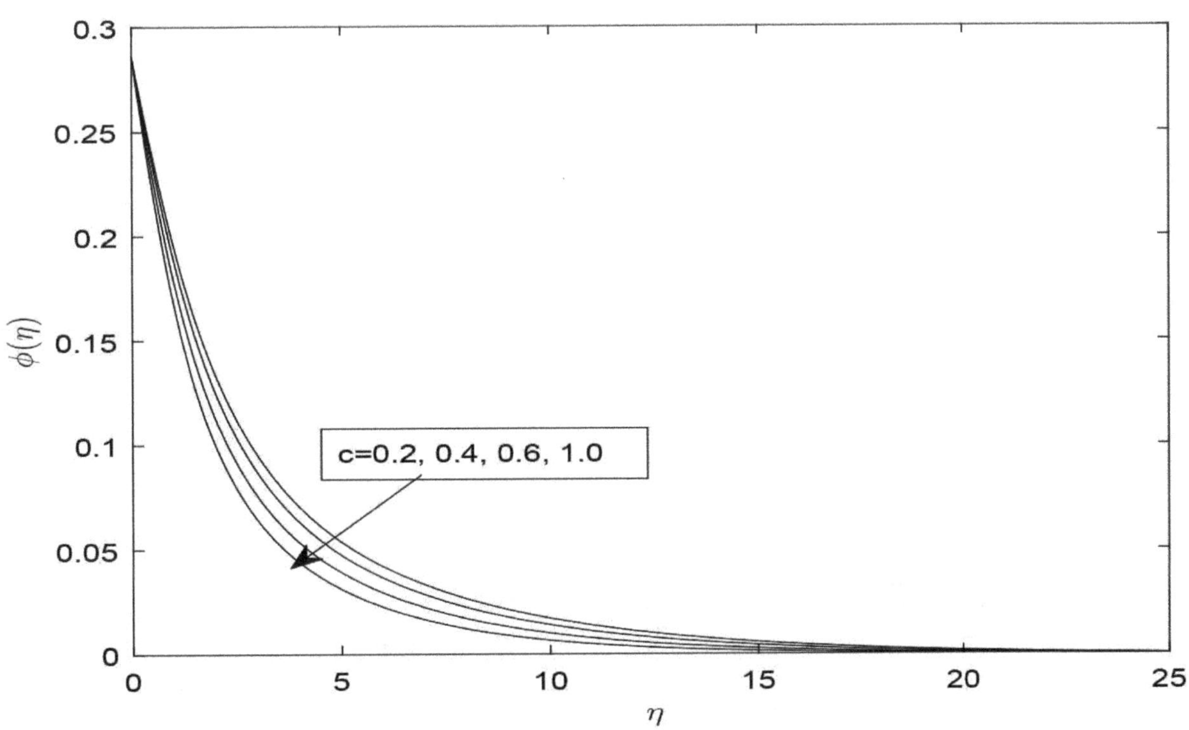

Figure 9. Effect of varying the stretching parameter on the temperature.

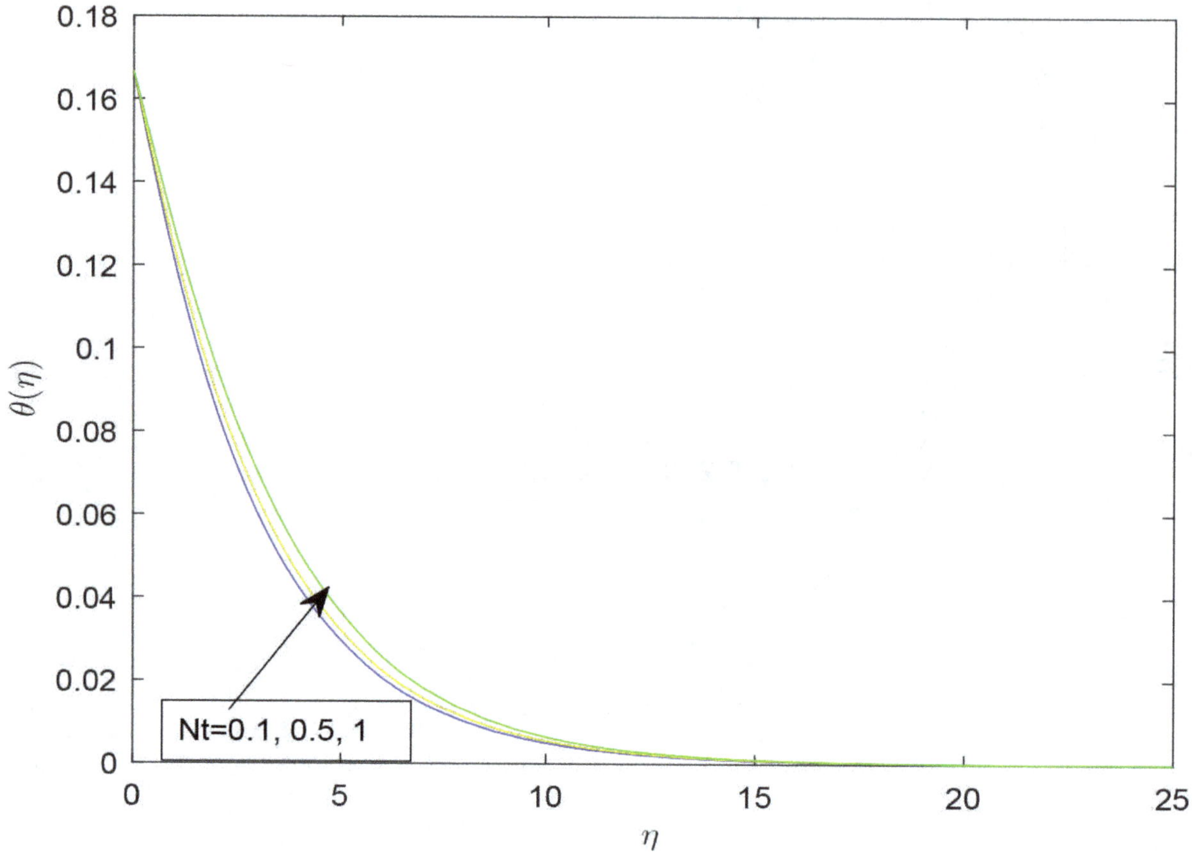

Figure 10. Effect of thermophoresis parameter on the temperature profiles.

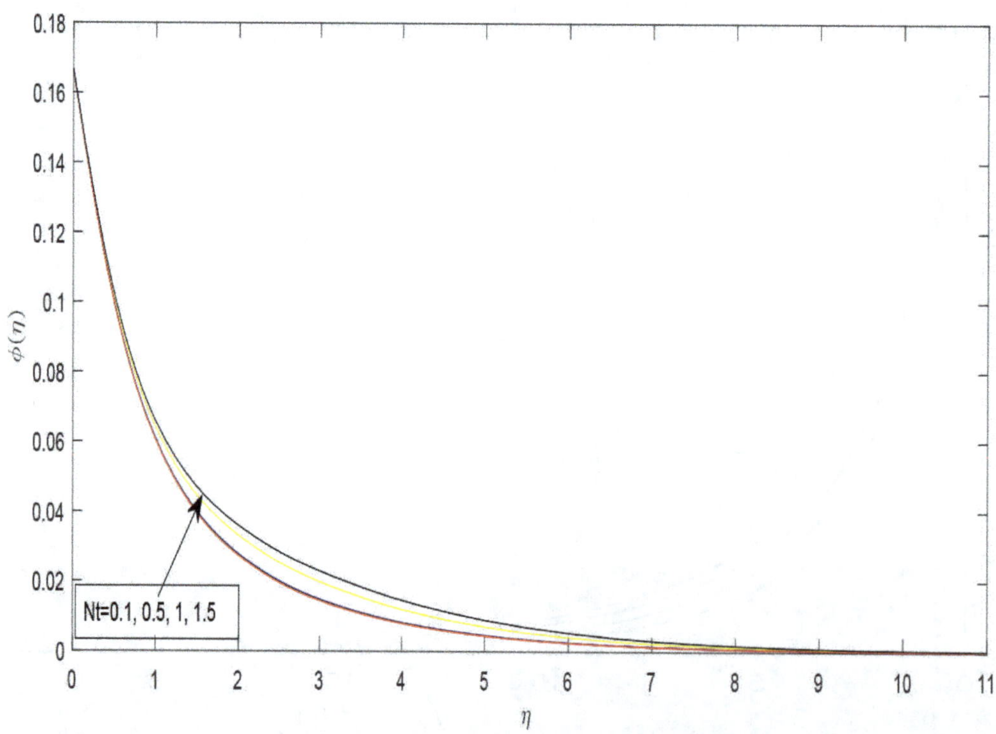

Figure 11. Effect of thermophoresis parameter on the concentration profiles.

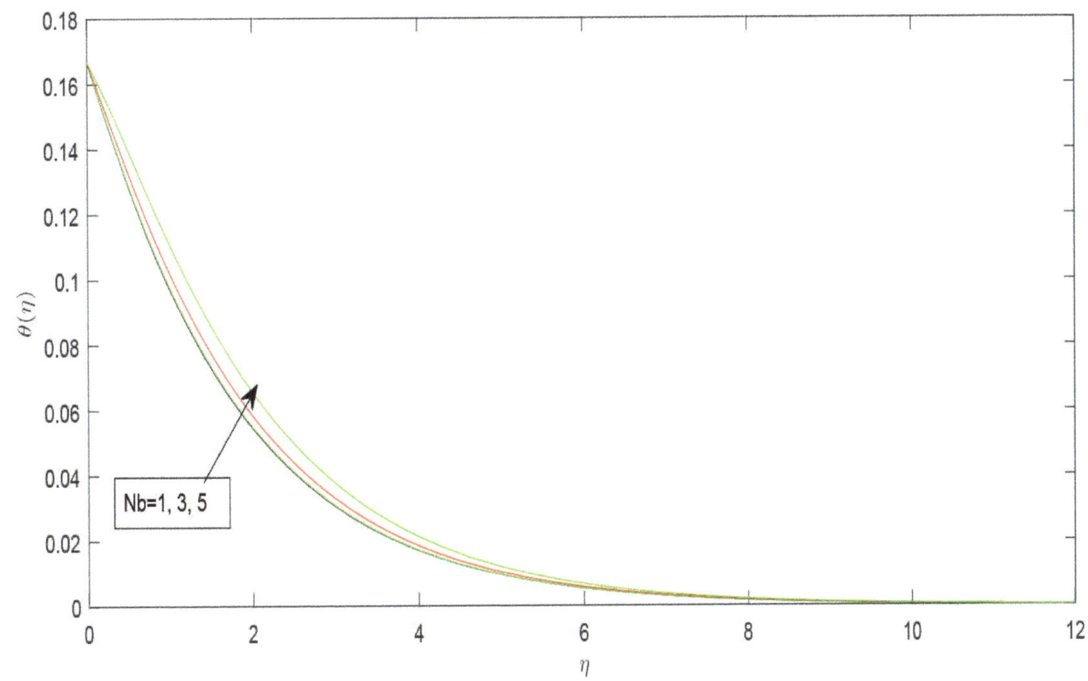

Figure 12. Effect of Brownian motion parameter on the temperature profiles.

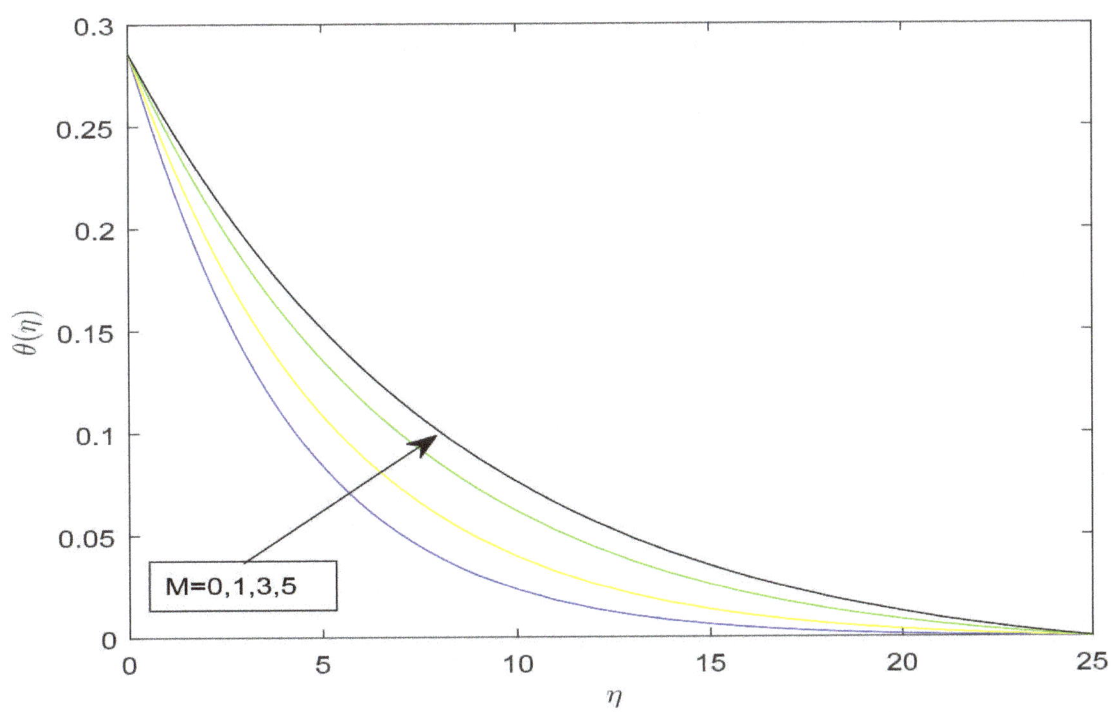

Figure 13. Effect of magnetic parameter on the temperature profiles.

the nanofluid increases with increases values of the permeability parameter. From **Figure 14,** we observe that the temperature profiles increase with the increasing values of the magnetic field parameter. **Figure 15** displays the effect of thermal radiation parameter R on the

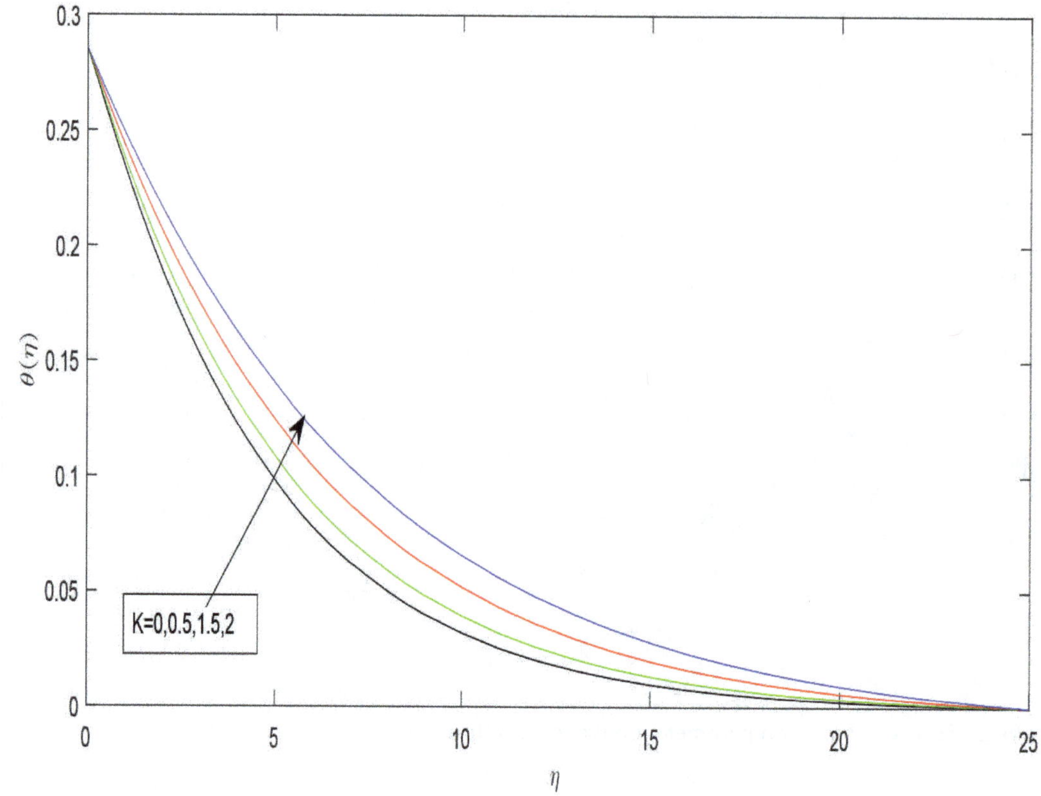

Figure 14. Influence of the permeability parameter on the nanofluid temperature.

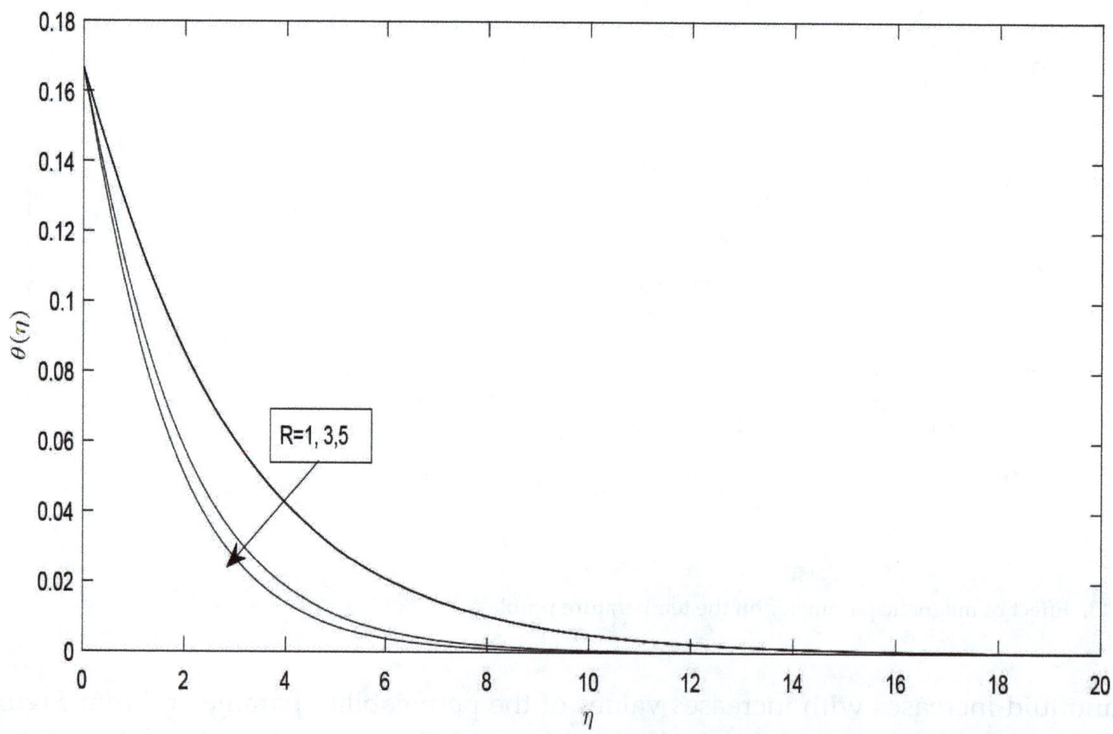

Figure 15. Influence of thermal radiation on the temperature.

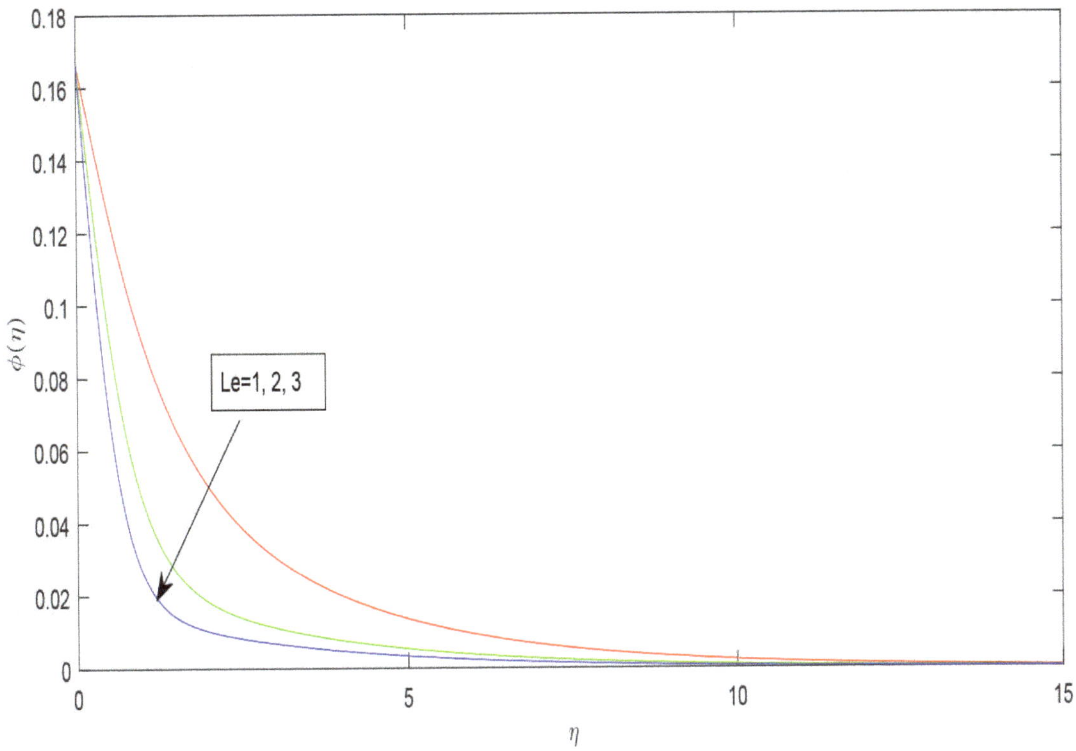

Figure 16. Influence of the Lewis number on the concentration.

temperature profiles. We observe in this figure that increasing the values of the thermal radiation produces a significant reduction in the thermal condition of the fluid flow.

Lastly, the effect of the Lewis number on the concentration profiles is depicted on **Figure 16.** Large values of the Lewis number implies increased values of the Schmidt number which results in the thinning of the solutal boundary layer.

6. Conclusion

A three-dimensional magnetohydrodynamic nanofluid, heat and mass transfer over a stretching surface with convective boundary conditions through a porous medium. The transformed governing equations are solved numerically using the spectral relaxation method. The accuracy of the SRM was validated against the MATLAB in-built bvp4c routine for solving boundary value problems. The following conclusions are driven from this study:

- The effect of increasing the magnetic field parameter is to reduce the momentum boundary layer there and to increase the thermal and solutal boundary layer thickness. The same effect on the flow characteristics is also experienced by increasing values of the stretching parameter (c).
- We observed that the local temperature rises as the Brownian motion, thermophoresis, permeability parameter and Biot numbers intensify. But opposite influences are observed when the values of the thermal radiation and stretching parameters increase.

- Increasing values of the Lewis number (*Le*) dimishes the concentration of the nanoparticles.
- The rise in the stretching ratio parameter increases the Nusselt and Sherwood number.
- Lastly, the Nusselt number decreases, while the Sherwood number increases as the Brownian motion and thermophoresis effects increase.

Nomenclature

a, b	positive constants
B_0	uniform magnetic field strength
B_{it}, B_{ic}	Biot numbers
c	stretching parameter
C	fluid concentration
C_f	skin friction coefficient
C_s	convective fluid concentration below the moving sheet
C_w	concentration on the wall
C_∞	free stream concentration
D_B	Brownian motion coefficient
D_T	thermopheric diffusion coefficient
f	dimensionless stream function
g	acceleration due to gravity
h_f	convective heat transfer coefficient
h_s	convective mass transfer
k_1	permeability of the porous medium
k	thermal conductivity
K	permeability parameter
Le	Lewis number
M	magnetic parameter
N_b	Brownian motion parameter
N_t	thermophoresis parameter
Nu	Nusselt number

Pr	Prandtl number
q_w, q_m	heat and mass fluxes at the surface
q_r	radiative heat flux
R	thermal radiation
Re	Reynolds number
Sh	Sherwood number
T	fluid temperature
T_f	convective fluid temperature
T_w	temperature
u, v, w	velocity components
x, y, z	Cartesian coordinates

Greek symbols

α	thermal expansion coefficient
ρ	fluid density
ν	kinematic viscosity
σ	electrical conductivity
θ	dimensionless temperature
ϕ	dimensionless concentration
τ	ratio of heat capacities
τ_{wx}, τ_{wy}	wall shears

Author details

Stanford Shateyi

Address all correspondence to: stanford.shateyi@univen.ac.za

Department of Mathematics, University of Venda, Thohoyandou, South Africa

References

[1] Crane, L. J. Flow past a stretching plate. Zeitschrift f ür Angewandte Mathematik und Physik. 1970; 21(4), 645–647.

[2] Gupta, P. S., Gupta, A. S. Heat and mass transfer on a stretching sheet with suction or blowing. The Canadian Journal of Chemical Engineering. 1977; 55, 744–746.

[3] Mohan Krishna, P., Sandeep, N., Sugunammai, V. Effects of radiation and chemical reaction of MHD convective flow over a permeable stretching surface suction and heat generation. Walaliak Journal of Science and Technology. 2015; 12,831–847.

[4] Sandeep, N., Sulochana, C. Dual solution for unsteady mixed convection flow of MHD micropolar fluid over a stretching/shrinking sheet with non-uniform heat source. International Journal of Engineering Science and Technology. 2015; 18, 1–8.

[5] Bhattacharya, K. Boundary layer stagnation point flow of Casson fluid and heat transfer towards a shrinking/stretching sheet. Front Heat Mass Transfer. 2013; 4, 1–9. http://dx.doi.org/10.5098/hmt.v4.2.3003.

[6] Sheikholeslami, M., Ellahi, R., Ashorynejod, H., Domairry, G., Hayat, T. Effects of heat transfer in flow of nanofluids over a permeable stretching wall in a porous medium. Journal of Computational and Theoretical Nanoscience. 2014; 11, 486–496.

[7] Hayat, T., Asad, S., Alsaed, A. Flow of Casson fluid with nanoparticles. Applied Mathematics and Mechanics. 2016; 37(4), 459–470.

[8] Ellahi, R., Aziz, S., Zeeshan, A. Non-Newtonian nanofluid flow through a porous medium between two coaxial cylinders with heat transfer and variable viscosity. Journal of Porous Media. 2013; 16, 205–216.

[9] Buongiorno, J. Convective transport in nanofluids. Journal of Heat Transfer. 2006; 128, 240–250.

[10] Zeeshan, A., Baig, M., Ellahi, R., Hayat, T. Flow of viscous nanofluid between the concentric cylinders. Journal of Computational and Theoretical Nanoscience. 2014; 11, 646–654.

[11] Shateyi, S., Prakash, J., A new numerical approach for MHD laminar boundary layer flow and heat transfer of nanofluids over a moving surface in the presence of thermal radiation. Boundary Value Problems. 2014; 2014, 2. http://www.boundaryvalueproblems.com/content/2014/1.

[12] Kuznetsov, A. V., Nield, D. A. Natural convective boundary-layer flow of a nanofluid past a vertical plate. International Journal of Thermal Sciences. 2010; 49, 243–247.

[13] Aziz, A., Khan, W. A., Natural convective boundary layer flow of a nanofluid past a convectively heated vertical plate. International Journal of Thermal Sciences. 2012; 53, 83–90.

[14] Makinde, O. D., Aziz, A., Boundary layer flow of a nanofluid past a stretching sheet with a convective boundary condition. International Journal of Thermal Sciences. 2011; 50, 1326–1332.

[15] Hayat, T., Aziz, A., Muhammad, T., Ahmad, B. On magnetohydrodynamic flow of second grade nanofluid over a nonlinear stretching sheet. Journal of Magnetism and Magnetic Materials. 2016; 408, 99–106.

[16] Zhao, Q., Xu, H., Tau, L., Raees, A., Sun, Q. Three-dimensional free bio-convection of nanofluid near stagnation point on general curved isothermal surface. Applied Mathematics and Mechanics. 2016; 37(4), 417–432.

[17] Sheikholeslami, M., Ganji, D. D. Heated permeable stretching surface in a porous medium using nanofluids. Journal of Applied Fluid Mechanics. 2014; 7(3), 535–542.

[18] Nayak, R. K., Bhattacharyya, S., Pop, I. Numerical study on mixed convection and entropy generation of Cu–water nanofluid in a differentially heated skewed enclosure. International Journal of Heat and Mass Transfer. 2015; 85, 620–634.

[19] Mabood, F., Das, K. Melting heat transfer on hydromagnetic flow of a nanofluid over a stretching sheet with radiation and second-order slip. European Physical Journal Plus. 2016; 131, 3.

[20] Naramgari, S., Sulochana, C. MHD flow of dusty nanofluid over a stretching surface with volume fraction of dust particles. Ain Shams Engineering Journal. 2016; 7, 709–716.

[21] Sandeep, N., Sulochana, C., Kumar, B. R. Unsteady MHD radiative flow and heat transfer of a dusty nanofluid over an exponentially stretching surface. Engineering Science and Technology, an International Journal. 2016; 19(1), 227–240.

[22] Sheikholeslami, M., Ashorynejad, H. R., Rana, P. Lattice Boltzmann simulation of nanofluid heat transfer enhancement and entropy generation. Journal of Molecular Liquids. 2016; 214, 86–95.

[23] Sheikholeslami, M., Ganji, D. D. Nanofluid convective heat transfer using semi analytical and numerical approaches: a review. Journal of the Taiwan Institute of Chemical Engineers. 2016; 65, 42–77.

[24] Sulochana, C., Ashwinkumar, G. P., Sandeep, N. Similarity solution of 3D Casson nanofluid flow over a stretching sheet with convective boundary conditions. Journal of the Nigerian Mathematical Society. 2016; 35, 128–141.

[25] Motsa, S. S., Dlamini, P. G., Khumalo, M. Solving hyperchaotic systems using the spectral relaxation method. Abstract and Applied Analysis. 2012; doi:10.1155/2012/203461.

6

Energy Transfer in Mixed Convection MHD Flow of Nanofluid Containing Different Shapes of Nanoparticles in a Channel Filled with Saturated Porous Medium

Aaiza Gul, Ilyas Khan and Sharidan Shafie

Abstract

Energy transfer in mixed convection unsteady magnetohydrodynamic (MHD) flow of an incompressible nanofluid inside a channel filled with a saturated porous medium is investigated. The walls of the channel are kept at constant temperature, and uniform magnetic field is applied perpendicular to the direction of the flow. Three different flow situations are discussed on the basis of physical boundary conditions. The problem is first written in terms of partial differential equations (PDEs), then reduces to ordinary differential equations (ODEs) by using a perturbation technique and solved for solutions of velocity and temperature. Four different shapes of nanoparticles inside ethylene glycol ($C_2H_6O_2$) and water (H_2O)-based nanofluids are used in equal volume fraction. The solutions of velocity and temperature are plotted graphically, and the physical behavior of the problem is discussed for different flow parameters. It is evaluated from this problem that viscosity and thermal conductivity are the dominant parameters responsible for different consequences of motion and temperature of nanofluids. Due to greater viscosity and thermal conductivity, $C_2H_6O_2$-based nanofluid is regarded as better convectional base fluid assimilated to H_2O.

Keywords: mixed convection, nanofluid, heat transfer, cylindrical-shaped nanoparticles, MHD flow, porous medium, analytical solutions

1. Introduction

1.1. Nanofluids

Thermal conductivity is the important thermophysical property that plays an essential role in heat transfer enhancement. The internal property of the material is depended on the nature of the fluids. Changing the nature of the fluids or materials can also be affected. In contrast, by applying external force, thermal convection can also be affected on the fluids. Thermal convections are depended on the nature of the fluids as well as on the fluid flow, fluid regime, geometry, etc. Heat transfer conventional fluids, such as ethylene glycol (EG), water, and oils, which are lubricant and kerosene oil have poor thermal conductivities compared to solids. On the other hand, solid particles exhibited rich thermal conductivities than those of conventional heat transfer fluids. The significant improvements in the thermal conductivity of the fluids, which are conventional base by adding up the nanosized particles, were considered by Choi [1] for the first time. The mixture of nanosized particles containing conventional-based fluids is usually known as nanofluids. More accurately, nanofluids are suspensions of nanosized particles in base fluids. The most ordinary nanoparticles used in nanofluids are oxides, carbides, and metals. Nanofluids are mainly used in electronic equipment, power generation, energy supply, production, and air conditioning. Vajjha and Das [2] investigated, for the first time, EG (60%) and water (40%) mixture that is a base fluid for the preparation of alumina (Al_2O_3), copper oxide (CuO), and zinc oxide (ZnO) nanofluids. At the equal temperature and concentration, CuO nanofluids have higher thermal conductivity than those of Al_2O_3 and ZnO nanofluids. Naik and Sundar [3] have prepared (CuO) nanofluids with 70% propylene glycol and 30% water. As estimated, they found that CuO nanofluids have rich thermal conductivity and viscosity as compared to the other fluids that are base. Recently, Mansur et al. [4] have studied magnetohydrodynamic (MHD) stagnation-point flow of nanofluids over a permeable sheet for stretching and shrinking cases. They used bvp4c program in MATLAB to obtain the numerical solutions and computed results for different parameters.

The quality of nanofluids in terms of heat transfer performance depends on the volume fraction and types of nanoparticles as well as also depends on the shapes of nanoparticles. Many researchers have used nanoparticles in spherical shapes. However, the nanofluids containing nonspherical shapes of nanoparticles have higher thermal conductivities as compared to spherical ones. Hence, nonspherical shapes of nanoparticles are most suitable to be used. Nanoparticles other than spherical shapes have been used in this research due to their above-mentioned properties. More closely, in this study we discuss four different types of nanoparticles, namely, cylinder, platelet, blade, and brick. Nanoparticles of nonspherical shapes with desirable properties are the main focus of present study. Recently, the development in the field of nanotechnology has shown that cylindrical-shaped nanoparticles are seven times more harmful than spherical-shaped nanoparticles in the deliverance of drugs to breast cancer. In the literature survey of nanofluids, we found that analytical studies for the suspension of a variety of shapes of nanoparticles containing EG and water-based fluids are not described yet. Although Timofeeva et al. [5] study the problem of Al_2O_3 nanofluids which are of different shaped nanoparticles, but they performed experimentally with theoretical modeling together in this study. More closely, they studied various shapes of Al_2O_3 nanoparticles in a base fluid mixture of EG and water of similar volumes. Improvement in the effective thermal conductivities due to particle shapes was mentioned by Timofeeva et al. Loganathan et al. [6] considered nanoparticles that are sphere shaped and analyzed radiation effects on an unsteady natural convection flow of nanofluids over an infinite vertical plate. They have found that the velocity of spherical silver (Ag) nanofluids is less than that of copper (Cu), titanium dioxide (TiO_2), and Al_2O_3 spherical nanofluids. Recently, Asma et al.

[7] found exact solutions for free-convection flow of nanofluids at ramped wall temperature by using five various kinds of spherical-shaped nanoparticles.

Sebdani et al. [8] studied heat transfer of Al_2O_3 water nanofluid in mixed convection flow inside a square cavity. Fan et al. [9] described mixed convection heat transfer in a horizontal channel filled with nanofluids. Tiwari and Das [10] and Sheikhzadeh et al. [11] investigated laminar mixed convection flow of a nanofluid in two-sided lid-driven enclosures. Furthermore, a magnetic field in nanofluids has diverse applications such as in the metallurgy and polymer industry where hydromagnetic techniques are used. Nadeem and Saleem [12] presented the unsteady flow of a rotating MHD nanofluid in a rotating cone in the presence of magnetic field. Al-Salem et al. [13] examined MHD mixed convection flow in a linearly heated cavity. The effects of variable viscosity and variable thermal conductivity on the MHD flow and heat transfer over a nonlinear stretching sheet was studied by Prasad et al. [14]. The problem of Darcy Forchheimer mixed convection heat and mass transfer in fluid-saturated porous media in the presence of thermophoresis was discussed by Rami et al. [15]. The effect of radiation and magnetic fields on the mixed convection stagnation-point flow over a vertical stretching sheet in a porous medium bounded by a stretching vertical plate was analyzed by Hayat et al. [16]. Some other studies on mixed convection flow of nanofluids are mentioned in Refs. [17–28]. Nanofluids due to the higher conduction or heat transfer rate of together with other several applications in the field of nanoscience have attracted the attention of researchers to perform future research. Altogether, several researchers are working experimentally, some of them are using numerical simulation. However, very limited research studies are obtainable on analytical side. Experimental research mainly focuses on the enhancement of heat transfer rate of nanofluids through thermal conduction. And this study also includes heat transfer through mixed convection. Several other efforts made on nanofluids are presented in Refs. [29–39].

In order to encounter the importance of MHD in nanofluids, Mansur et al. [4] explored the MHD stagnation point flow of nanofluids over a stretching/shrinking sheet with suction. Colla et al. [21] conducted water-based nanofluids characterization, thermal conductivity and viscosity measurements, and correlation. Abareshi et al. [40] investigated fabrication, characterization, and measurement of thermal conductivity of nanofluids. Borglin et al. [41] examined experimentally the flow of magnetic nanofluids in porous media. MHD effect on nanofluid with energy and hydrothermal behavior between two collateral plates was presented by Sheikholeslami et al. [42]. Sheikholeslami et al. [43] studied forced convection heat transfer in a semiannulus under the influence of a variable magnetic field. Some other research studies on electrically conducting nanofluids are mentioned in Refs. [44–50].

In view of the above literature, the existing research is concerned with the radiative heat transfer in mixed convection MHD flow of different shapes of Al_2O_3 in EG base nanofluid in a channel filled with saturated porous medium. The foremost focus of this study is the importance of cylindrical-shaped nanofluids on several flow parameters. The uniform constant magnetic field is applied at 90° to the flow and the nanofluids are supposed electrically conducting. No slip condition is taken at the boundary walls of the channel. Three different flow cases on the basis of appropriate boundary conditions are explored. Both of the boundary walls of the channel are at rest in the first case, while motion in the fluids is induced due to buoyancy force and external pressure gradient. In the second case, the right wall of the channel is oscillating in its own plane whereas in the third case, both of the bounding walls of the channel are set into oscillatory motions. The analytical solutions of velocity and temperature are obtained by using a perturbation technique. Graphs are plotted, and the physical behavior of the problem is discussed for different parameters of interest. Nusselt number and skin friction are also computed.

2. Derivation and solutions of governing equations

Consider flow in oscillating form of an incompressible nanofluid inside vertical porous channel. Flow of the nanofluids is supposed electrically conducting under the influence of uniform constant magnetic field. The uniform magnetic field is applied perpendicular to the direction of the flow. Reynolds number is supposed small to ignore the effect of evoked magnetic field. The external electric and electric fields due to polarization of charges are taken zero in order to ignore the influence of electric force in Lorentz force. It is assumed that at time $t = 0$, the flow is at constant temperature θ_0. Right boundary of the channel $(y = d)$ is maintained at constant temperature θ_{wR}, while the left boundary $(y = 0)$ has a uniform temperature θ_{wL}. Coordinate axis is considered where flow of fluids is moving in the x.axis direction, while the y-axis is taken perpendicular to the velocity of the flow direction.

After derivation of governing equations for quantity of motion and energy under the assumption of Boussinesq approximation are as follow:

$$\rho_{\text{nf}} \frac{\partial u}{\partial t} = -\frac{\partial p}{\partial x} + \mu_{\text{nf}} \frac{\partial^2 u}{\partial y^2} - \left(\sigma_{\text{nf}} B_0^2 + \frac{\mu_{\text{nf}}}{\kappa_1} \right) u + (\rho\beta)_{\text{nf}}\, g(\theta - \theta_0), \tag{1}$$

$$(\rho c_p)_{\text{nf}} \frac{\partial \theta}{\partial t} = \kappa_{\text{nf}} \frac{\partial^2 \theta}{\partial y^2} - \frac{\partial q}{\partial y}, \tag{2}$$

where $u = u(y, t)$, $\theta = \theta(y, t)$, ρ_{nf}, μ_{nf}, σ_{nf} , $\kappa_1 > 0$, $(\rho\beta)_{\text{nf}}$, g, $(\rho c_p)_{\text{nf}}$, κ_{nf} q specify the fluid velocity, temperature, density, dynamic viscosity, electrical conductivity, permeability of the porous medium, thermal expansion coefficient, gravitational acceleration, heat capacitance, thermal conductivity of nanofluids, and radiative heat flux in the x-direction. $-\partial p/\partial x = \lambda \exp{(i\omega t)}$ represents the pressure gradient of the flow in oscillatory form where ω is the oscillation parameter and λ is the amplitude of oscillation.

The Hamilton and Crosser [17] model for thermal conductivity and the Timofeeva et al. [5] model for calculating dynamic viscosity of nanofluids are used in order to encounter spherical and other than spherical shapes of nanoparticles inside nanofluids. From these models:

$$\mu_{\text{nf}} = \mu_f(1 + a\phi + \phi^2 b), \tag{3}$$

$$\frac{\kappa\text{nf}}{\kappa_f} = \frac{\kappa_s + (m-1)\kappa_f + (n-1)(\kappa_s - \kappa_s)\phi}{\kappa_s + (m-1)\kappa_f - (\kappa_s - \kappa_f)\phi}, \tag{4}$$

where κ_f and κ_s are the thermal conductivities of the base fluid and solid nanoparticles, respectively. The density ρ_{nf}, thermal expansion coefficient $(\rho\beta)_{nf}$, heat capacitance $(\rho c_p)_{nf}$ and thermal conductivity σ_{nf}, of nanofluids are derived by using the relations given by [6, 7]

$$(\rho c_p)_{nf} = (1-\phi)(\rho c_p)_f + \phi(\rho c_p)_s, \rho_{nf} = (1-\phi)\rho_f + \phi\rho_s,$$

$$(\rho\beta)_{nf} = (1-\phi)(\rho\beta)_f + \phi(\rho\beta)_s, \kappa_{nf} = \alpha_{nf}(\rho c_p)_{nf},$$

$$\sigma_{nf} = \sigma_f\left[1 + \frac{3(\sigma-1)\phi}{(\sigma+2)-(\sigma-1)\phi}\right], \sigma = \frac{\sigma_s}{\sigma_f}, \tag{5}$$

where ϕ is the volume fraction of the nanoparticles, ρ_f and ρ_s are the density of the base fluid and solid nanoparticles, respectively, μ_f is the dynamic viscosity of the base fluid, $(c_p)_f$ and $(c_p)_s$ denote the specific heat at constant pressure corresponding to the base fluid and solid nanoparticles and σ_f and σ_s are electrical conductivities of base fluids and solid nanoparticles. The total term (ρc_p) is known as heat capacitance.

Here a and b are shape constants and different for different shape of nanoparticles as presented in **Table 1** [5], μ_f, κ_f, and κ_s are the dynamic viscosity, thermal conductivity of the base fluid, and thermal conductivity of solid nanoparticles, respectively. The empirical shape factor n defined in Eq. (4) is equal to $n = 3/\Psi$, where Ψ is the sphericity. Sphericity given in Hamilton and Crosser model is the ratio of surface area of the sphere to the surface area of real particle with equal volumes. The values of sphericity for different shapes of nanoparticles are given in **Table 2** [5].

Model	Platelet	Blade	Cylinder	Brick
a	37.1	14.6	13.5	1.9
b	612.6	123.3	904.4	471.4

Table 1. Constants a and b empirical shape factors.

Model	Platelet	Blade	Cylinder	Brick
Ψ	0.52	0.36	0.62	0.81

Table 2. Sphericity Ψ for different shapes nanoparticles.

In Eqs. (1) and (2), the $(\rho c_p)_{nf}$, ρ_{nf}, $(\rho\beta)_{nf}$, κ_{nf}, and σ_{nf} of nanofluids are used from the relations given by Asma et al. [7], as

The temperature θ_0 and θ_w of both walls of the channel is assumed high and both walls are emitting radiations [20]. Therefore, radiative heat flux as a function of temperature is given by

$$-\frac{\partial q}{\partial y} = 4\alpha_0^2(\theta - \theta_0), \tag{6}$$

where α_0 is the radiation absorption coefficient.

Using Eq. (6) into Eq. (2) gives

$$(\rho c_p)_{\text{nf}}\frac{\partial \theta}{\partial t}=\kappa_{\text{nf}}\frac{\partial^2 \theta}{\partial y^2}-4\alpha_0^2(\theta-\theta_0), \tag{7}$$

In order to convert dimensional partial differential equations (PDEs) into dimensionless PDEs, the following nondimensional variable are introduced

$$\begin{aligned} &x^*=\frac{x}{d},\ y^*=\frac{y}{d},\ u^*=\frac{u}{U_0},\quad t^*=\frac{tU_0}{d},\ \theta^*=\frac{\theta-\theta_0}{\theta_w-\theta_0},\\ &p^*=\frac{d}{\mu U_0}p,\quad \omega^*=\frac{d\omega_1}{U_0}, \end{aligned} \tag{8}$$

After dimensionalization (dropping for convenience) Eqs. (1) and (7) give

$$\begin{aligned} &\left[(1-\phi)+\phi\frac{\rho_s}{\rho_f}\right]\text{Re}\frac{\partial u}{\partial t}=\lambda\exp(i\omega t)+(1+a\phi+b\phi^2)\frac{\partial^2 u}{\partial y^2}-\\ &\left[1+\frac{3(\sigma-1)\phi}{(\sigma+2)-(\sigma-1)\phi}\right]M^2u-\frac{1}{K}(1+a\phi+b\phi^2)+\left[(1-\phi)+\frac{\phi(\rho\beta)_s}{(\rho\beta)_f}\right]Gr\theta, \end{aligned} \tag{9}$$

$$\frac{Pe}{\lambda_n}\left[(1-\phi)+\phi\frac{(\rho c_p)_s}{(\rho c_p)_f}\right]\frac{\partial \theta}{\partial t}=\frac{\partial^2\theta}{\partial y^2}+\frac{N^2}{\lambda_n}\theta, \tag{10}$$

where

$$\begin{aligned} &Re=\frac{U_0 d}{v_f},M^2=\frac{\sigma_f B_0^2 d^2}{\mu_f},K=\frac{\kappa_1}{d^2}Gr=\frac{g\beta_f d^2(\theta_w-\theta_0)}{v_f U_0},\ Pe=\frac{(\rho c_p)_f dU_0}{\kappa_f},\\ &\lambda_n=\frac{\kappa_{nf}}{\kappa_f}=\frac{\kappa_s+(m-1)\kappa_f+(n-1)(\kappa_s-\kappa_s)\phi}{\kappa_s+(m-1)\kappa_f-(\kappa_s-\kappa_f)\phi},N^2=\frac{4d^2\alpha_0^2}{\kappa_f}. \end{aligned}$$

Here *Re*, *M*, *Gr*, *Pe*, and *N* denote the Reynolds number, magnetic parameter, the thermal Grashof number, the Peclet number, and the radiation parameter, respectively. Three different flow cases are considered in order to solve Eqs. (9) and (10). These are as follow:

2.1. Case I: Stationary walls of the channel

In this case, the gap between the two plates of the channel is denoted by d and both plates are assumed stationary at $y=0$ and $y=d$. The flow of nanofluids is unidirectional and moving with velocity in the x-axis. Both plates of the channel are maintained at constant and uniform temperature θ_w and θ_0. Thus, the appropriate boundary conditions are

$$u(0,\ t)=0,\quad u(d,\ t)=0, \tag{11}$$

$$\theta(0,\ t)=\theta_0,\ \ \theta(d,\ t)=\theta_w. \tag{12}$$

After reducing Eqs. (11) and (12) into dimensionless form, we get

$$u(0,\ t) = 0;\quad u(1,\ t) = 0,\ \ t > 0, \tag{13}$$

$$\theta(0,\ t) = 0;\ \ \theta(1,\ t) = 1;\quad t > 0. \tag{14}$$

Further simplification of Eqs. (9) and (10) after conversion of dimensional PDEs, we get

$$d_0 \frac{\partial u}{\partial t} = \lambda \exp(i\omega t) + \phi_2 \frac{\partial^2 u}{\partial y^2} - h_0^2 u + d_1 \theta, \tag{15}$$

$$e_0 \frac{\partial \theta}{\partial t} = \frac{\partial^2 \theta}{\partial y^2} + e_1 \theta, \tag{16}$$

where

$$d_0 = \phi_1 Re, \phi_1 = (1-\phi) + \phi \frac{\rho_s}{\rho_f}, \phi_2 = (1 + a\phi + b\phi^2), h_0^2 = \phi_5 M^2 + 1/K,$$

$$\phi_5 = \left[1 + \frac{3(\sigma-1)\phi}{(\sigma+2) - (\sigma-1)\phi}\right], d_1 = \phi_3 Gr, \phi_3 = (1-\phi) + \phi \frac{(\rho\beta)_s}{(\rho\beta)_f}, {e_0}^2 = \frac{Pe\phi_4}{\lambda_n},$$
$$\phi_4 = \left[(1-\phi) + \phi \frac{(\rho c_p)_s}{(\rho c_p)_f}\right], {e_1}^2 = \frac{N^2}{\lambda_n}.$$

Perturb-type solutions are supposed to convert PDEs in Eqs. (15) and (16) under appropriate boundary conditions, and in Eqs. (11) and (12) into ordinary differential equations (ODEs) are

$$u(y,\ t) = [u_0(y) + \varepsilon \exp(i\omega t)\ u_1(y)], \tag{17}$$

$$\theta(y,\ t) = [\theta_0(y) + \varepsilon \exp(i\omega t)\ \theta_1(y)]. \tag{18}$$

Using Eqs. (17) and (18) into Eqs. (15) and (16), we obtain the following system of ordinary differential equations

$$\frac{d^2 u_0(y)}{dy^2} - h_1^2 u_0(y) = d_2 \theta_0(y), \tag{19}$$

$$\frac{d^2 u_1(y)}{dy^2} - h_2^2 u_1(y) = -\frac{\lambda}{\phi_2}, \tag{20}$$

$$\frac{d^2 \theta_0(y)}{dy^2} - e_1^2 \theta_0(y) = 0, \tag{21}$$

$$\frac{d^2 \theta_1(y)}{dy^2} + h_3^2 \theta_1(y) = 0, \tag{22}$$

where

$$h_1 = \sqrt{\frac{{h^2}_0}{\phi_2}},\ d_2 = \frac{d_1}{\phi_2},\ h_2 = \sqrt{\frac{h_0^2 + i\omega d_0}{\phi_2}},\ h_3 = \sqrt{e_1 - i\omega e_0}.$$

The associated boundary conditions (Eqs. (13) and (14)) are reduce to

$$u_0(0) = 0; \;\; u_0(1) = 0, \tag{23}$$

$$u_1(0) = 0; \;\; u_1(1) = 0, \tag{24}$$

$$\theta_0(0) = 0; \;\; \theta_0(1) = 1, \tag{25}$$

$$\theta_1(0) = 0; \;\; \theta_1(1) = 0. \tag{26}$$

Perturb solution for temperature in Eqs. (21) and (22) under appropriate boundary conditions (Eqs. (25) and (26)) give

$$\theta_0(y) = \frac{\sin(e_1 y)}{\sin(e_1)}, \tag{27}$$

$$\theta_1(y) = 0. \tag{28}$$

Temperature of nanofluids, Eq. (18) using Eqs. (27) and (28) gives

$$\theta(y,t) = \theta(y) = \frac{\sin(e_1 y)}{\sin(e_1)}. \tag{29}$$

Perturb solution for temperature in Eqs. (19) and (20), using Eq. (27) under appropriate boundary conditions, Eqs. (23) and (24) yield

$$u_0(y) = c_1 \sinh(h_1 y) + c_2 \cosh(h_1 y) + \frac{d_2}{(e_1^2 + h_1^2)} \frac{\sin(e_1 y)}{\sin(e_1)}. \tag{30}$$

$$u_1(y) = c_3 \sinh(h_2 y) + c_4 \cosh(h_2 y) + \frac{\lambda}{h_2^2 \phi_2}. \tag{31}$$

Here c_1, c_2, c_3 and c_4 are arbitrary constants given by

$$c_1 = -\frac{d_2}{\sinh(h_1)(e_1^2 + h_1^2)}, c_2 = 0, c_3 = \frac{\lambda}{h_2^2 \phi_2} \frac{1}{\sinh(h_2)} (\cosh(h_2) - 1), c_4 = -\frac{\lambda}{h_2^2 \phi_2}. \tag{32}$$

Final velocity for nanofluids, substituting Eqs. (30)–(32) into Eq. (17), we obtain

$$u(y,\, t) = -\frac{d_2 \sinh(h_1 y)}{(e_1^2 + h_1^2)\sinh(h_1)} + \frac{d_2 \sin(e_1 y)}{(e_1^2 + h_1^2)\sin(e_1)}$$

$$+\varepsilon \exp(i\omega t) \begin{bmatrix} \dfrac{\lambda(\cosh(h_2) - 1)\sinh(h_2 y)}{h_2^2 \phi_2 \sinh(h_2)} \\ + \dfrac{\lambda}{h_2^2 \phi_2} \left(1 - \cosh(h_2 y)\right) \end{bmatrix}. \tag{33}$$

2.2. Case 2: right plate of the channel is oscillating in its own plane

In this case, the right wall of the channel ($y = d$) is locate into oscillatory motion, while on other hand, the left wall ($y = 0$) is taken as stationary. The first boundary condition is the same, while the second boundary condition in dimensionless form modifies to

$$u(1,\ t) = H(t)\varepsilon \exp(i\omega t);\ \ t > 0, \tag{34}$$

where $H(t)$ is the Heaviside step function.

By using the same procedure as in Case-1, and the velocity solution is obtained as

$$u(y,\ t) = -\frac{d_2 \sinh(h_1 y)}{\sinh(h_1)(e_1^2 + h_1^2)} + \frac{d_2}{(e_1^2 + h_1^2)} \frac{\sin(e_1 y)}{\sin(e_1)}$$

$$+\varepsilon \exp(i\omega t) \begin{bmatrix} \dfrac{\sinh(h_2 y)}{\sinh(h_2)} \left(H(t) + \dfrac{\lambda}{(h_2^2 \phi_2)} \left(\cosh(h_2) - 1\right) \right) \\ -\dfrac{\lambda}{(h_2^2 \phi_2)} \cosh(h_2 y) + \dfrac{\lambda}{(h_2^2 \phi_2)} \end{bmatrix}. \tag{35}$$

2.3. Case 3: both plates of the channel are oscillating in its own plane

In this case, both plates of the channel are chosen into oscillatory motions. The dimensionless form of the boundary conditions is

$$u(0,\ t) = u(1,\ t) = H(t)\varepsilon \exp(i\omega t);\ \ t > 0. \tag{36}$$

The resulting expression for velocity is obtained as

$$u(y,\ t) = -\frac{d_2 \sinh(h_1 y)}{\sinh(h_1)(e_1^2 + h_1^2)} + \frac{d_2}{(e_1^2 + h_1^2)} \frac{\sin(e_1 y)}{\sin(e_1)}$$

$$+\varepsilon \exp(i\omega t) \begin{bmatrix} \dfrac{\sinh(h_2 y)}{\sinh(h_2)} \begin{pmatrix} H(t)\left(1 - \cosh(h_2 y)\right) + \\ \dfrac{\lambda}{(h_2^2 \phi_2)} \left(\cosh(h_2) - 1\right) \end{pmatrix} \\ + \left\{ H(t) - \dfrac{\lambda}{(h_2^2 \phi_2)} \right\} \cosh(h_2 y) + \dfrac{\lambda}{(h_2^2 \phi_2)} \end{bmatrix}. \tag{37}$$

2.4. Evaluation of Nusselt number and skin-friction

The dimensionless derivations for Nusselt number and skin-frictions are evaluated from Eqs. (29), (33), (35) and (37) as follows:

$$Nu = \frac{e_1}{\sin(e_1)}, \tag{38}$$

$$\begin{aligned}\tau_1 = \tau_1(t) = &-\frac{d_2 h_1}{(e_1^2 + h_1^2)\sinh(h_1)} + \frac{d_2 b_1}{(e_1^2 + h_1^2)\sin(e_1)} \\ &+\varepsilon \exp(i\omega t)\left[\frac{\lambda(\cosh(h_2) - 1)}{h_2\phi_2 \sinh(h_2)}\right].\end{aligned} \tag{39}$$

$$\begin{aligned}\tau_2 = \tau_2(t) = &-\frac{d_2 h_1}{(e_1^2 + h_1^2)\sinh(h_1)} + \frac{d_2 e_1}{(e_1^2 + h_1^2)\sin(e_1)} \\ &+\varepsilon \exp(i\omega t)\left[\frac{h_2}{\sinh(h_2)}\left\{H(t) + \frac{\lambda}{(h_2\phi_2)}\left(\cosh(h_2) - 1\right)\right\}\right],\end{aligned} \tag{40}$$

$$\begin{aligned}\tau_3 = \tau_3(t) = &-\frac{d_2 h_1}{(e_1^2 + h_1^2)\sinh(h_1)} + \frac{d_2 b_1}{(e_1^2 + h_1^2)\sin(e_1)} \\ &+\varepsilon \exp(i\omega t)\left[\frac{h_2}{\sinh(h_2)}\left\{\begin{array}{l} H(t)\left(1 - \cosh(h_2)\right) \\ + \frac{\lambda}{(h_2^2\phi_2)}\left(\cosh(h_2) - 1\right)\end{array}\right\}\right].\end{aligned} \tag{41}$$

3. Graphical consequences and depiction

In this segment, graphical consequences are figured and debated. Similarly, influence of the radiation effect on heat transfer in mixed convection MHD flow of nanofluids inside a channel filled with a saturated porous medium is explored. Three different flow cases on the basis of appropriate boundary conditions are examined. Four dissimilar shapes of Al_2O_3 as solid nanoparticles, which are cylinder, platelet, brick, and blade, are dangling into conventional base fluids, ethylene glycol and water. The physical performance of the exhibited graphs is conferred for various embedded parameters. The numerical values of constants a and b (called shape factors) are chosen from **Table 1**, and sphericity Ψ is given in **Table 2**. It should be acclaimed that a and b coefficients vary highly with particle shape. The numerical values for the various shapes of nanoparticles (platelet, blade, cylinder and brick) at equal volumes are presented in **Tables 1–3**.

The geometry of the problem is shown in **Figure 1**. The impact of dissimilar shapes of alumina Al_2O_3 nanoparticles on the motion of ethylene glycol-based nanofluids is represented in **Figure 2**. It is viewed that the blade shape of alumina Al_2O_3 nanoparticles inside ethylene glycol-based nanofluids has the uppermost velocity chased by brick, platelet, and cylindrical-shaped nanoparticles in ethylene glycol-based nanofluids. The impact of the shapes on the motion of nanofluids is because of the strong subjection of viscosity on particle shapes for $\phi < 0.1$. It is exonerated from the present consequences that the elongated shaped nanoparticles like cylinder and platelet have larger viscosities as assimilated to nanofluids comprising square-shaped nanoparticles like brick and blade. The acquired consequences agree well with the experimental consequences predicted by Timofeeva et al.. A very small divergence is perceived in the present research, where the cylindrical-shaped nanoparticles have the larger viscosity, whereas from the experimental findings perceived by Timofeeva et al., the platelet has the larger viscosity. Timofeeva et al. had assimilated their consequences with the Hamilton and Crosser model and found that their consequences were equivalent with the Hamilton and Crosser model. In the current work, the model of Hamilton and Crosser is applied and found that analytical consequences in this research also identical to the experimental consequences of Timofeeva et al.

Model	**$\rho(Kgm^{-3})$**	**$c_P(kg^{-1}K^{-1})$**	**$k(Wm^{-1}K^{-1})$**	**$\beta \times 10^{-5}(K^{-1})$**	**σ (s/m)**
H_2O	997.1	4179	0.613	21	5.5×10^{-6}
$C_2H_6O_2$	1.115	0.58	0.1490	6.5	1.07×10^{-6}
Cu	8933	385	401	1.67	59.6×10^{6}
TiO_2	4250	686.2	8.9528	0.9	2.6×10^{6}
Ag	10,500	235	429	1.89	6.30×10^{7}
Al_2O_3	3970	765	40	0.85	$1.07 \times 10_{-6}$
Fe_3O_4	5180	670	9.7	0.5	12.7

Table 3. Thermophysical properties of water and nanoparticles.

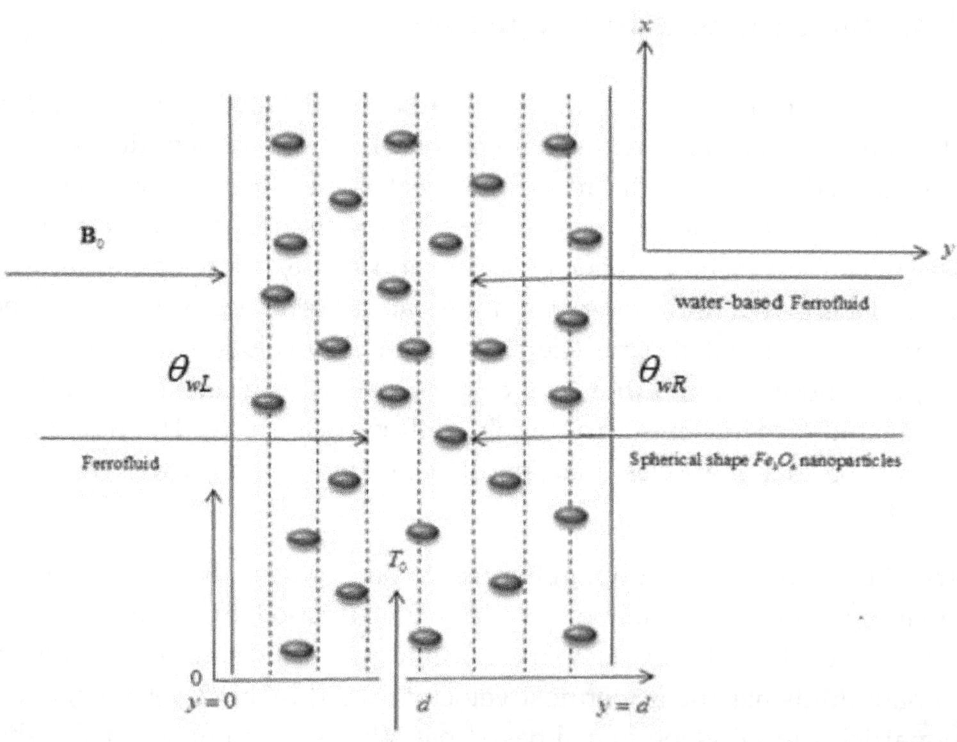

Figure 1. Physical model and coordinates system.

Figure 3 shows the result of dissimilar shapes of alumina Al_2O_3 nanoparticles on the motion of water H_2O-based nanofluids. It is surely noticed that the cylindrical-shaped alumina Al_2O_3

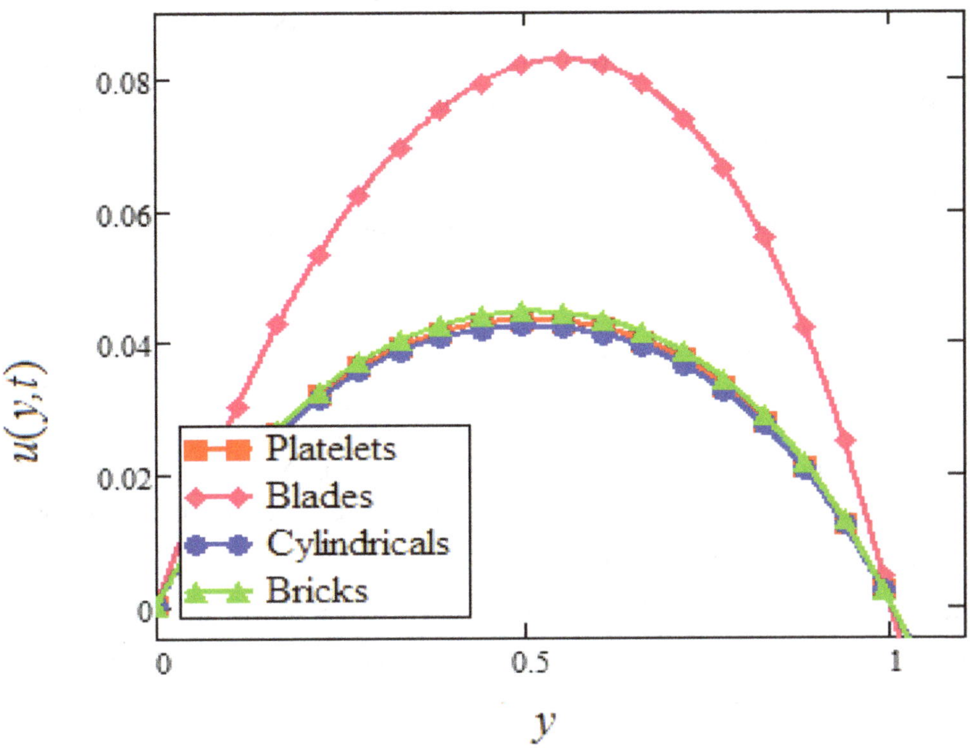

Figure 2. Velocity profiles for different shapes of Al_2O_3 nanoparticles in EG-based nanofluids when $Gr = 0.1$, $N = 0.1$, $M = 1$, $\lambda = 1$, $K = 1$, $t = 5$, $\phi = 0.04$, $\omega = 0.2$.

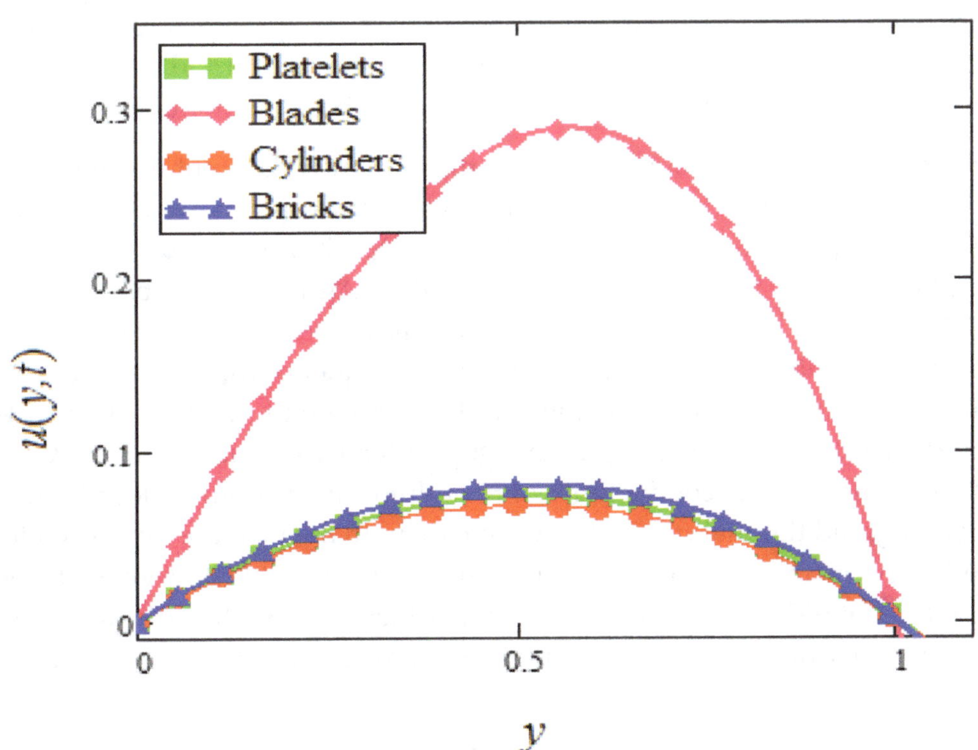

Figure 3. Velocity profiles for different shapes of Al_2O_3 nanoparticles in water-based nanofluids when $Gr = 0.1$, $N = 0.1$, $M = 1$, $\lambda = 1$, $K = 1$, $t = 5$, $\phi = 0.04$, $\omega = 0.2$.

nanoparticles in ethylene glycol-based nanofluids have the uppermost velocity chased by platelet, brick, and blade. Thus, in accordance with Hamilton and Crosser model, solution of elongated and thin shaped particles (high shape factor m) should have larger thermal conductivities, if the ratio of k_{nf}/k_f is higher than 100. It is also predicted by Colla et al. that the thermal conductivity and viscosity increase with the increase of particle absorption due to which motion of nanofluids decreases. For that reason, the cylindrical-shaped alumina Al_2O_3 nanoparticles have the higher thermal conductivity chased by platelet, brick, and blade. Timofeeva et al. depicted the conclusion that, when the sphericity of nanoparticles is less than 0.6, the negative concession of heat flow resistance at the solid-liquid interface increases much faster than the particle shape concession. Thus, the inclusive thermal conductivity of solution starts decreasing less than sphericity of 0.6. However, it is increasing in the case of Hamilton and Crosser model because of the only concession of particle shape parameter m. Nevertheless, flow in this work is one-directional and one-dimensional; therefore, the negative concession of heat flow resistance is ignored. Timofeeva et al. reported the model $k_{\mathrm{nf}}/k_f = 1 + (c_k^{\mathrm{shape}} + c_k^{\mathrm{surface}})\,\phi$, for calculating the thermal conductivity of nanoparticles. In accordance with this model, c_k^{shape} and c_k^{surface} coefficients reflecting concessions to the effective thermal conductivity because of particle shape (positive influence) and because of surface resistance (negative effect), respectively. Particle shape coefficient c_k^{shape} was also derived from the Hamilton and Crosser's equation (1962).

A comparability of alumina Al_2O_3 in $C_2H_6O_2$-based nanofluids with alumina Al_2O_3 in H_2O.based nanofluids is shown in **Figure 4**. It is viewed that the motion of H_2O-based nanofluids is larger than the motion of $C_2H_6O_2$-based nanofluids. The thermal conductivity and viscosity of $C_2H_6O_2$- and H_2O-based nanofluids are also perceived by the Hamilton and Crosser model for equal value of ϕ. This consequence shows that $C_2H_6O_2$-based nanofluids have larger thermal conductivity and viscosity than H_2O-based nanofluids.

The influence of different solid nanoparticles on the motion of different nanofluids is shown in **Figure 5**. From this figure, it is notable that cylindrical-shaped Al_2O_3 in $C_2H_6O_2$-based nanofluids has the uppermost motion chased by Fe_3O_4, TiO_2, Cu, and Ag in $C_2H_6O_2$-based nanofluids. This indicates that cylindrical-shaped silver Ag in $C_2H_6O_2$-based nanofluids has the higher viscosity and thermal conductivity assimilated to Cu, TiO_2, Fe_3O_4, and Al_2O_3 in $C_2H_6O_2$-based nanofluids. One can view from this consequence that cylindrical-shaped Ag in $C_2H_6O_2$-based nanofluids has better quality fluids assimilated to Fe_3O_4 cylindrical-shaped in $C_2H_6O_2$-based nanofluids. This consequence is supported by the Hamilton and Crosser model that the viscosity and thermal conductivity of nanofluids are also influenced by nanoparticles ϕ, i.e., the viscosity and thermal conductivity increase with the increase in ϕ. Therefore, motion decreases with the increase in ϕ. This figure further shows that the viscosity of Al_2O_3 in $C_2H_6O_2$-based nanofluids at ϕ is below 0.1, which increases nonlinearly with nanoparticles suspension. This consequence is found similar to the experimental consequence predicted by Colla et al.

Different ϕ of nonspherical cylindrical-shaped alumina Al_2O_3 nanoparticles on the motion of alumina Al_2O_3 in $C_2H_6O_2$-based nanofluids is exhibited in **Figure 6**. It is view from this figure that with the increase of ϕ the motion of nanofluids is decreased. Due to this reason, the

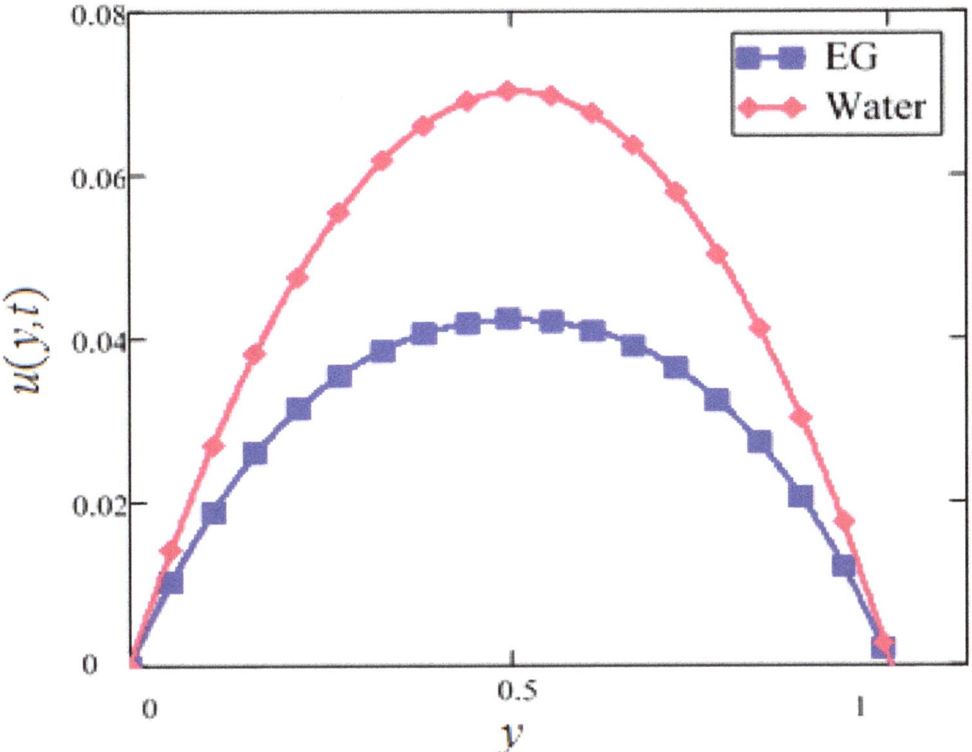

Figure 4. Comparison of velocity profiles of Al_2O_3 in EG and water-based nanofluids when $Gr = 0.1$, $N = 0.1$, $M = 1$, $\lambda = 1$, $K = 1$, $t = 5$, $\phi = 0.04$, $\omega = 0.2$.

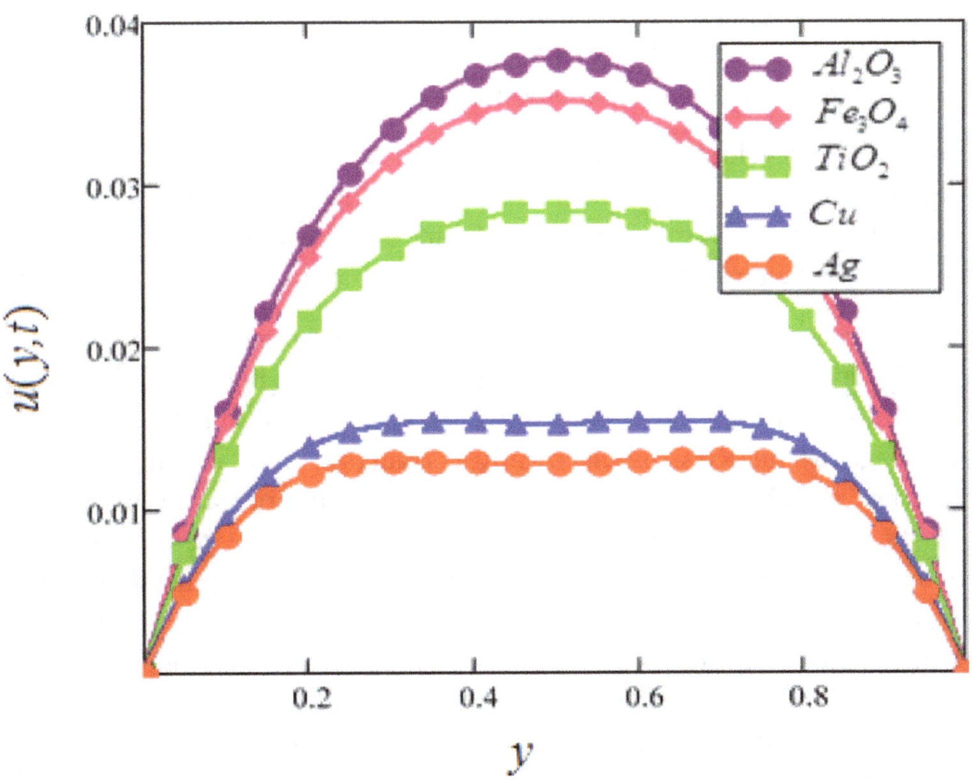

Figure 5. Velocity profiles of different nanoparticles in EG-based nanofluids when $Gr = 0.1$, $N = 0.1$, $M = 1$, $\lambda = 1$, $K = 1$, $t = 5$, $\phi = 0.04$, $\omega = 0.2$.

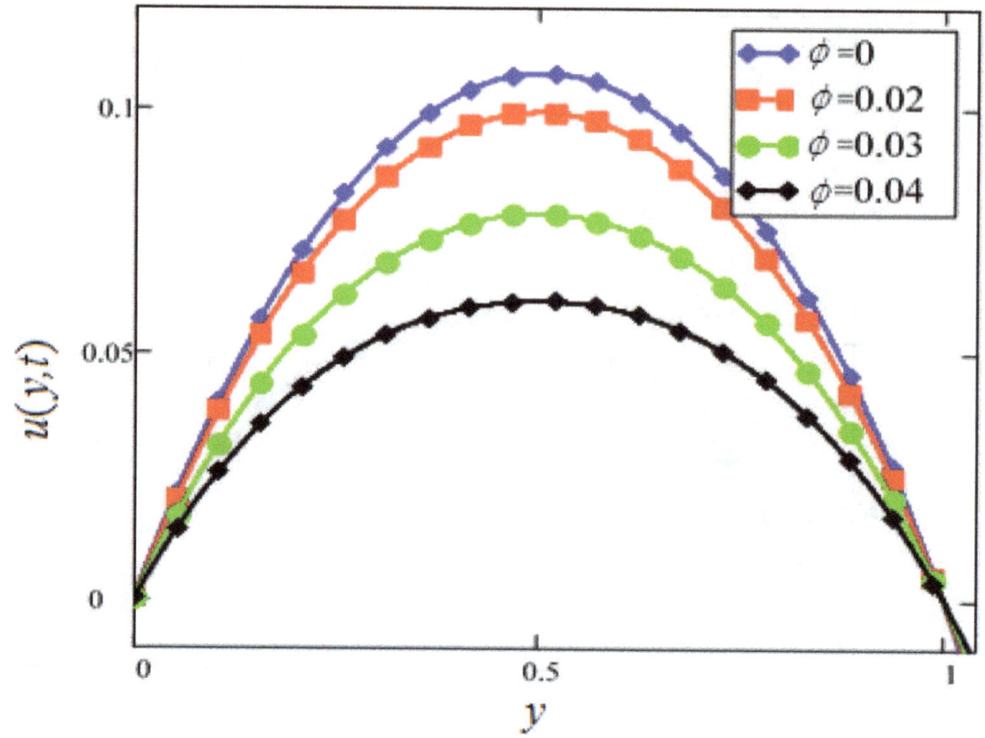

Figure 6. Velocity profiles for different values of ϕ of Al_2O_3 in EG-based nanofluids when $Gr = 0.1,\ \ N = 0.1,\ \ M = 1,$ $\lambda = 1,\ \ K = 1,\ \ \ t = 5,\ \ \ \omega = 0.2.$

nanofluid becomes more viscous with the increase of ϕ, which governs to decrease the motion of nanofluids. The thermal conductivity of nanofluids also increases with the increase of ϕ. The experimental research by Colla et al. also supports these results.

Consequences for various values of radiation parameter N of Al_2O_3 in H_2O-based nanofluids are displayed in **Figure 7**. It is observed that motion increases with the increase of N. This consequence agrees well with the consequence reported by Makinde and Mhone. Physically, this means that with the increase of N, the amount of heat energy transfers to the fluids also increases.

The graphical outcomes of the motion of nanofluids for several values of magnetic parameter M of alumina Al_2O_3 in H_2O-based nanofluids are exhibited in **Figure 8**. Increasing magnetic parameter, M, results in the decrease of the motion of the alumina nanofluids. Increasing perpendicular magnetic field on the electrically conducting fluid imparts to a resistive force called Lorentz force, which is identical to drag force, and upon increasing the value of magnetic parameter, M, the drag force rises which has the tendency to reduce the motion of the nanofluid. The resistive force is maximum near the plates of the channel and minimum in the middle of the plates. Therefore, motion on the alumina nanofluids is maximum in the middle of the plates and minimum at the plates. The motion of alumina nanofluids for various numerical values of Grashof number, Gr of Al_2O_3 in H_2O-based nanofluids is shown in **Figure 9**. It is concluded that an increase in Grashof number, Gr, governs to an increase in the motion of alumina nanofluids. An increase in Grashof number, Gr, rises temperature of alumina in ethylene glycol nanofluids, which govern to an increase in the upward buoyancy force. Therefore, motion in alumina nanofluids increases with Gr, because of the increment of buoyancy force. **Figure 10** shows

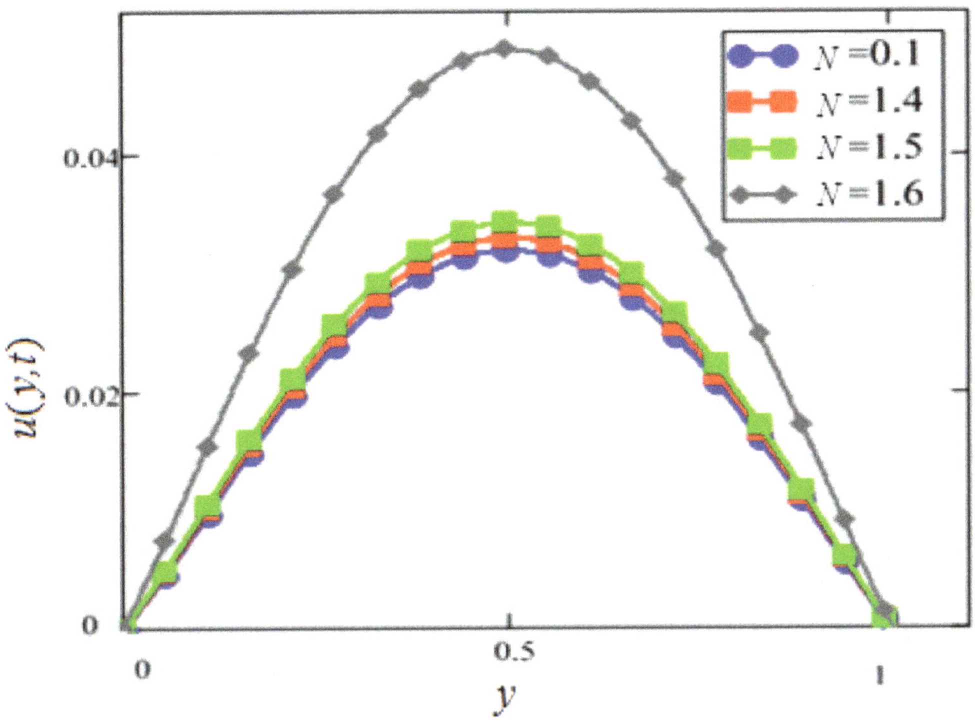

Figure 7. Velocity profiles for different values of N of Al_2O_3 in EG-based nanofluids when $Gr = 0.1$, $M = 1$, $\lambda = 1$, $K = 1$, $t = 5$, $\phi = 0.04$, $\omega = 0.2$.

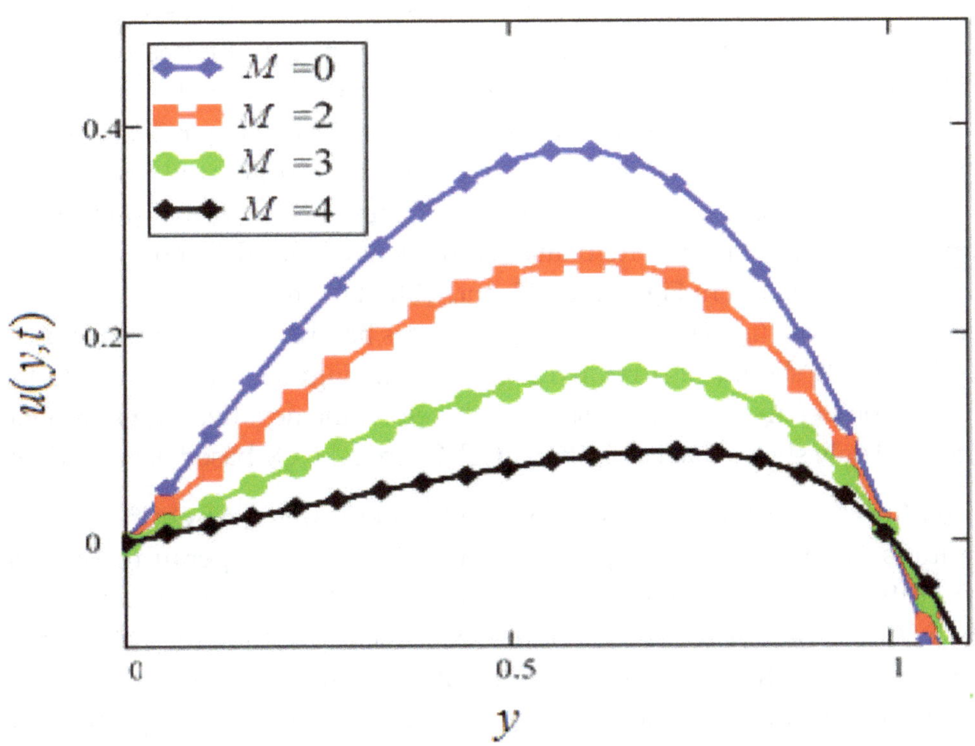

Figure 8. Velocity profiles for different values of M of Al_2O_3 in EG-based nanofluids when $Gr = 0.1$, $N = 0.1$, $\lambda = 1$, $K = 1$, $t = 5$, $\phi = 0.04$, $\omega = 0.2$.

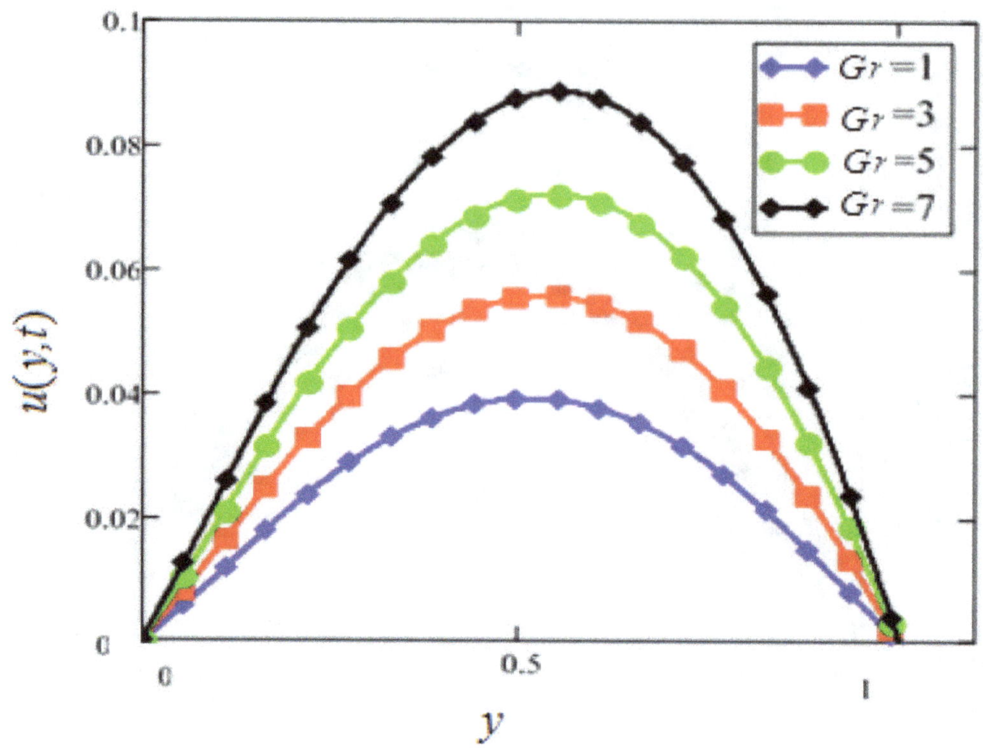

Figure 9. Velocity profiles for different values of Gr of Al_2O_3 in EG-based nanofluids when $N = 0.1,\ M = 1,\ \lambda = 1,\ K = 1,\ t = 10,\ \phi = 0.04,\ \omega = 0.2$.

permeability parameter K. It is view that motion of Al_2O_3 in $C_2H_6O_2$-based nanofluids increases with increasing values of permeability parameter K of Al_2O_3 in $C_2H_6O_2$-based nanofluids because of the small friction force. Similarly, increasing K reduces the nanofluids friction within the plate walls and motion of alumina nanofluids enhances. In the second case, **Figures 11–18** show the flow condition when the left wall is oscillating in its own plane and the right wall is stationary. Under the last conditions, when both plates of the channel are oscillating in their planes (**Figures 19–26**). From all these graphs, we found that they are qualitatively identical but different quantitatively to **Figures 2–10**. It can also viewed from **Figures 11, 13,** and **14** that the motion of Al_2O_3 in $C_2H_6O_2$-based nanofluids at the right wall ($y = 1$) is not equal to zero. It is because of the reason of oscillating right plate of the channel. It should also be investigated from **Figures 20, 24,** and **26** that only the magnitude of motion of nanofluids is considered; therefore, the negative sign in motion is ignored in these figures and only shows that the motion of Al_2O_3 in $C_2H_6O_2$-based nanofluids is reduced.

The influence of dissimilar particle shapes on the temperature of Al_2O_3 in H_2O and $C_2H_6O_2$-based nanofluids is presented in **Figures 27** and **28**. The temperature of both types of nanofluids is different for dissimilar shapes because of the various viscosity and thermal conductivity of these nanoparticles. It should be concluded that the influence of thermal conductivity increases with the increase of temperature. However, the viscosity decreases with the increase of temperature. It is view that an elongated shape of nanoparticles inside H_2O-based nanofluids like cylinder and platelet has small temperature due to the larger viscosity and thermal conductivity while the blade shape of nanoparticles has the uppermost temperature because of least viscosity and thermal conductivity. The brick shape nanoparticles suspended fluids are lowest in the temperature range; however, it has low viscosity. This is

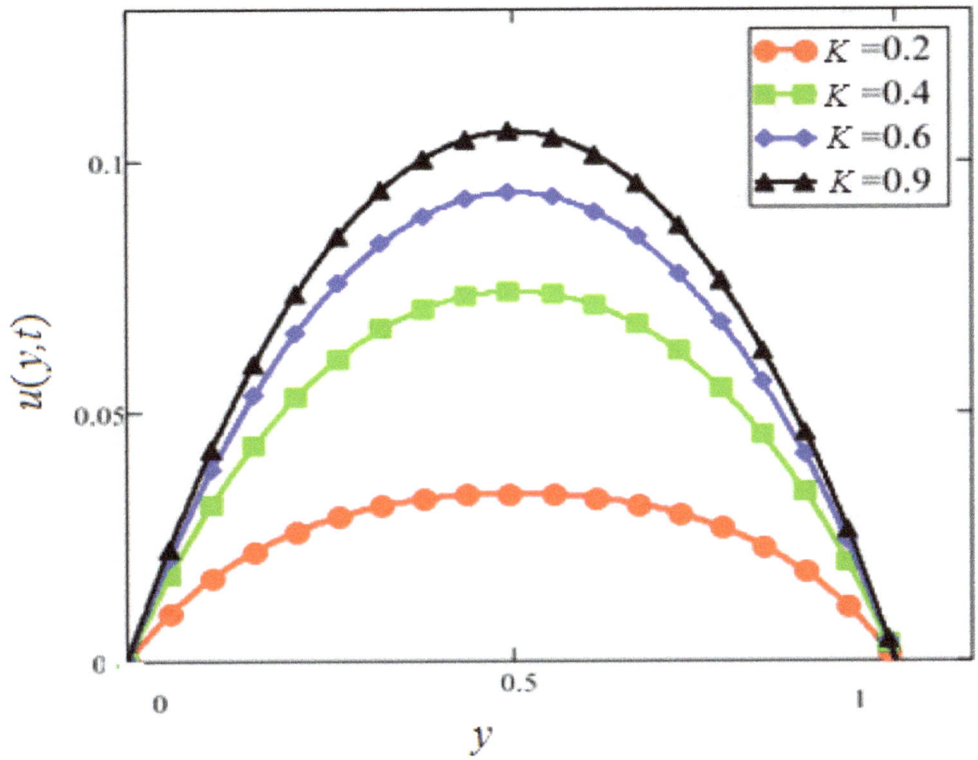

Figure 10. Velocity profiles of different values of K of Al_2O_3 in EG-based nanofluids when $Gr = 0.1,\ \ N = 0.1,\ \ M = 1,$ $\lambda = 1,\ \ t = 5,\ \ \phi = 0.04,\ \ \omega = 0.2.$

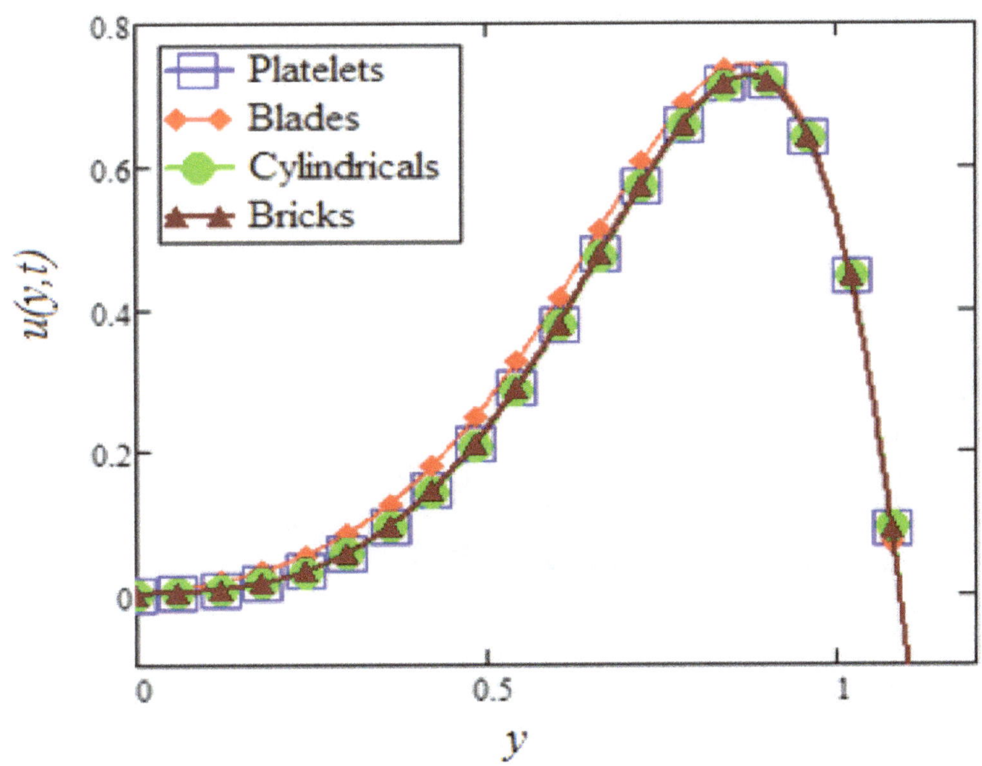

Figure 11. Velocity profiles for different shapes of Al_2O_3 nanoparticles in EG-based nanofluids when $Gr = 0.1,\ \ N = 0.1,$ $M = 1,\ \ \lambda = 1,\ \ K = 1,\ \ t = 5,\ \ \phi = 0.04,\ \omega = 0.2.$

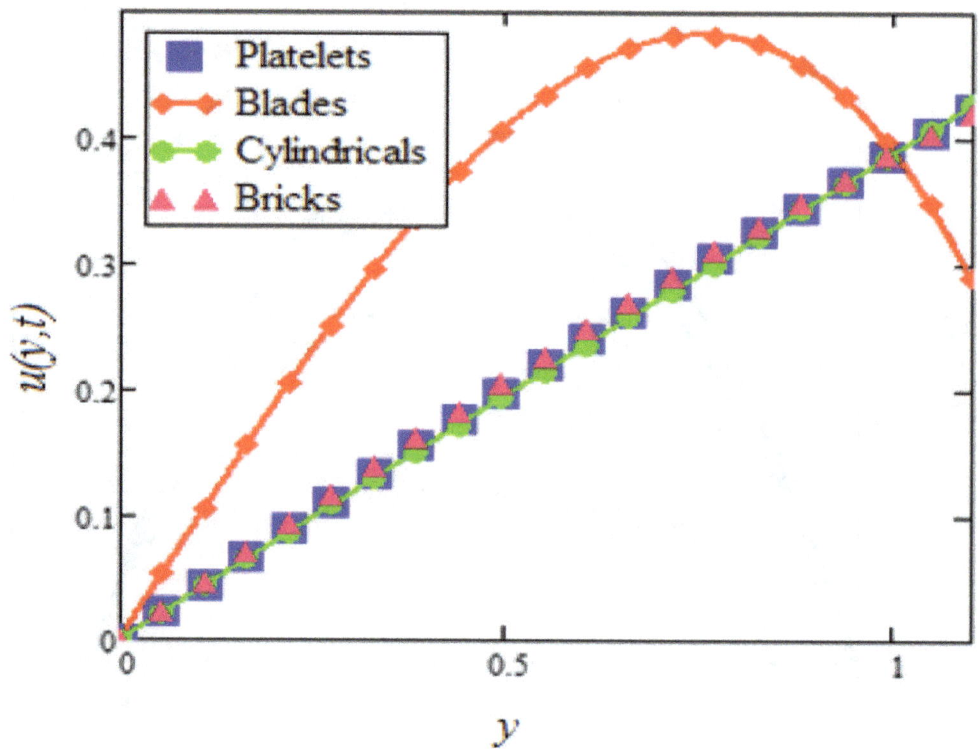

Figure 12. Velocity profiles for different shapes of Al_2O_3 nanoparticles in water-based nanofluids when $Gr = 0.1$, $N = 0.1$, $M = 1$, $\lambda = 1$, $K = 1$, $t = 5$, $\phi = 0.04$, $\omega = 0.2$.

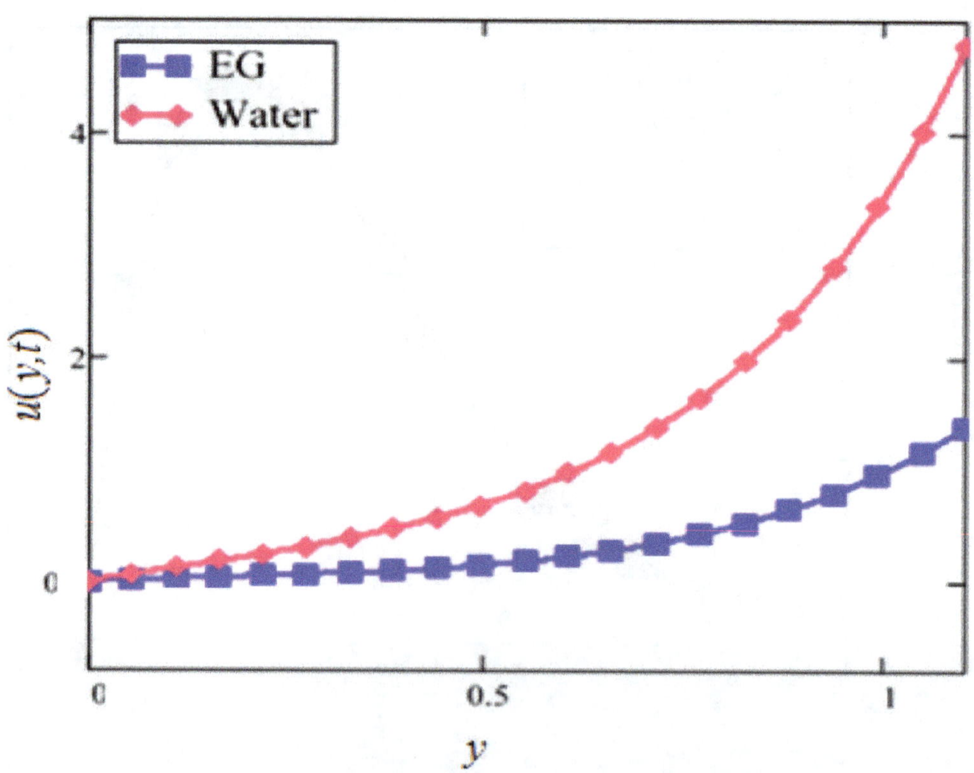

Figure 13. Comparison of velocity profiles of Al_2O_3 in EG- and water-based nanofluids when $Gr = 0.1$, $N = 0.1$, $M = 1$, $\lambda = 1$, $K = 1$, $t = 5$, $\phi = 0.04$, $\omega = 0.2$.

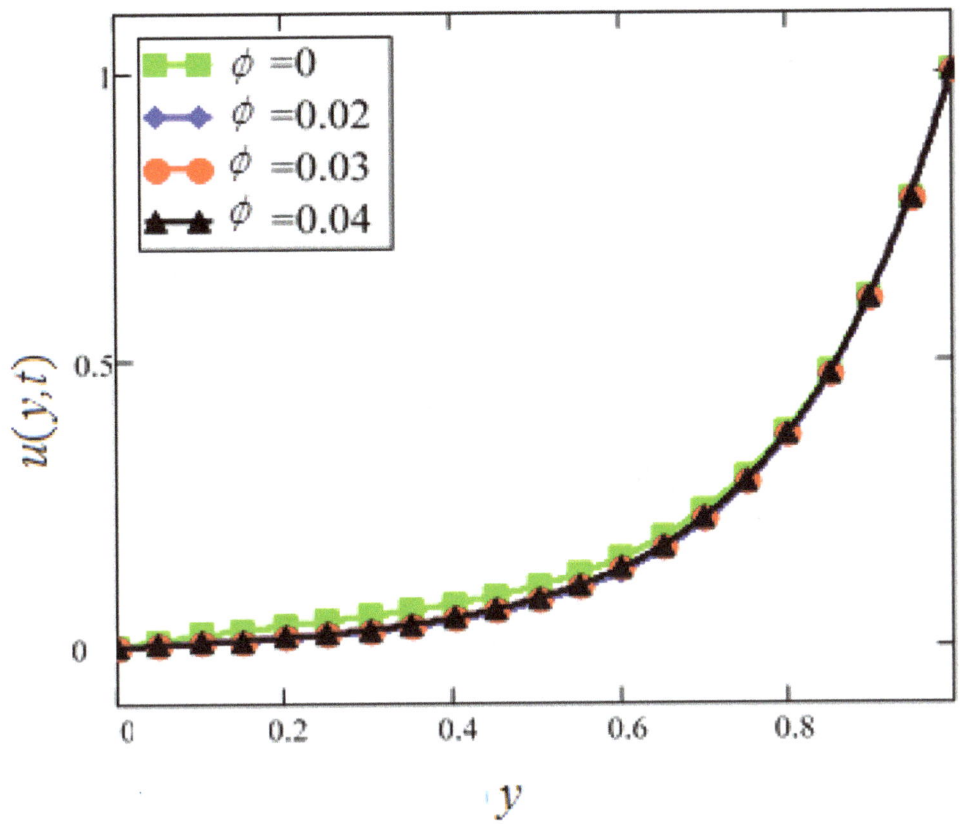

Figure 14. Velocity profiles for different values of ϕ of Al_2O_3 in EG-based nanofluids when $Gr = 0.1,\ N = 0.1,\ M = 1,\ \lambda = 0.01,\ K = 0.3,\ t = 10,\ \omega = 0.2$.

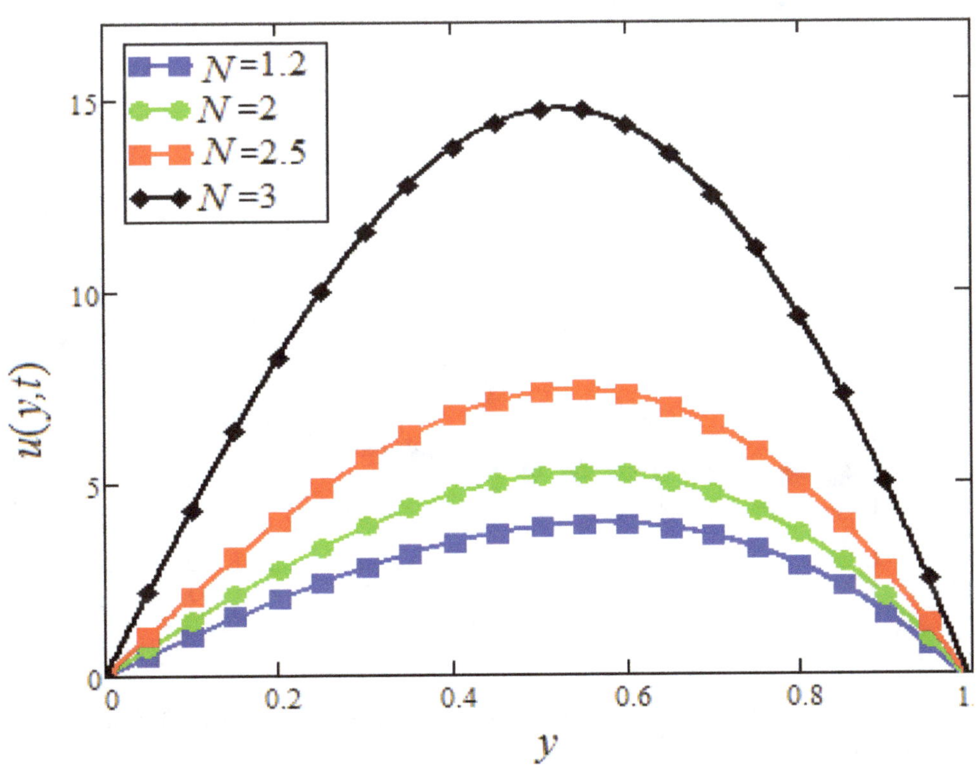

Figure 15. Velocity profiles for different values of N of Al_2O_3 in EG-based nanofluids when $Gr = 1,\ M = 1,\ \lambda = 0.01,\ K = 0.2,\ t = 10,\ \phi = 0.04,\ \omega = 0.2$.

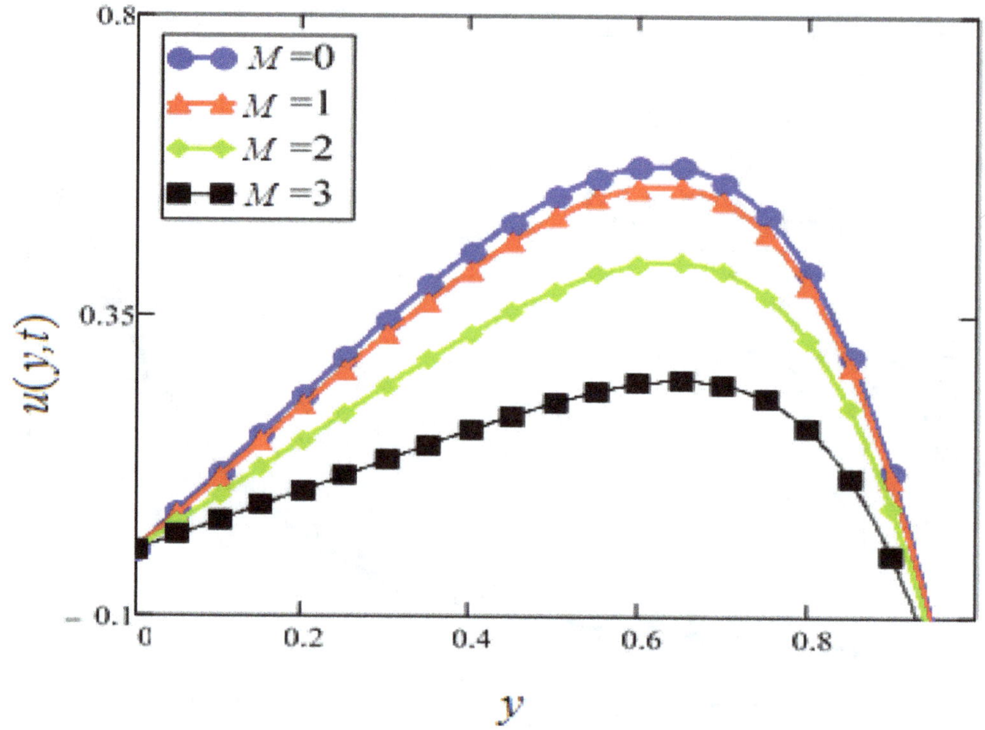

Figure 16. Velocity profiles for different values of M of Al_2O_3 in EG-based nanofluids when $Gr = 1$, $N = 0.1$, $\lambda = 0.001$, $K = 1$, $t = 10$, $\phi = 0.04$, $\omega = 0.2$.

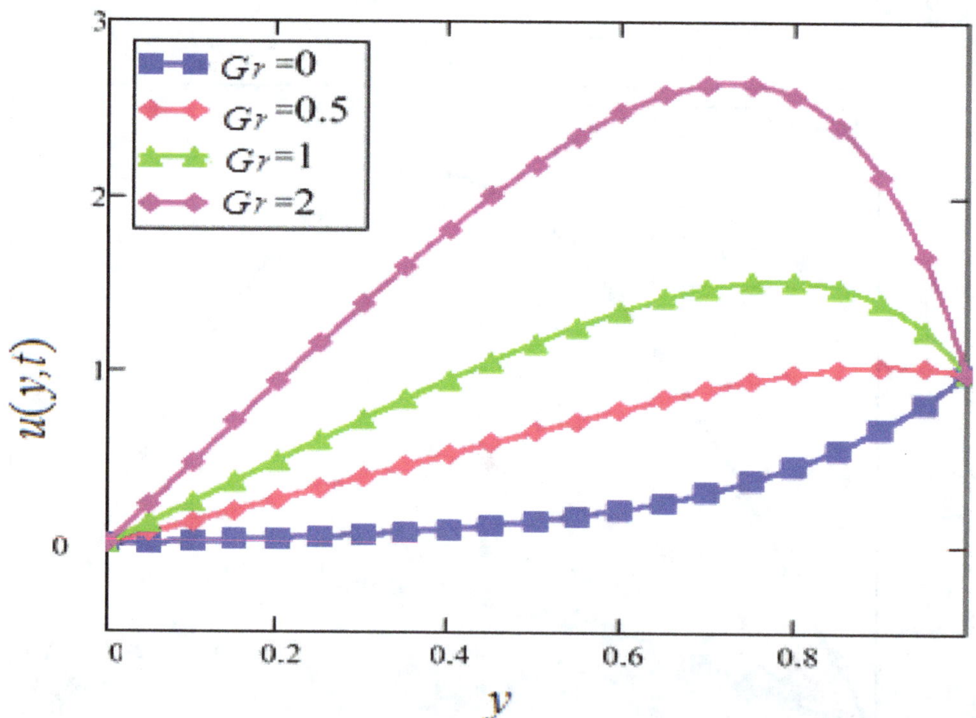

Figure 17. Velocity profiles for different values of Gr of Al_2O_3 in EG-based nanofluids when $N = 0.1$, $M = 1$, $\lambda = 0.01$, $K = 0.2$, $t = 10$, $\phi = 0.04$, $\omega = 0.2$.

because of the shear thinning behavior with temperature of nanofluids. Moreover, cylindrical-shaped nanofluids also show shear thinning behavior. However, the effect is less dominant. All the other dissimilar shapes like platelet and blade show Newtonian behavior and independence of viscosity on shear rate. This shear thinning behavior is also analyzed experimentally by Timofeeva et al.

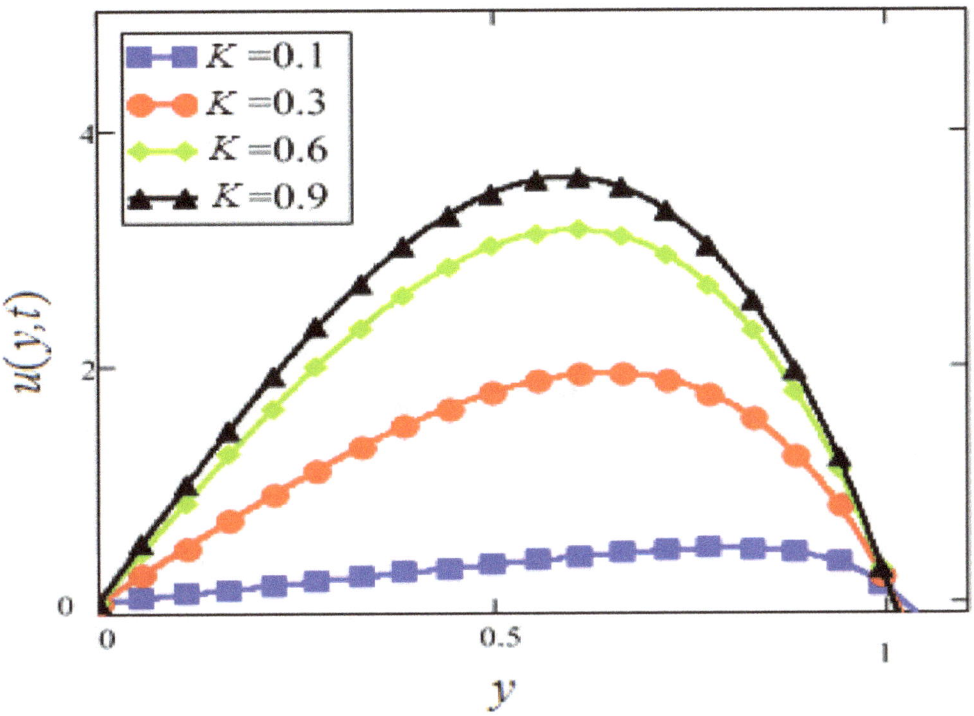

Figure 18. Velocity profiles for different values of K of Al_2O_3 in EG-based nanofluids when $Gr = 0.1,\ N = 0.1,\ M = 1,\ \lambda = 0.01,\ t = 10,\ \phi = 0.04,\ \omega = 0.2$.

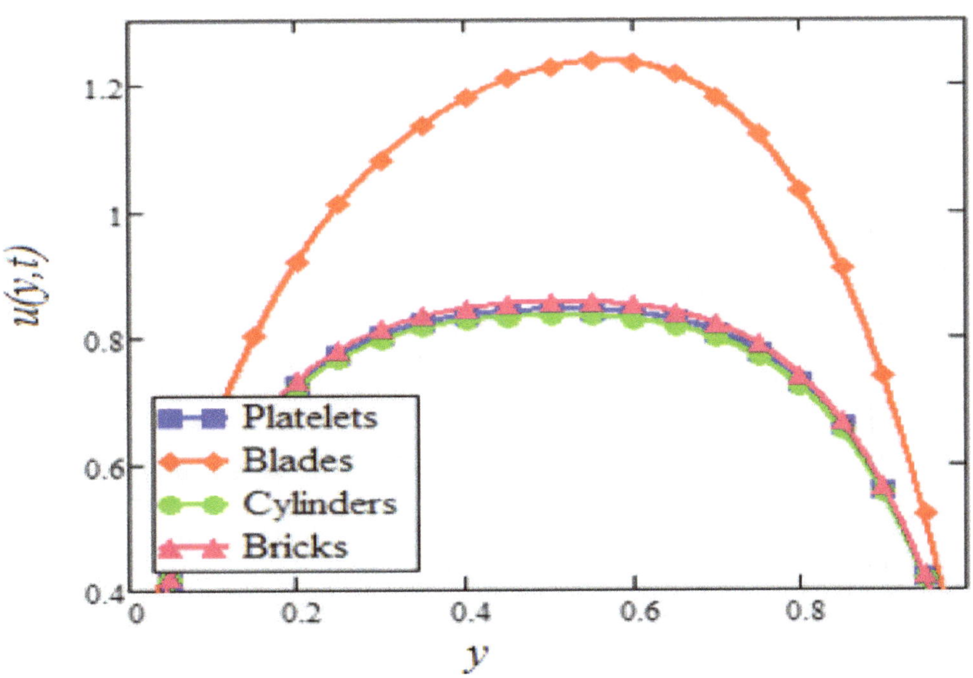

Figure 19. Velocity profiles for different shapes of Al_2O_3 nanoparticles in EG-based nanofluids when $Gr = 1,\ N = 0.1,\ M = 1,\ \lambda = 0.01,\ K = 1,\ t = 10,\ \phi = 0.04,\ \omega = 0.2$.

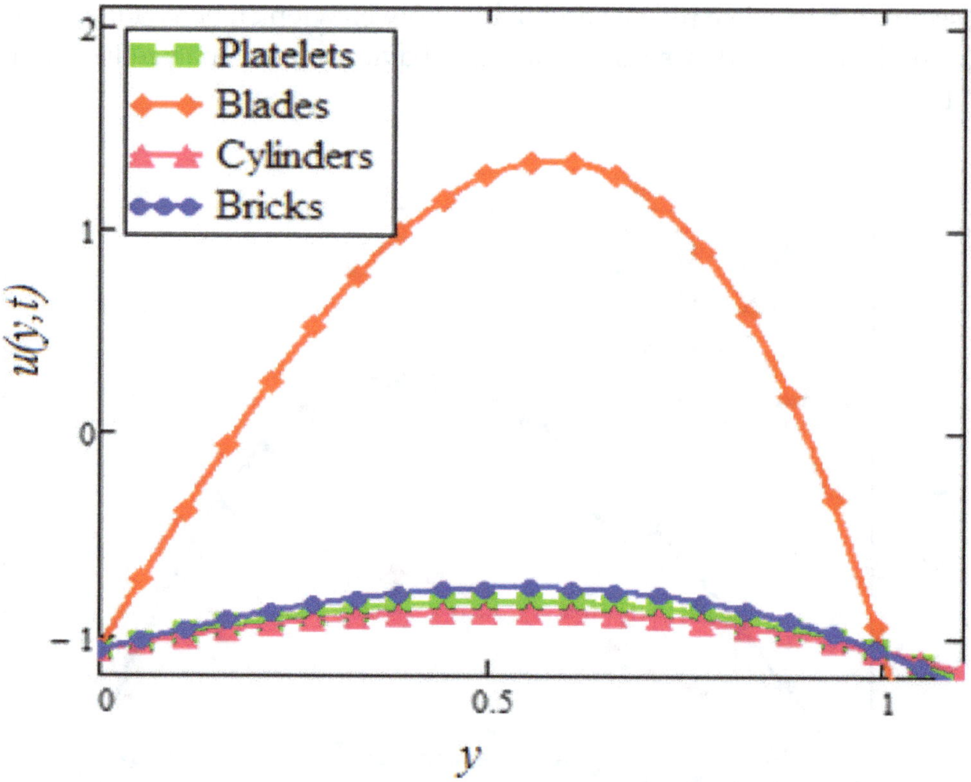

Figure 20. Velocity profiles for different shapes of Al_2O_3 nanoparticles in water-based nanofluids when $Gr = 1$, $N = 0.1$, $M = 1$, $\lambda = 0.01$, $K = 1$, $t = 10$, $\phi = 0.04$, $\omega = 0.2$.

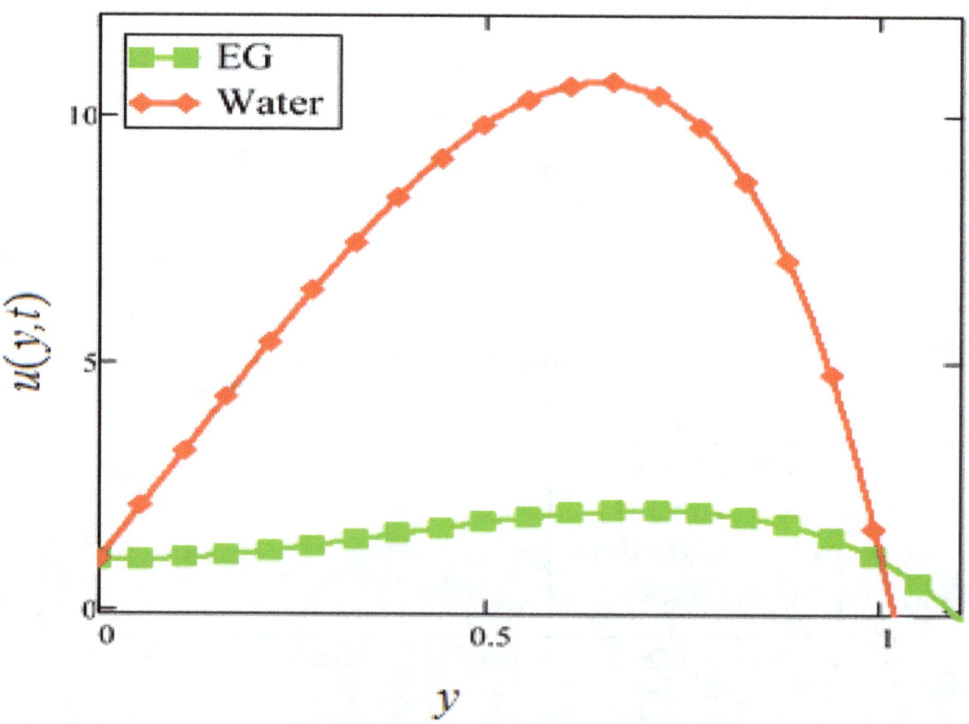

Figure 21. Comparison of velocity profiles of Al_2O_3 in EG- and water-based nanofluids when $Gr = 1$, $N = 0.1$, $M = 1$, $\lambda = 0.01$, $K = 1$, $t = 10$, $\phi = 0.04$, $\omega = 0.2$.

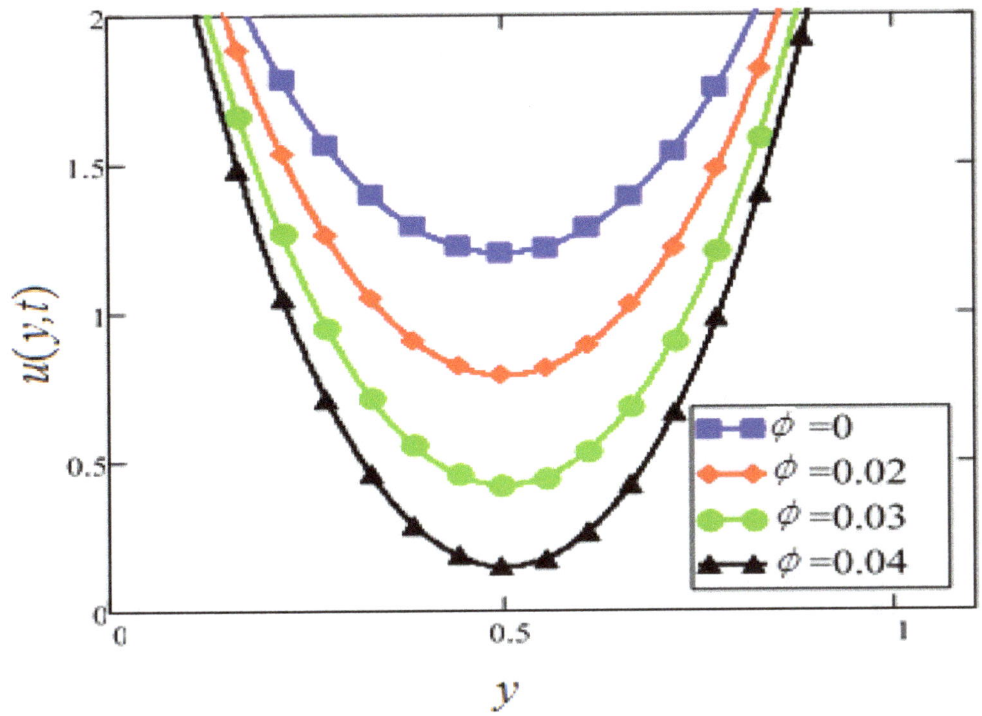

Figure 22. Velocity profiles for different values of ϕ of Al_2O_3 in EG-based nanofluids when $Gr = 1,\ \ N = 0.1,\ \ M = 1,$ $\lambda = 0.01,\ \ K = 1,\ \ \ t = 10,\ \ \ \omega = 0.2.$

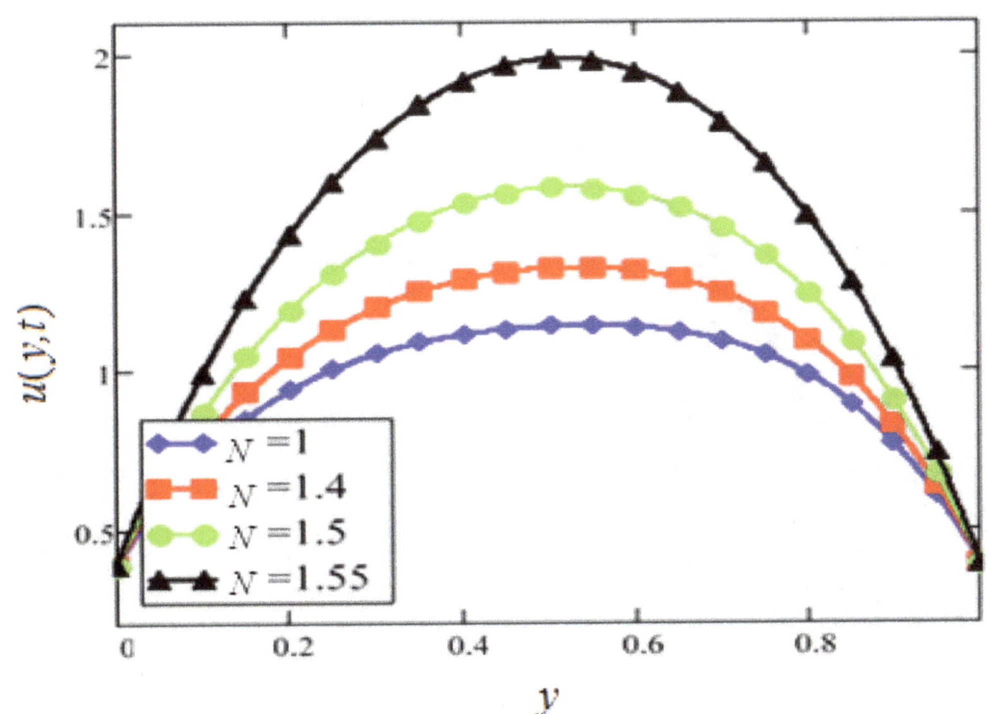

Figure 23. Velocity profiles for different values of N of Al_2O_3 in EG-based nanofluids when $Gr = 1,\ \ M = 1,\ \ \lambda = 0.01,$ $K = 1,\ \ \ t = 10,\ \ \ \phi = 0.04,\ \omega = 0.2.$

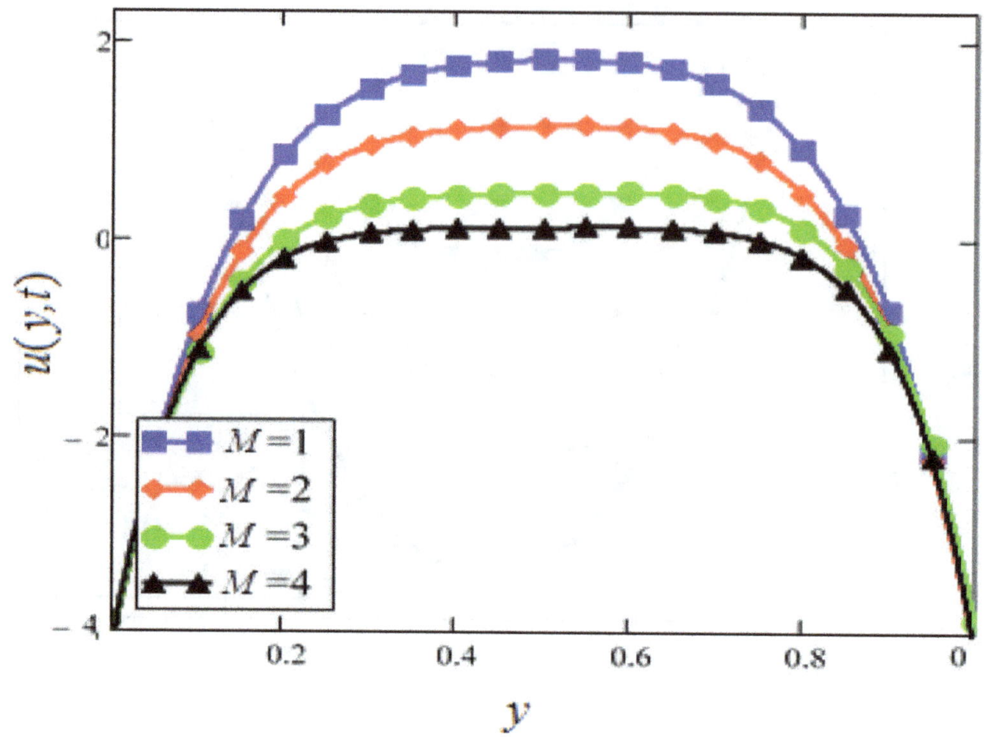

Figure 24. Velocity profiles for different values of M of Al_2O_3 in EG-based nanofluids when $Gr = 1$, $M = 1$, $\lambda = 0.01$, $K = 1$, $t = 10$, $\phi = 0.04$, $\lambda = 0.01$, $\omega = 0.2$.

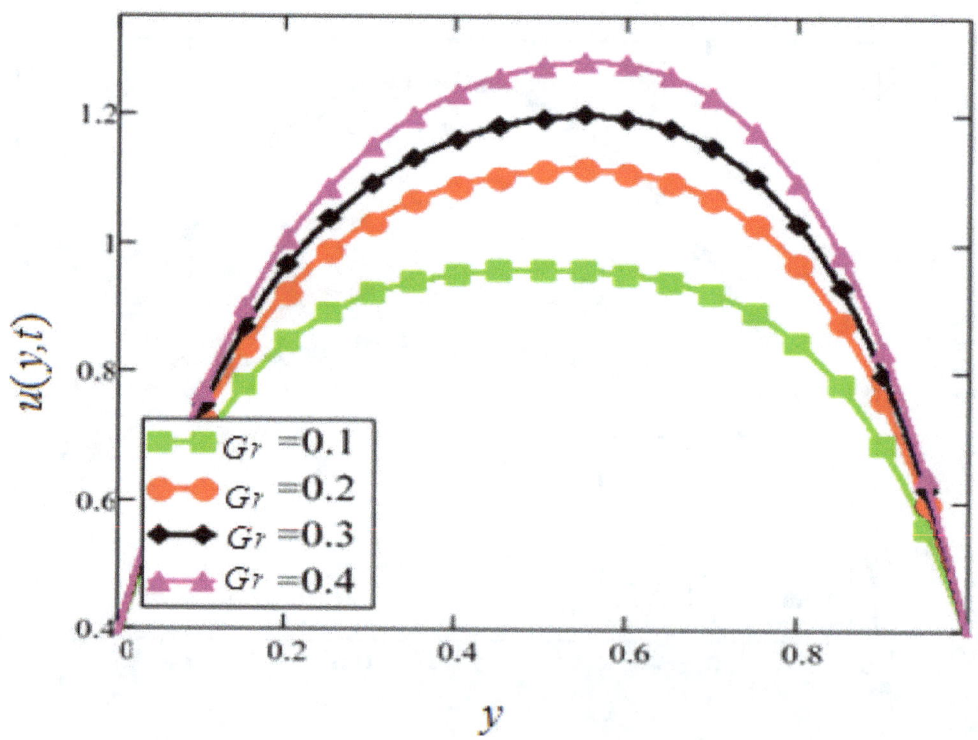

Figure 25. Velocity profiles for different values of Gr of Al_2O_3 in EG-based nanofluids when $Gr = 1$, $M = 1$, $\lambda = 0.01$, $K = 1$, $t = 10$, $\phi = 0.04$, $\omega = 0.2$.

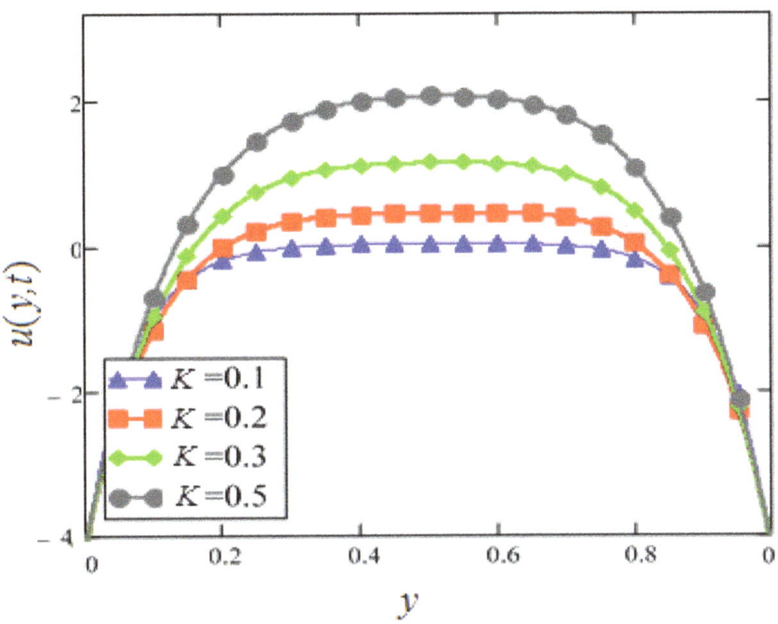

Figure 26. Velocity profiles for different values of K of Al_2O_3 in EG-based nanofluids when $Gr = 1$, $M = 1$, $\lambda = 0.01$, $K = 1$, $t = 10$, $\phi = 0.04$, $\omega = 0.2$.

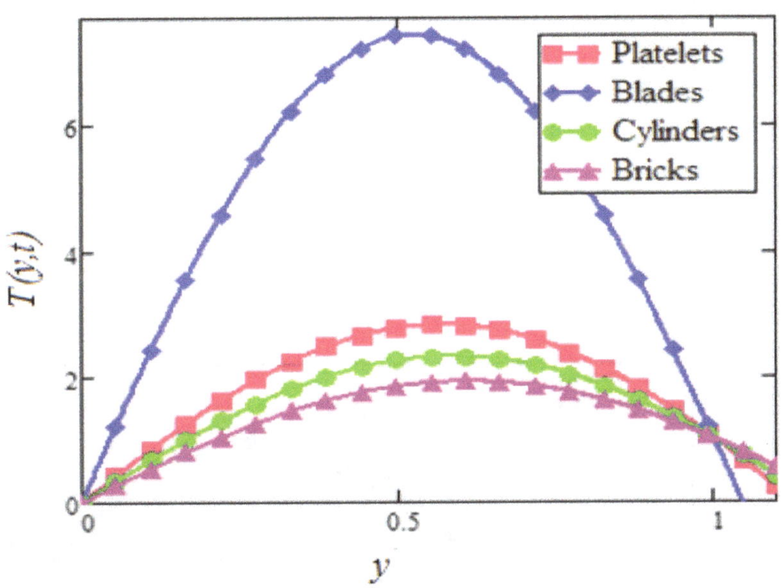

Figure 27. Temperature profiles for different shapes of Al_2O_3 nanoparticles in EG-based nanofluids when $N = 1.5$, $t = 1$.

A comparatively study of Al_2O_3 in H_2O- and $C_2H_6O_2$-based nanofluids is displayed in **Figure 29**. It is analyzed that both nanofluids are temperature dependent and the variation is found at the same rate for both types of nanofluids. This means that the influence of temperature of nanofluids on the thermal conductivity and viscosity of two types of base nanofluids may cause at the same rate. **Figure 30** exhibits the influence of ϕ on the temperature of Al_2O_3 in $C_2H_6O_2$-based nanofluids. It is evaluated that with the increase of ϕ temperature of the nanofluid increases because of the shear thinning nature. The viscosity of cylindrical-shaped nanoparticles inside water and $C_2H_6O_2$-based nanofluids shows shear thinning nature at the highest suspension. This was also experimentally investigated by Timofeeva et al.

The graphical consequences of temperature of nanofluids for several values of N are displayed in **Figure 31**. It is indicated that the temperature of the cylindrical-shaped alumina Al_2O_3 nanoparticles in $C_2H_6O_2$-based nanofluids shows larger oscillation with the increase of N. It is evaluated in the solution of the problem that temperature of the alumina nanofluids is oscillating and the influence of oscillation is increase with the increase of N. The increasing N means cooler or dense nanofluids or reduce the influence of energy transport to the nanofluids. The cylindrical-shape nanofluids have temperature dependent viscosity because of the shear thinning nature.

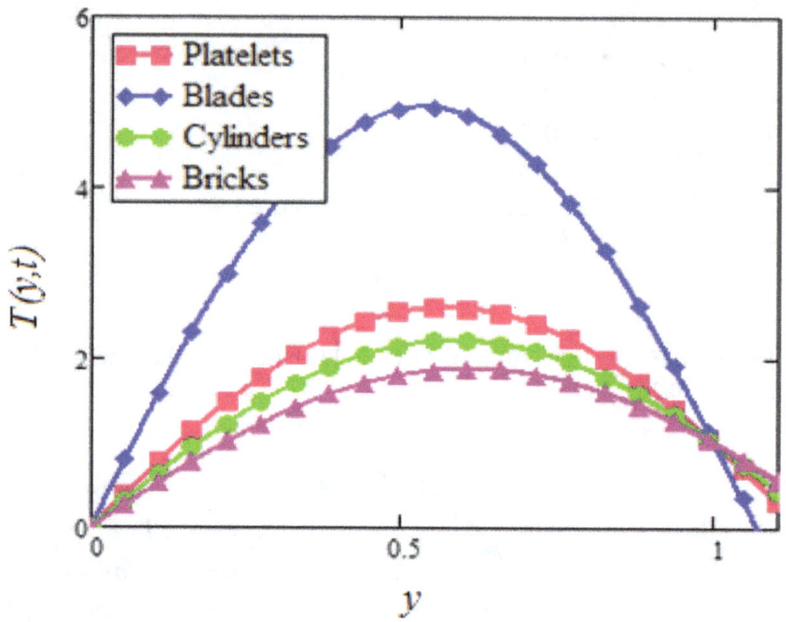

Figure 28. Temperature profiles for different shapes of Al_2O_3 nanoparticles in water-based nanofluids when $N = 1.5$, $t = 1$.

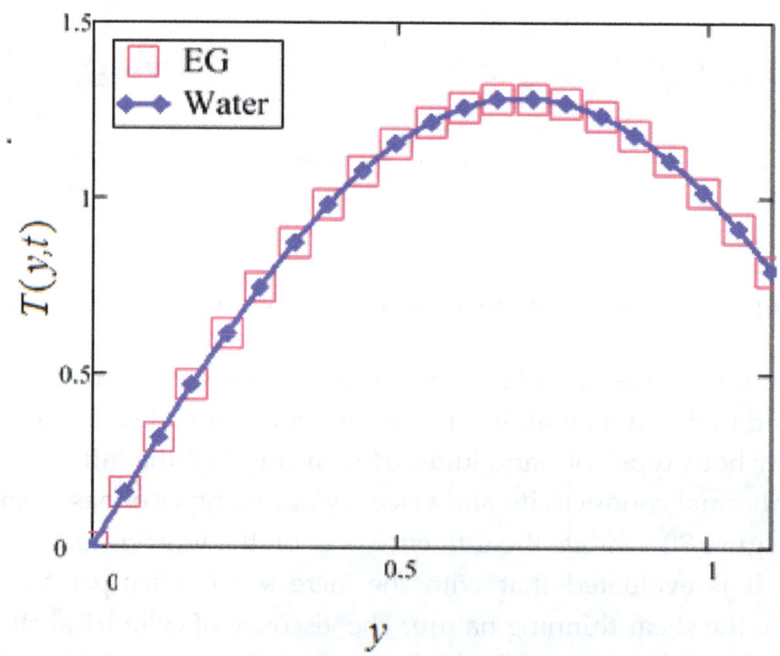

Figure 29. Comparison of temperature profiles of Al_2O_3 in EG- and water-based nanofluids when $N = 1.5$, $t = 1$.

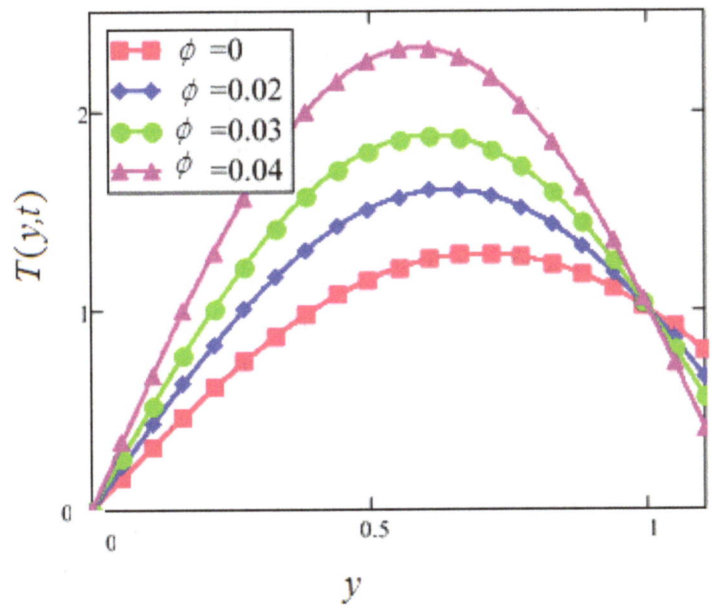

Figure 30. Temperature profiles for different values of ϕ of Al_2O_3 in EG-based nanofluids when $N = 1.5,\ t = 1$.

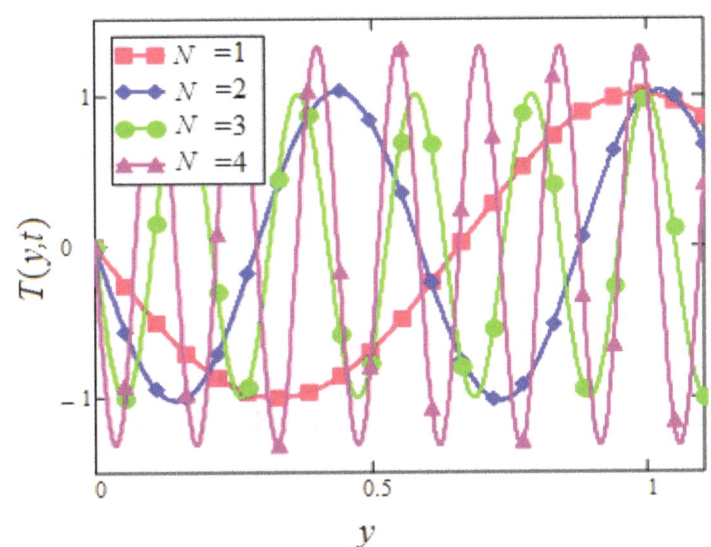

Figure 31. Temperature profiles for different values of N of Al_2O_3 in EG-based nanofluids when $t = 1$.

4. Conclusions

In this chapter, the influence of radiative heat transfer on mixed convection MHD flow of different shapes of Al_2O_3 in $C_2H_6O_2$, and H_2O base nanofluids in a channel filled with a saturated porous medium is analyzed. The two plates of the channel at finite distance with nonuniform wall temperature are chosen in a vertical direction under the influence of a perpendicular magnetic field. The governing PDEs are solved by the perturbation method for three different flow cases, and analytic solutions are evaluated. The influence of the dissimilar shapes of nanoparticles, namely, platelet, blade, cylinder, and brick of the same volume, on the motion of nanofluids and temperature of nanofluids is examined with different consequences. An elongated shape of nanoparticles inside base fluids like cylinder and platelet results in greater viscosity at the equal volume fraction due to structural limitation of rotational and transitional Brownian motion. The shear thinning nature of cylinder and blade shape of nanoparticles inside H_2O and $C_2H_6O_2$ is also investigated in this research. Viscosities and thermal conductivities of nanofluids are viewed depending on nanoparticle shapes, suspension of volume fraction, and base fluid of solid nanoparticles. The concluded results are as follow:

1. The motion of nanofluid decreases with the increase of volume fraction of nanoparticles because of the increment in viscosity and thermal conductivity.
2. The motion of $C_2H_6O_2$ base nanofluid is found smaller than H_2O base nanofluid because the viscosity of base fluid influences the Brownian movement of the nanoparticles.
3. Elongated nanoparticles inside water and $C_2H_6O_2$ like cylinder and platelet shapes have smaller motion as assimilated to blade and brick shapes of nanoparticles because of greater viscosity.
4. The motion of the nanofluid decreases with the enhancement of magnetic parameter because of the increase of the resistive type drag force that has the tendency to reduce the motion of the nanofluids.
5. The motion of the nanofluid also decreases with the increases of thermal Grashof number. The increase of Gr increases the temperature of nanofluids, which governs to an increase in the upward buoyancy force. Therefore, the motion in nanofluids increases with increasing Gr because of the increment of buoyancy force.

Acknowledgements

The authors are grateful to the reviewers for their excellent comments to improve the quality of the present article. The authors would also like to acknowledge the Research Management Center-UTM for the financial support through vote numbers 4F109 and 03J62 for this research.

Conflicts of interest

The authors declare that they have no conflicts of interest.

Nomenclature

Roman letters	
a, b	Constants depend on shape of nanoparticles
$\mathbf{B}_0$	Applied magnetic field
B_0	Magnitude of applied magnetic field
$\mathbf{B}$	Total magnetic field
$\mathbf{b}$	Induced magnetic field
$(c_p)_s$	Heat capacity of solid nanoparticles
$(c_p)_f$	Heat capacity of base fluids

Roman letters	
$(c_p)_{nf}$	Heat capacity of nanofluids
$\mathbf{D}$	Rate of strain tensor
d_p	Diameter of solid nanoparticles
$\mathbf{E}$	Total electric field
e	Internal energy per unit volume
exp	Exponential function
$\mathbf{F}$	Force
f	Function of temperature and volume fraction, etc.
Gr	Thermal Grashof number
g	Gravitational acceleration
H(.)	Heaviside function
$\mathbf{H}$	Total momentum of the system
$\mathbf{I}$	Identity tensor
$\mathbf{i}$	Cartesian unit vector in the x-direction
$\mathbf{J}$	Current density
$\mathbf{J}\times\mathbf{B}$	Lorentz force
$\mathbf{j}$	Cartesian unit vector in the y-direction
K	Dimensionless permeability parameter
k_s	Thermal conductivity of solid nanoparticles
k_f	Thermal conductivity of base fluids
k_{nf}	Thermal conductivity of nanofluids
k_b	Boltzmann constant
$\mathbf{k}$	Cartesian unit vector in the z-direction
M	Magnetic parameter
m	Mass of the flow of fluids
N	Radiation parameter
Nu	Nusselt number
n	Empirical shape factors
p	Pressure
p_h	Hydrostatic pressure
p_d	Dynamic pressure
Pe	Peclet number
Q	Heat generation parameter
	Radiant flux vector
q_r	Magnitude of radiant heat flux
$\mathbf{q}''$	Heat conduction per unit area
q''	Magnitude of heat conduction per unit area

Roman letters	
Re	Reynolds number
r_c	Radius of gyration
S	Surface of the control volume
$\mathbf{T}$	Cauchy stress tensor
T	Temperature
t	Time
u	Velocity in x-direction
U_0	Reference velocity
$\nabla\mathbf{V}$	Dyadic tensor
$\mathbf{V}$	Velocity vector field
V	Control volume
V	Magnitude of velocity
v_0	Constant velocity in y-direction
$\mathbf{W}$	Work done
Greek letters	
ρ_s	Density of solid nanoparticles
ρ_f	Density of base fluids
ρ_{nf}	Density of nanofluids
β_s	Volumetric coefficient of thermal expansion of solid nanoparticles
β_f	Volumetric coefficient of thermal expansion of base fluids.
β_{nf}	Volumetric coefficient of thermal expansion of nanofluids
μ_s	Dynamic viscosity of solid nanoparticles
μ_f	Dynamic viscosity of base fluids
μ_{nf}	Dynamic viscosity of solid nanofluids
β	Modeling function
φ	Volume fraction of solid nanoparticles
$\boldsymbol{\Psi}$	Sphericity
ω	Oscillating parameter
ε	Perturbed parameter
λ	Williamson parameter
α_0	Mean absorption coefficient
σ_{nf}	Electrical conductivity of nanofluids
μ_m	Magnetic permeability
∇	Delta function
τ_1	Skin friction
$\boldsymbol{\tau}$	Viscous stress tensor

Author details

Aaiza Gul[1], Ilyas Khan[2]* and Sharidan Shafie[1]

*Address all correspondence to: ilyaskhanqau@yahoo.com

1 Department of Mathematical Sciences, Faculty of Science, Universiti Teknologi Malaysia, Skudai, Malaysia

2 Basic Engineering Sciences Department, College of Engineering Majmaah University, Majmaah, Saudi Arabia

References

[1] S. U. S. Choi, Enhancing thermal conductivity of fluids with nanoparticle, ASME FED, 66 (1995), 99–105.

[2] R. S. Vajjha and D.K. Das, Experimental determination of thermal conductivity of three nanofluids and development of new correlations, International Journal of Heat and Mass Transfer, 52 (2009), 4675–4682.

[3] M. T. Naik and L. S. Sundar, Investigation into thermophysical properties of glycol based CuO nanofluid for heat transfer applications, World Academy of Science Engineer Technology, 59 (2011), 440–446.

[4] S. Mansur, A. Ishak, and I. Pop, The magnetohydrodynamic stagnation point flow of a nanofluid over a stretching/shrinking sheet with suction, PLoS ONE, 10 (3) (2015), e0117733.

[5] E. V. Timofeeva, R. L. Jules, and S. Dileep, Particle shape effect on thermophysical properties of alumina nanofluids, Journal of Applied Physics, 106 (2009), 014304.

[6] P. Loganathan, P. N. Chand, and P. Ganesan, Radiation effects on an unsteady natural convection flow of a nanofluids past an infinite vertical plate, World Scientific Publishing Company, 8 (2013), 1350001–1350010.

[7] K. Asma, I. Khan, and S. Sharidan, Exact solutions for free convection flow of nanofluids with ramped wall temperature, The European Physical Journal Plus, 130 (2015), 57–71.

[8] S. Sebdani, M. Mahmoodi, and S. Hashemi, Effect of nanofluid variable properties on mixed convection in a square cavity, International Journal of Thermal Sciences, 52 (2012), 112–126.

[9] T. Fan, H. Xu, and I. Pop, Mixed convection heat transfer in horizontal channel filled with nanofluids, International Journal of Springer Plus, 34 (2013), 339–350.

[10] R. K. Tiwari and M. K. Das, Heat transfer augmentation in a two-sided lid-driven differentially heated square cavity utilizing nanofluids, International Journal of Heat and Mass Transfer, 50 (2007), 9–10.

[11] G. A. Sheikhzadeh, N. Hajialigol, M. E. Qomi, and A. Fattahi, Laminar mixed convection of Cu-water nano-fluid in two sided lid-driven enclosures, Journal of Nanostructures, 1 (2012), 44–53.

[12] S. Nadeem and S. Saleem, Unsteady mixed convection flow of nanofluid on a rotating cone with magnetic field, Apply Nanoscience, 4 (2014), 405–414.

[13] K. Al-Salem, H. F. Oztop, I. Pop, and Y. Varol, Effect of moving lid direction on MHD mixed convection in a linearly heated cavity, International Journal of Heat and Mass transfer, 55 (2012), 1103–1112.

[14] K. V. Prasad, K. Vajravelu, and P. S. Datti, The effect of variable fluid properties on the MHD flow and heat transfer over a non-linear stretching sheet, International Journal of Thermal Science, 49 (2010), 603–610.

[15] J. Y. Rami, A. Fawzi, and F. Abu-Al-Rub, Darcy-Forchheimer mixed convection heat and mass transfer in fluid saturated porous media, International Journal of Numerical Methods for Heat & Fluid Flow, 11 (2011), 600–618.

[16] T. Hayat, Z. Abbas, I. Pop, and S. Asghar, Effect of radiation and magnetic field on the mixed convection stagnation-point flow over a vertical stretching sheet in a porous medium, International Journal of Heat and Mass Transfer, 53 (2010), 466–474.

[17] R. L. Hamilton and O. K. Crosser, Thermal conductivity of heterogeneous two-component systems, Journal of Industrial & Engineering Chemistry Fundamentals, 1 (1962), 187–191.

[18] M. Turkyilmazoglu, Unsteady convection flow of some nanofluids past a moving vertical flat plate with heat transfer, Journal of Heat Transfer, 136 (2014), 031704–031711.

[19] A. Zeehan, R. Ellahi, and M. Hassan, Magnetohydrodynamic flow of water/ethylene glycol based nanofluids with natural convection through porous medium, European Physical Journal Plus, 129 (2014), 261.

[20] O. D. Makinde and P. Y. Mhone, Heat transfer to MHD oscillatory flow in a channel filled with porous medium, Romanian Journal of Physics, 50 (2005), 931–938.

[21] L. Colla, L. Fedele, M. Scattolini, and S. Bobbo, Water-based Fe_2O_3 nanofluid characterization: thermal conductivity and viscosity measurements and correlation, Advances in Mechanical Engineering, 2012 (Article ID 674947), 8.

[22] A. S. Noreen, M. Raza, and R. Ellahi, Influence of induced magnetic field and heat flux with the suspension of carbon nanotubes for the peristaltic flow in a permeable channel, Journal of Magnetism and Magnetic Materials, 381(2015), 405–415.

[23] R. Ellahi, S. Aziz, and A. Zeeshan, Non Newtonian nanofluids flow through a porous medium between two coaxial cylinders with heat transfer and variable viscosity, Journal of Porous Media, 16 (3) (2013), 205–216.

[24] A. S. Noreen, M. Raza, and R. Ellahi, Interaction of nanoparticles for the peristaltic flow in an asymmetric channel with the induced magnetic field, The European Physical Journal – Plus, 129 (2014), 155–167.

[25] M. Sheikholeslami and R. Ellahi, Three dimensional mesoscopic simulation of magnetic field effect on natural convection of nanofluid, International Journal of Heat and Mass Transfer, 89 (2015), 799–808.

[26] R. Ellahi, M. Hassan, and A. Zeeshan, Study on magnetohydrodynamic nanofluid by means of single and multi-walled carbon nanotubes suspended in a salt water solution, IEEE Transactions on Nanotechnology, 14 (4) (2015), 726–734.

[27] M. Sheikholeslami and R. Ellahi, Simulation of ferrofluid flow for magnetic drug targeting using Lattice Boltzmann method, Journal of Zeitschrift Fur Naturforschung A, Verlag der Zeitschrift für Naturforschung, 70 (2015), 115–124.

[28] M. Sheikholeslami, Effect of uniform suction on nanofluid flow and heat transfer over a cylinder, Journal of the Brazilian Society of Mechanical Sciences and Engineering, 37 (6) (2015), 1623–1633.

[29] R. Ellahi, M. Hassan, and A. Zeeshan, Shape effects of nanosize particles in Cu-H2O nanofluid on entropy generation, International Journal of Heat and Mass Transfer, 81 (2015), 449–456.

[30] M. Sheikholeslami, D. D. Ganji, M. Y. Javed, and R. Ellahi, Effect of thermal radiation on magnetohydrodynamics nanofluid flow and heat transfer by means of two phase model, Journal of Magnetism and Magnetic Materials, 374 (2015), 36–43.

[31] S. Rashidi, M. Dehghan, R. Ellahi, M. Riaz, and M. T. Jamal-Abad, Study of stream wise transverse magnetic fluid flow with heat transfer around an obstacle embedded in a porous medium, Journal of Magnetism and Magnetic Materials, 378 (2015), 128–137.

[32] S. A. Noreen, M. Raza, and R. Ellahi, Influence of heat generation and heat flux in peristalsis with interaction of nanoparticles, The European Physical Journal Plus, 129 (2014), 185.

[33] R. Ellahi, M. Hassan, and S. Soleimani, A study of natural convection heat transfer in a nanofluid filled enclosure with elliptic inner cylinder, International Journal for Numerical Methods for Heat and Fluid Flow, 24 (8) (2014), 1906–1927.

[34] M. Sheikholeslami, R. Ellahi, H. R. Ashorynejad, G. Domairry, and T. Hayat, Effects of heat transfer in flow of nanofluids over a permeable stretching wall in a porous medium, Computational and Theoretical Nanoscience, 11 (2) (2014), 486–496.

[35] M. SheikholeslamI, M. G. Bandpy, R. Ellahi, and A. Zeeshan, Simulation of CuO-water nanofluid flow and convective heat transfer considering Lorentz forces, Journal of Magnetism and Magnetic Materials, 369 (2014), 69–80.

[36] S. A. Noreen, S. U. Rahman, R. Ellahi, and S. Nadeem, Nano fluid flow in tapering stenosed arteries with permeable walls, International Journal of Thermal Sciences, 85 (2014), 54–61.

[37] R. Ellahi, The effects of MHD and temperature dependent viscosity on the flow of non-Newtonian nanofluid in a pipe: analytical solutions, Applied Mathematical Modeling, 37 (2013), 1451–1467.

[38] R. Ellahi, M. Raza, and K. Vafai, Series solutions of non-Newtonian nanofluids with Reynolds' model and Vogel's model by means of the homotopy analysis method, Mathematical and Computer Modelling, 55 (2012), 1876–1891.

[39] X. B. Wang, P. L. Zhou, and F. X. Peng, A fractal model for predicting the effective thermal conductivity of liquid with suspension of nanoparticles, International Journal of Heat and Mass Transfer, 46 (14) (2003), 2665–2672.

[40] M. Abareshi, E. K. Goharshadi, S. M. Zebarjad, H. K. Fadafan, and A. Youssefi, Fabrication, characterization and measurement of thermal conductivity of Fe_3O_4 nanofluids, Journal of Magnetism and Magnetic Materials, 322(24) (2010), 3895–3901.

[41] S. E. Borglin, G. J. Moridis, and C. M. Oldenburg, Experimental studies of the flow of ferrofluid in porous media, Transport in Porous Media, 41 (1) (2000), 61–80.

[42] M. Sheikholeslami, N. S. Akbar, and M. T. Mustafa, MHD effect on nanofluid with energy and hydrothermal behavior between two collateral plates: application of new semi analytical technique, Thermal Science, (2015), 95–95. doi:10.2298/TSCI150228095S.

[43] M. Sheikholeslami, K. Vajravelu, and M. M. Rashidi, Forced convection heat transfer in a semi annulus under the influence of a variable magnetic field, International Journal of Heat and Mass Transfer, 92 (2016), 339–348.

[44] M. Sheikholeslami and M. M Rashid, Non-uniform magnetic field effect on nanofluid hydrothermal treatment considering Brownian motion and thermophoresis effects, Journal of the Brazilian Society of Mechanical Sciences and Engineering, 38 (4) (2016), 1171–1184.

[45] M. Sheikholeslami, M. M. Rashidi, and D. D. Ganji, Effect of non-uniform magnetic field on forced convection heat transfer of water nanofluid, Computer Methods in Applied Mechanics and Engineering, 294 (2015), 299–312.

[46] M. Sheikholeslami and M. M. Rashidi, Ferrofluid heat transfer treatment in the presence of variable magnetic field, The European Physical Journal Plus, 130 (6) (2015), 1–12.

[47] M. Sheikholeslami and S. Abelman, Two-Phase simulation of nanofluid flow and heat transfer in an annulus in the presence of an axial magnetic field, IEEE Transactions on Nanotechnology, 14 (3) (2015), 561–569.

[48] M. Sheikholeslami and A. J. Chamkha, Flow and convective heat transfer of a ferro-nanofluid in a double-sided lid-driven cavity with a wavy wall in the presence of a variable magnetic field, Numerical Heat Transfer, Part A: Applications 69 (10) (2016), 1186–1200.

[49] M. Sheikholeslami, M. M. Rashidi, T. Hayat, and D. D. Ganji, Free convection of magnetic nanofluid considering MFD viscosity effect, Journal of Molecular Liquids, 218 (2016), 393–399.

[50] M. Sheikholeslami, M. G. Bandpy, R. Ellahi, M. Hassan, and S. Soleimani, Effects of MHD on Cu–water nanofluid flow and heat transfer by means of CVFEM, Journal of Magnetism and Magnetic Materials, 349 (2014), 188–200.

7

Problems Faced while Simulating Nanofluids

Adil Loya

Abstract

Problems are faced when something is already been adopted for a considerable amount of time–here the problem that is discussed is related with nanofluids. The nanofluids have been considered for different engineering applications since last three decades; however, the work on its simulation has been started since last two decades. With the time, nanofluid simulations are increasing as compared to experimental testing. Researchers conducting nanofluid simulations do find difficulties and problems while trying to simulate this system. In addition to this, most of the time researchers are unaware of some basic problems and they find themselves stuck in relentless difficulties. Most of the time, these problems are very basic and can waste a lot of useful time of a research. Therefore, this chapter introduces some fundamental problems which a researcher can find while simulating nanofluids and with a simple way of dealing with it. Moreover, the chapter withholds lots of information regarding the way to design and to model a nanofluid system. Not only this, it also tends to elaborate the nanofluid simulation methodology in a precise manner. Moreover, the literature shows that nanofluid simulation has gained high consideration since last two decades, as experimental techniques are out of reach for everyone. In addition to experimental techniques, they are expensive, time-consuming and require high skills. However, it seems the simulation is picking pace with the due time and is considerably being adopted by the expertise dealing with nanofluids. This opens a high prospect of simulating nanofluids in future. Nevertheless, it seems there will be user-friendly software to conduct nanofluid simulations. Finally, issues and their resolution have also been conveyed which is the main aspect of this topic.

Keywords: nanoparticles, nanofluids, molecular dynamics, simulations, problems

1. Introduction

Couple of decades back, nanofluid research was mostly conducted using experimental techniques. With time, as the computational power acquired drastic developments, new algorithms were designed, and therefore, today, we have got sophisticated software and mathematical models to solve and simulate the nanofluid environment.

1.1. Background knowledge

Nanofluids comprises of two constitutes, i.e. Nano comes from nanoparticles and fluid comes from base fluid. The need of combining nanoparticles with fluid was necessary for enhancing the properties of the base fluid. Addition of nanoparticles to the base fluid helps in altering and optimizing properties such as physiochemical [1], thermo-physical [2], rheological [2–4], etc.; to give a new composite performance. The initial mixing of nanofluids can be dated back to the time of the US choi in 1995, he was the first one to form nanofluid at Argonne laboratories USA [5, 6]. He used the nanofluid for optimization of thermal conductivity. Since then there have been several experimental studies over the thermal conductivity analysis of different nanoparticles in various base fluids [5, 7–9].

By looking at thermal conductivity improvement, other researchers came up with different ideas and formulations for utilization of this technique in various fields of science. Today, nanofluids are being used in biological, pharmaceuticals and medicine [10], engineering [7], lubrication industries [11, 12]. The major work on experimental side in all these industries has been carried out; however, these experiments of nanofluid require high skilled labour and expensive equipment. Furthermore, material purchase and characterization are costly. Due to this, researchers and industrialists working with nanofluids are trying to develop a model that can replicate mechanisms dealing with nanoparticle and fluid interactions. However, this subject is wide and requires huge expertise to deal with.

Currently, as the computational power has enhanced to a level where people are finding it easy to simulate and replicate systems within their personal computers, it is now becoming quite manageable task to simulate nanofluids. But the task is not as simple as it seems, it requires a lot of understanding of physiochemical interactions with thermo-physical boundary conditions. There are many algorithms and mathematical models to be considered. As the number of these models and algorithms increases, higher the computational power is required for solving. Nevertheless, the endless applications and usage makes it convincible for an end-user to adopt this creativity, as it enables one to understand the process and makes it visually quantifiable.

Before moving forward, it is necessary to understand some basic theory that is behind the dispersion of nanoparticles within a certain fluid.

1.1.1. Theory behind dispersion of nanoparticle

Dispersion of nanoparticles is a process in which they are dispersed in a medium like fluid. These fluids are of different grades such as biological, aerospace, automotive and buffering solutions. According to the kinetic theory of molecules, as the molecule interacts with other molecule, it starts to generate some heat due to kinetic molecular movement of the particle. This movement is accountable for the dispersion of nanoparticles in different fluids; thereby, this model causes anomalous increase in the heat transfer of the nanofluids. Furthermore, using this model, four major effects produced by nanoparticles dispersion can be explained i.e. (a) Brownian motion of nanoparticle, (b) liquid layering at liquid particle interface, (c) nature of heat transport between nanoparticles and (d) the clustering effect of nanoparticles in fluid. These factors are responsible for inducing random motion within particle and liquid layers, and this phenomenon is Brownian motion. During the interaction between nanoparticle and fluid, heat is evolved, causing nanoparticles to cluster and agglomerate. These mechanisms have already been replicated by various researchers for analysing properties such as; (a) rheological, (b) thermo-physical and (c) physiochemical as mentioned in Section 1.2.

1.2. Applications

There are various applications in the area of nanofluid simulation. Currently, nanofluid simulation is being applied for analysing the rheological properties of nanofluid environment, which is useful for biological, oil and gas, lubrication and chemical industries. Now, by the help of simulation, it is possible to test those undesirable conditions that could not be tested before, such as testing viscosity at low and very high temperatures. Properties of ideal nanofluid can be tested and their results can also be validated using autocorrelation functions for satisfaction.

The use of molecular dynamics has enabled us to test and quantify thermo-physical quantities of nanofluid at obnoxious level. The chemical interactions that were complicated to understand from the real interface, now it has become straightforward to know how the atoms of fluid and nanoparticle interacts together, nevertheless, Brownian dynamics is more appreciably demonstrated and visualized. Having this all, analysing different properties of fluid and nanoparticle interaction, now it is easy to know other parameters such as specific heat [13], total energy, bond formation at molecular level, chemical interactions, etc. [14]. Furthermore, various effects that could not be judged by experimental testing can now easily be known such as the effect of liquid layering on thermal conductivity as investigated by Li et al. [15]. Particle effect on thermal conductivity analysis can now be determined as carried out by Lu and Fan [16]. Nevertheless, effect of surfactant addition in nanofluid system can also be tested using molecular dynamics, which can better tell about the chemical interaction and aggregation dynamics within this system as conveyed by Mingxiang and Lenore [17]. Rudyak also succeeded in showing that by changing nanoparticle size and shape effects the viscosity [18]. Therefore, by looking at the vast applications of nanofluid simulation, it is necessary to know some overview about how these simulations can easily be conducted.

2. The literature review

2.1. Need of simulations over experiments

Simulations are being preferred over experimental practices in the twenty-first century. As experiments require a lot of man power and material, which is costly and time-consuming, therefore, researchers are favouring simulations, as it saves material, money and time. With the advancement in computational technology, simulations are being approached to replicate the nanofluids. Simulations are not an old technique, and it has got a firm ground. Currently, the area of simulation to replicate the real phenomena of dispersion is through the intermediate stages. Before moving to simulations, it is important to understand dispersion and interaction mechanism of nanoparticles with fluids. For this, the major phenomena that is used for dispersion is Brownian motion, which is an important aspect that controls the random factor of nanoparticle dispersion.

2.2. Simulations of nanofluids

Nowadays, the necessity of using simulation techniques is increasing due to its cost-effectiveness and time-saving capabilities. Simulations for nanofluids are mostly referred to as molecular dynamics simulation (MDS). However, before MDS, researchers adopted theoretical and numerical calculation method for computing thermo-physical quantities. Earlier theoretical formation, related to MDS research, has not established a strong hold position for replicating the mechanism of heat transfer, rheology and thermo-physics involved for nanofluid dispersion. This is because several researchers had modelled system using various assumptions rather using a definite formulation. This creates ambiguity in collecting results; however, they were well utilized for initial prediction of thermal transfer properties of nanofluid at the cost of wide inaccuracies. Experimental results that are representing actual system sometime are way off from the ideal method, in addition to this, researchers apply various differential equations for equating the system to realistic results as possible.

These methods are single-phase and two-phase methods [19] of nanofluid heat convection. They are still being used for predicting several properties related to heat transfer, convection and conduction within nanofluid systems [19–21]. Now these two methods are being embedded in computation fluid dynamic and molecular dynamics for heat transfer analysis [21]. The single-phase method of heat convection in nanofluid is an old method and is good for initial prediction of the thermal properties of nanofluid; however, the second-phase method is costlier as it requires higher computing power. In addition to the second-phase method, it is quite versatile as its prediction is in higher accuracy to the experimental results. Numerical approach simulates the nanofluid system using classical thermodynamics principles, which is more close to the single-phase model. Different correlations are applied to estimate the imbalance between the heat propagation values from actual to the ideal system. Physical interaction kinetics involved in real nanofluid system are not mimicked. This is why the real prediction is hard to achieve by this approach; moreover, two-phase fluid heat transfer involves higher mathematical complexity, which requires high computational power for general analysis of nanofluid heat transfer, rheology and thermo-physical quantities.

It was investigated by Sergis Antonis that due to not standardizing the procedure of nanofluid preparation diversifies accuracy of the experimental results obtained [2]. In this respect, MDS comes in to play, as it helps in simulating both nanoparticle and fluid particle system in one single domain, enabling us to mimic reaction kinetics of both materials in one single domain. However, these simulations require high computational power for simulating the system as it involves kinetic molecular movement of different atoms. Initially, MDS involved heat transfer within a nanofluid system in which it did not involve analysis with respect to the geometrical features or spherical with no surface texture. It used to be simple analysis in a uniform and homogeneous system. Earlier, properties of SiO_2 nanoparticles were calculated using Stillinger-Weber [22] and later fluid particles were represented by L-J potential.

There are two different dispersion prospects of MDS i.e. (1) non-equilibrium MDS (NEMD) and (2) equilibrium MDS (EMD). The macroscopic MDS mimics the molecular interactions between different molecules of various elements; in compound or ionic form. These different thermo-physical types of interactions of molecular dynamic quantities can be tailored and analysed by true boundary conditions. These boundary conditions are related to the physical settings, chemical interactions, charges, viscosity of the system and motion exhibition of particles. The interaction between the molecules is exhibited by Brownian motion as this mimics the random forces in the system. The system relies on different algorithms behind the scene to design a virtual nanoparticles dispersion in fluid. Furthermore, this is because the interaction kinetics of nanofluid system adhere with nanoparticle surface interacting with the surrounding fluid; this involves exchange of energy, surface tension between two, orientation of nanoparticle, surface energy, bonding configuration, nanoparticle dynamics and kinematics (including nanoparticle spin), liquid layering between nanoparticle and fluid molecule, and diffusion rate.

To explain the trajectories and velocities of a fluidic system, it is necessary to adopt a hydrodynamic framework. Computer simulations for mimicking trajectory of hydrodynamic dispersion of a dispersed particle in a fluid system was used by Ermak [23]. Nevertheless, Ermak and McCammon [24] work was more focused on the hyrdrodynamically concentrated system. The hydrodynamical system exhibited that the inter-particle distance is much greater than the range of hydrodynamic interactions. However, by implementation of Brownian dynamics by Ermak gave highly concurrent results with the experimental values achieved. The hydrodynamics of the system display combinations of Coulomb interactions; i.e. long range interactions as well as the Vander Waal interactions; short range interactions. Furthermore, the dynamics of the system is more convincing after applying the Derjaguin, Landau, Verwey and Overbeek (DLVO) [25] theory/factor in the system to mimic the charges and to enhance the realistic intermolecular attractions and repulsions.

Currently, there are different nanoparticles being considered for various applications. Therefore, for simulating nanofluids, modelling the nanoparticle is important, for that nanoparticle structure, shape and its properties should be known.

Subsequently, the mimicking of interaction potentials; i.e. using force fields such as embedded atom method (EAM), COMPASS, universal, etc; and the other forces between the atoms and molecules, the velocity verlet theorem is implemented. The velocity verlet theorem is a time-dependent movement of the atoms from one position to another using an algorithm for defining the movement, which is based on Brownian dynamics (BD). In addition to this, velocities or movements of atoms are controlled using thermal ensembles i.e. canonical (NVT), grand canonical (ΔPT), isobaric and isothermal (NPT) and micro canonical (NVE). These ensembles support in conducting thermal and physical perturbation to change the dynamical position of the atoms and molecules within a desired system. This causes the system to move to an un-equilibrium state. After starting and moving from an un-equilibrium state, the system is then equilibrated for convergence to equilibrium state. Finally, by this convergence, the system acquires stability of temperature and physical quantity fluctuations. However, this convergence is an iterative process for which time steps are varied to achieve the real convergence results [26, 27].

Currently, there are various simulation of nanofluids, for example; CuO, TiO_2 and CeO_2 nanoparticle dispersion in water [3, 4]; furthermore, there are also studies of dispersing nanoparticles in hydrocarbons [28]. By having two different simulation strategies, a perspectives and robust methodology can be formulated. As these simulations are performed on two different types of fluids i.e. polar and non-polar, so a concurrent methodology for both fluids can be deduced. Furthermore, up to the date, investigators have carried out various researches on nanofluid MDS, in addition to this, last two decades of work has been cumulated in **Figure 1**. Following are the details of their work in the field of nanofluid simulations.

In 1998, Malevanets and Kapral [29] formulated a method for computing complex fluidic systems using H theorem, which helped in solving hydrodynamics equations and transport coefficients. Colloidal model and random stochastic movement algorithm was established using Brownian dynamics which was formulated by Lodge and Heyes [30].

Francis W. Starr investigated effect of glass transition temperature on the bead spring polymer melts with a nanoscopic particle. He found that the surface interaction dominates due to nanoparticle diffusion within the melted polymeric system [31].

Simulation of chemical interactions was also carried out, and the bond length and structural orientation was noted for Silica nanoparticles in poly ethylene oxide (PEO) oligomer system. By this study, Barbier et al. concluded that the silica nanoparticles influence structural properties of PEO up to two to three layers [32].

Mingxiang and Lenore worked on hydrocarbon surfactant in an aqueous environment with a nanoparticle diffused within this system. It was observed from interactions that the agglomeration created between water molecules and surfactant was independent of nanoparticle i.e. it does not matter whether it is present or not [17].

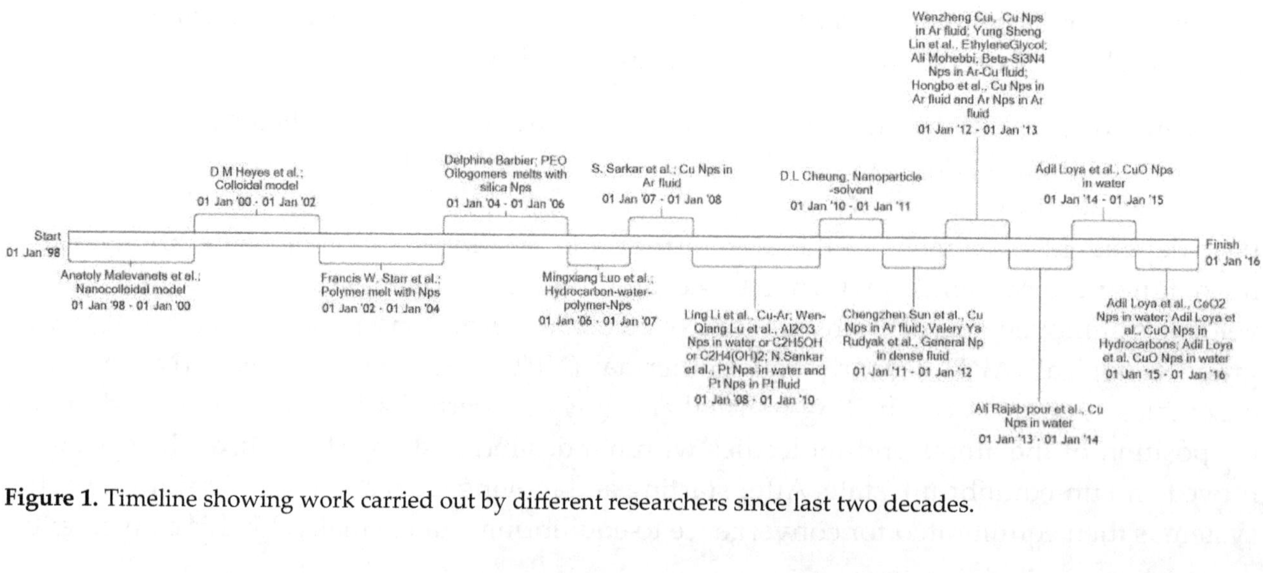

Figure 1. Timeline showing work carried out by different researchers since last two decades.

Sarkar and Selvam designed a nanofluid system of Cu nanoparticle and Argon as basefluid, for this, he used EAM potential and Green Kubo technique to find the thermal conductivity of this system. He examined that the periodic oscillation existed due to the heat fluxes imposed by Leonard Jones (L-J) potential [9].

Li et al. later worked on similar system of Cu nanoparticle with Ar base fluid; however, they investigated Brownian dynamics induces a thin layer around a particle, giving a hydrodynamic effect to the particle dispersion [33].

Lu and Fan investigated thermo-physical quantities of Alumina nanoparticles dispersed in water and concluded that the particle volume fraction and size effects the viscosity and thermal conductivity [16].

Sankar et al. examined and formulated an algorithm for calculating metallic nanoparticle thermal conductivity in fluid. They articulated that the volume fraction of nanoparticles and temperature of the system effects the overall thermal conductivity [8].

Moreover, Cheung carried out research on L-J nanoparticles within solvent and quantified that the detachment energy decreases as the nanoparticle solvent attraction rises [1].

Sun et al. devised a technique using EMD using Green Kubo method to find the effective thermal conductivity of the Cu nanoparticles in Ar liquid. It was found that there was a linear increase in the effective thermal conductivity of shearing nanofluid due to micro-convection [34].

Rudyak and Krasnolutskii later on worked on Aluminium and Lithium nanoparticles with liquid Ar and suggested that the size and material of nanoparticle considerably effects the viscosity [18].

Lin Yun Sheng et al. also detected increment in thermal conductivity by Cu nanoparticle dispersion in Ethylene glycol fluid. In this study, he used Green kubo formulation for finding thermal conductivity using NEMD [35]. Furthermore, Mohebbi investigated a method to calculate thermal conductivity of nanoparticles in fluid using a non-periodic boundary conditions with EMD and NEMD [14].

Kang H et al. carried out work on coupling factor between nanoparticle of Copper and Ar as base fluid, his investigations suggest that coupling factor is proportional to the volume concentration of particles, nevertheless, he also suggested the that there is no effect of temperature change from 90 to 200 K on coupling factor [36].

Rajabpour et al. investigated the specific heat capacity of Cu nanoparticles within water and he found that the specific heat capacity of this system decreases by increasing the volume fraction of particles in base fluid [13].

Loya et al. initiated work on CuO nanoparticles dispersion in water focusing on the change of viscosity due to temperature increase, he figured that temperature increment decreases the viscosity of nanofluid as also initially predicted using experimental testing [37].

In addition to above, further rheological analysis of CuO nanoparticles in straight chain alkanes [28] and water [4] and CeO_2 in water [3] was carried out by Loya et al. For conducting these simulations, molecular dynamics was used and studies provided highly accurate results of viscosity to experimental findings.

Finally, after knowing the perspective of nanofluid simulation, a simple and general way is deduced for researcher, industrialist and their co-worker in Section 2.3.

2.3. Mimicking different properties of nanofluids using simulation

Several studies about simulation work were reported on the diffusion of polymeric, ionic and mineral nanoparticles [38–40]. An example of this is calcite nanoparticles. These have been simulated in water for salt molecular dynamics for thermal energy storage nanofluidic simulations [38]. Simulations such as these are mostly conceiving diffusions of the polymeric nanoparticles or di-block polymers represented by spheres. The major diffusion phenomena that have been implemented on the nanoparticle or the polymer dispersion is with the help of BD, targeting the random motion of the particles in a solvent or any solution system. Some further surveys show that one of the best simulations for the dispersion of the metal oxide nanoparticle in the water system was carried out using the DPD potential [41–43]. This potential has the power to disperse nanoparticles as well as replicating the phenomena of the BD [44]. DPD was first carried out on nano-water systems by Hooggerbrugge and Koelman [44, 45]. Moreover, the work was carried out by Español and Warren for implementing the DPD technique using statistical mechanics. DPD technique imparts stochastic phenomena on particle dynamics [46]. This is how BD was integrated into DPD technique. However, the random forces will only be in pairwise interaction since DPD at the same time imparts the hydrodynamic effect on the system. Many studies of DPD for complex fluidic systems [41–43] show that the dispersion of nanoparticles in water exhibits complex properties and to simulate this, initial selection of boundary conditions are important to replicate the real scenario. Thereby, the best way to simulate is to acquire the boundary conditions of the existing experimental system and then use a molecular dynamic simulator to further implement it [47]. The considerations of boundary conditions are particle sizes, force field for particle-to-particle interactions, solvent in which the particles will be diffused, and physiochemical nature of the system [48, 49]. Within the simulation system, force field plays an important role since it provides charges on atoms for interaction. The force field is a mathematical parameter that governs the energies and potentials between interactive atoms. The physiochemical settings of the system refer to the thermal, chemical and physical properties of the system such as initial temperature settings, charges and dynamics. Finally, the temperature is controlled using different ensembles.

3. Methodology

3.1. Simulation strategy

The nanofluid interactions are carried out at molecular level. Therefore, by keeping this in mind to conduct nanofluid simulations, it is necessary to have a simulation technique which allows us to do simulation at molecular level. Hence, the technique use for this is molecular dynamics and package that is focused through this chapter is Large-scale Atomic/Molecular Massively Parallel Simulator (LAMMPS). Furthermore, how to approach this is mentioned in the next section of this chapter i.e. Section 3.1.1.

3.1.1. Approach

The simulation of nanoparticle dispersion is related to the MDS. For this, the software or the package that needs to be selected was based on the criteria of the conditions that were needed to be simulated, and the flexibility was a major concern for the applicability of different systems. The LAMMPS can be a best molecular dynamics package for simulating the nanofluidic system. This is the code generated by the Sandia Laboratories by Plimpton [50]. This molecular dynamics software has high viability over other available software like Montecarlo and Gromacs.

After selection of the MD package, to simulate a desired system with realistic features, it is highly vital to know and understand initial boundary conditions. These initial conditions for a dispersion of nanoparticles are related to charges within the system for interaction, molecular bonding, forces of attraction i.e. Vander Waal or electrostatic coulombs interactions, forcefields, pair potentials (i.e. molecular mechanics constants) and molecular weight. To perform MD simulation, initial boundary conditions are major and fundamental parameters to devise actual dynamics that exist in a real system. After setting the initial parameters, velocity of the system is equilibrated and ensembles are applied to mimic the real thermo-physical conditions.

In addition to above, after setting all the boundary conditions related to chemical and thermo-physical parameters, the system is then equilibrated for certain time steps. Simulations are processed until converging results are obtained as that of the actual system. Over here, "time step" is the major dependent factor. This accounts for equilibrating the kinetics of the system that takes place; i.e. movement of system from an un-equilibrated state to equilibrium conditions. The above explained method has been compressed and illustrated using a flowchart for better understanding as shown in **Figure 2**.

After suggesting how to approach and initiate your work for simulation of nanofluids, it is also important to know the briefed-out details about the steps like force field, pair potentials, ensembles, etc.

After setting up the atoms in a coordinate system using a molecular modelling software, then force field is applied on the system (i.e. Universal, COMPASS, OPLS, etc.) by this atomic charges and bond configurations are setup. These force fields are interlinked with pair potentials (such as DPD, BD, Smoothed Particle Hydrodynamic, LJ, etc.), they are parameters which are used to describe vibrational and oscillation settings between two different atoms. Finally, ensembles are applied on the molecular dynamic system for equilibrating the actual thermal settings for example NVT, NPT, NPH, etc.

3.1.2. Techniques and tools

As of now, it is known from the previous sections that to simulate and perform MDS it is necessary to know techniques and tools that can be beneficial for use and executing the work.

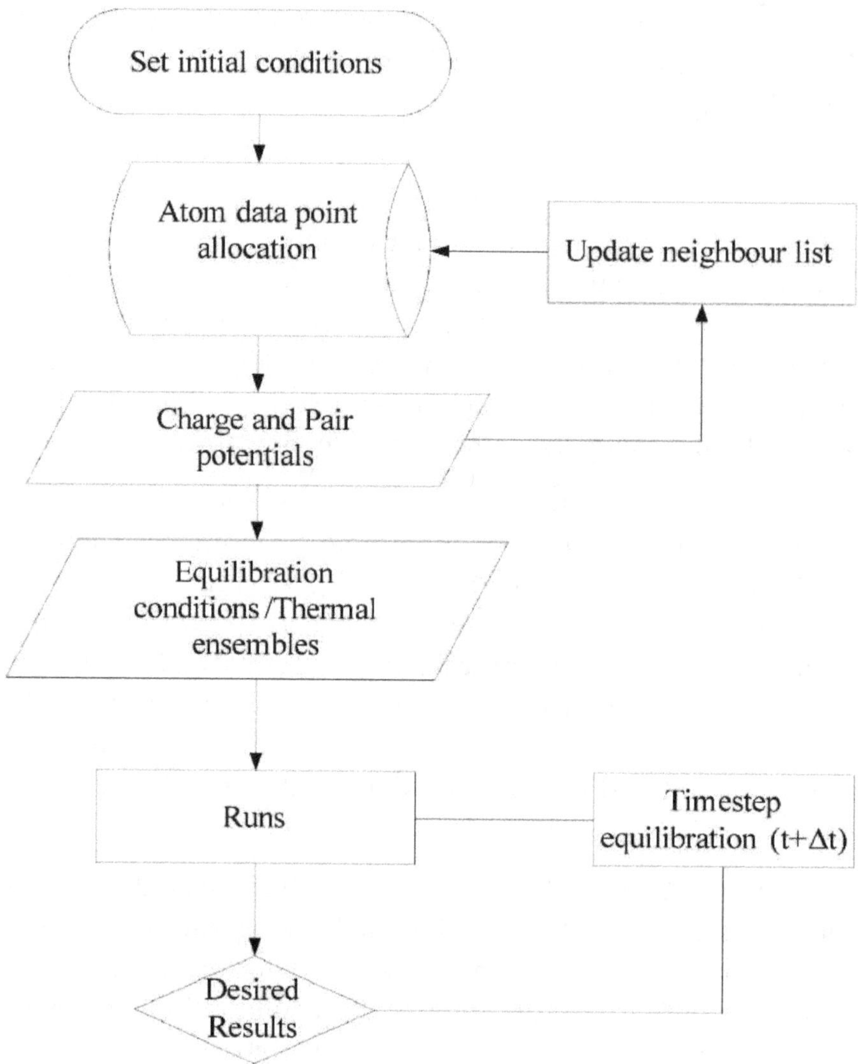

Figure 2. Flow chart of molecular dynamics simulation [26].

Today, there are several tools and ways to perform this; however, still researchers are unsure about "what are the clear steps for conducting nanofluid simulations using molecular dynamics?" Therefore, through this section, a brief and concise way is illustrated and conveyed for better and easier understanding for people working under the horizon of nanofluid simulations. These steps are as follow:

a. Firstly, for creating nanofluid simulation system, it is required to setup a nanoparticle and fluid, then combine them together, for which material studio is the best software for designing a nanoparticle. Now, the nanoparticles can be inserted and replicated in a box containing fluid particles, however, this may be tedious for bigger systems. Therefore, it is suggested to use Packmol after creating the Protein Data Bank (PDB) file from material studio and then create an input script for Packmol to replicate the system with as many particles and fluid molecules as per required. This software automatically packs up the overall molecular arrangement with in a confined imaginary box.

b. As the nanofluid system is set up, now an input data file is required for LAMMPS software, this can be generated by using the PDB file and converting it to required .CAR and .COR format using Material studio. Before conversion do not forget to implement charges on the atoms of the nanoparticle and fluid molecules for this Discover module of the Material studio software can be used. After conversion to .CAR and .COR, use msi2lmp package provided with LAMMPS for converting the file to a LAMMPS readable input.

c. Once the LAMMPS readable input file is generated now use "read data" command for LAMMPS to read this file during the simulation execution.

Finally, the data quantification, visualizing the effects and properties that can be analysed have been jotted below in different sections.

3.1.3. Data quantification

Now, the data obtained by using different compute commands can be quantified on MATLAB or Excel. MATLAB initially requires more time for developing its script for computing the mathematical problem or graphs. However, on a long run, it does save time. Whereas, excel is easy going but requires more time for plotting graph each time you feed the new data.

MATLAB scripting helps in formulating the work in a precise manner, and digitalise the work with high quality publishing of the data for journal publications. However, MATLAB requires good command over the MATLAB scripting and functions. By using MATLAB, it is easy to apply discrete as well as continuous algorithms and equations for refining and optimization of results. Furthermore, it helps in applying the regression on the noisy data for refinement.

In Excel, similar stuff is possible as in MATLAB, but in excel, it is quite complicated as you need to apply macros. These days the computation of MATLAB can be computed in parallel mode; again for excel, it is quite difficult. However, for graphical representation of data, excel is quite versatile.

Vice versa both tools have their own benefits over each other; it depends totally on a user-friendliness with certain software. In addition to excel, to compute or establish complex calculations, it will be required to interlink its macros with visual basic scripting, which is under a developer's tool library, mostly hidden from newbies.

3.1.4. Visualizing the effects

After the successful execution of simulation, you will get dump files from LAMMPS, here a software that can read LAMMPS trajectories can be used for reading the file and visualizing it. For which Visual Molecular Dynamic (VMD) can be used. However, OVITO is also a good software for visualizing your trajectories.

The results generated by OVITO are represented as small spheres merged together to form a particular system representation, as shown in **Figure 3**, i.e. of a CuO-water nanofluid system.

In the similar way for showing how the VMD gives visual output is shown in **Figure 4**. It is similar to that of OVITO, however, VMD has capability of representing the trajectories in the form of molecular structure. This gives an extra possibility for researchers working in the area of Biochemistry, pharmacy, drug delivery and biomedical to represent and observe the chemical kinetics in real-time, i.e. how one atom reacts and interacts with another atom within a confined system.

3.1.5. Properties that can be analysed

Some properties and parameters can directly be analysed using VMD using trajectories dump files. VMD has option for analysing the radial distribution function (RDF) and mean square displacement (MSD), they indicate about the agglomeration and dispersion rate, respectively.

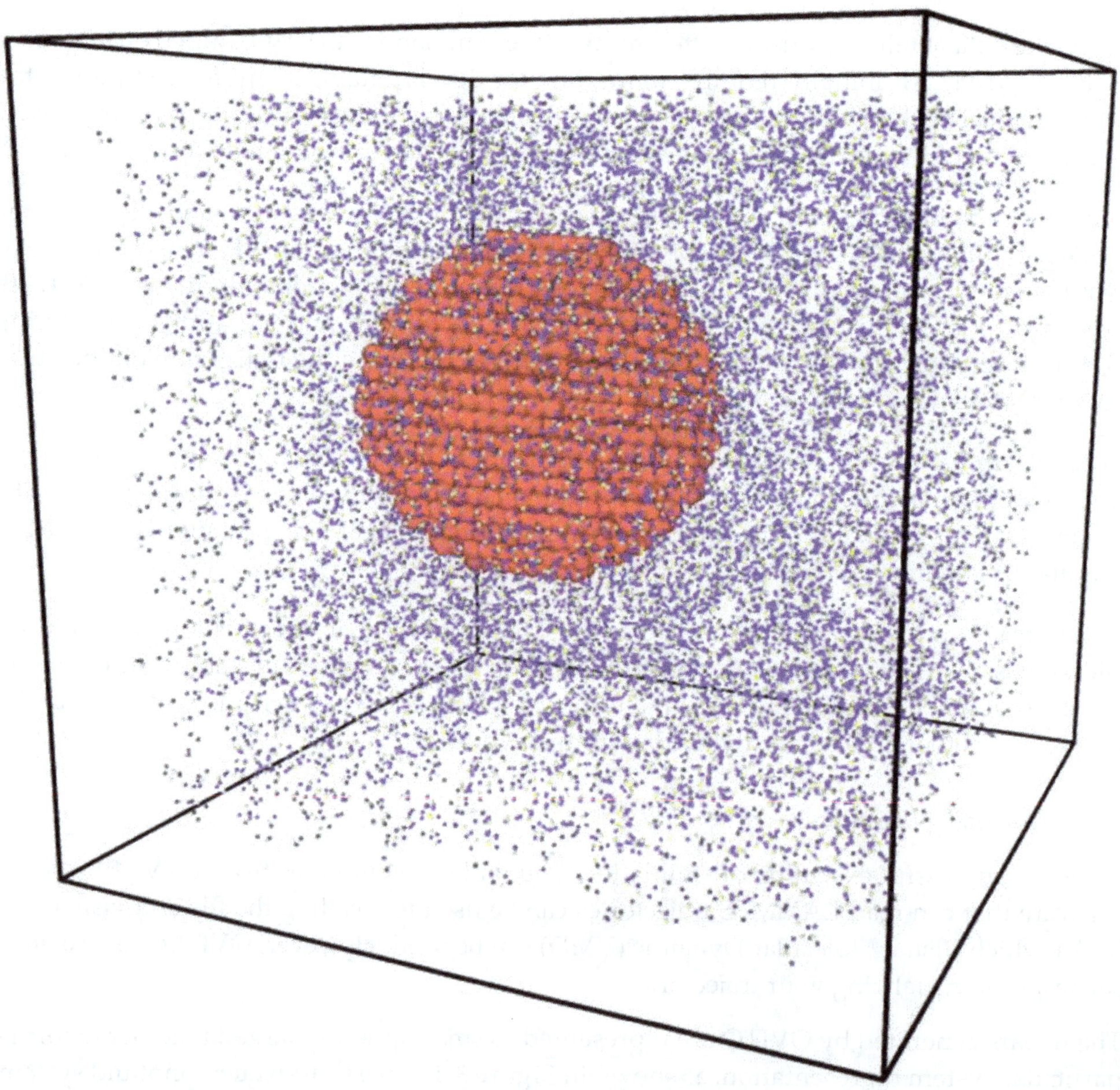

Figure 3. Representation of OVITO output of molecular dynamics of CuO nanoparticles in water system [4].

Figure 4. Visual output showing two CuO nanoparticles in a water-based nanofluid.

When nanofluids are concerned the major parameters or properties researcher are interested to investigate are viscosity, thermal conductivity, specific heat capacity, thermal diffusivity, diffusion coefficient, total energy, heat loss, etc. To find these properties LAMMPS provide versatile options to compute what you require, using different algorithms or previously established techniques. Currently, main concerned variables out of above mentioned ones are viscosity, diffusion coefficient and thermal conductivity. Therefore, in the next section, we will discuss about how to validate and quantify your results obtained from the simulation.

4. Validation and quantification of results

To validate the three major properties mentioned in Section 3.1.5, it is required to know initial experimental results, however, sometime it is hard to obtain those results as some simulation condition cannot be tested, either due to lack of experimental device or it is not possible to meet the boundary conditions as setup over the simulation platform.

Now, in this case, the best way is to analyse using autocorrelation function; which is a time series modelling of a function of a variable dependent on time fluctuation. Let us take the case of viscosity, as it is related with shearing stress, there are shear forces acting between the layers of molecular interaction causing pressure function to be induced. This pressure function is dependent on stress due to shearing force. If this stress is analysed using the function of time, this becomes stress tensor. This stress tensor is used for analysing stresses exiting between the molecular layers. Therefore, this is known as stress autocorrelation function (SACF). The SACF accounts for the stresses imparted on the system due to the diffusion of molecules and intermolecular kinetics; i.e. molecular stresses caused by attraction and repulsion of molecules. During the intermolecular kinetics drag is created between the molecular layers, this drag is due to the effect of shearing forces. Ultimately as the system is equilibrated, it shows unstable response of the SACF, however, as it approaches stability the SACF starts to converge to a monotonic level, which satisfies that the viscosity analysed is acceptable.

In the similar manner, thermal conductivity is quantified, but here instead of stress and shear forces, heat is considered. Therefore, this is known as heat autocorrelation function (HACF), which quantifies or validates the thermal conductivity obtained is satisfactory.

In addition to HACF and SACF for thermal conductivity and viscosity, respectively, for diffusion coefficient, velocity autocorrelation function is used for its quantification. As diffusion coefficient is measured by taking the slope of the MSD. So to quantify and validate it, displacement with respect to time i.e. velocity can be used.

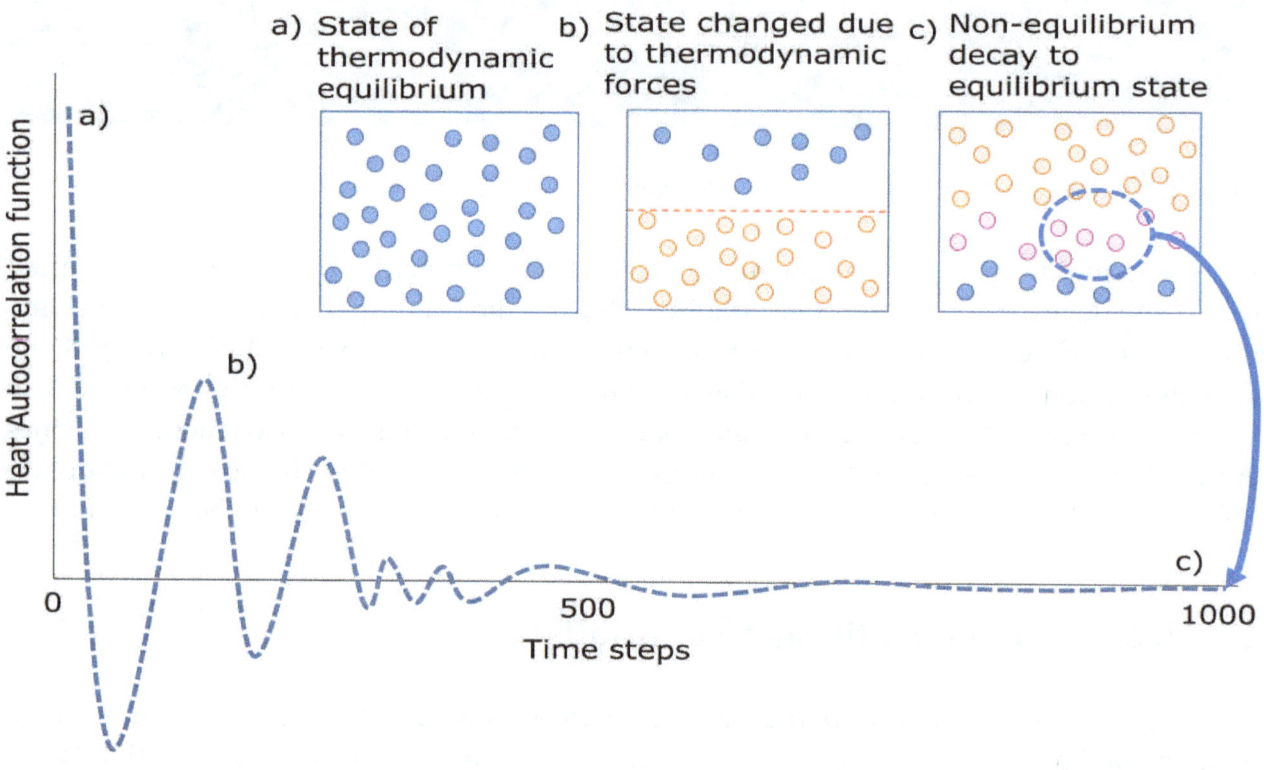

Figure 5. Autocorrelation output gained by running a molecular dynamics simulations [26].

The accuracy of results equilibrated for measuring the viscosity and thermal conductivity of a system can be justified in a better way with the estimation of heat autocorrelation function and stress autocorrelation function as show in **Figure 5**. The graphical result in **Figure 5** explains the process of the integration of non-equilibrated system to equilibration.

At step (a), the system starts with a thermodynamic equilibrium, but the system is not at equilibrium state. At step (b), the thermodynamic conditions are changed due to implementation of thermal ensemble so the system tends to go towards equilibrium. At step (c), the non-equilibrium system moves to equilibrated level of convergence at this level the system satisfies the convergences. This process is followed during the equilibration of the thermo-physical quantities, the convergence time steps depend on the volume and quantity of the atoms in that system. For the larger system, large amount of computational power and time step will be required for convergence.

5. Discussion

5.1. Problems faced for simulating nanofluids

So far the topic has been conveying the techniques, approach and method for carrying out nanofluid simulations. Moreover, there has been no data available for the expertise to know what are the problems faced when these simulations are conducted, number of questions can arise, for example, (1) Till what level, computational power can support our simulations? (2) Is there any other way out rather than this? (3) How larger systems can be simulated? etc.

Therefore, to answer these questions, it is necessary to understand the material and knowledge given before, however, as the number of atoms are increased within a nanofluid system the molecular dynamics demonstrates sluggish performance due to less computational capabilities i.e. either central processing unit (CPU) power or graphic processing unit (GPU). Furthermore, it is not just simulation that need to be carried out but for the data quantification, the data that are gathered requires huge memory for storage. Thereby, requiring the random access memory (RAM) and hard disk drive (HDD) to be large enough to store the required data easily [51].

After hardware issues, the second set of problems faced by nanofluid simulation is the use of multiple software for designing, modelling, processing and visualization, which needs a lot of understanding of computer for a new geek. Furthermore, if this all is combined in one package, this can marvellously save time and money for purchasing different software for data acquisition. It is slightly known at the moment that there are few software in market for helping in simulating nanofluid; however, academia is not yet aware of it due to less versatility such as Medea and Scienomics MAPS.

One of the major problem is that, people of twenty-first century like working using graphical user interface (GUI), as it is easy and you can do everything by just clicks rather than using complicated commands, however, most of the molecular dynamics package are used on Linux operating system, moreover, commands are used for computing and feeding the data for computation.

In addition to high computing power, it should be known that before attempting to simulate large scale molecular dynamics (i.e. with more than 0.1 million atoms), it is required to have parallel processing enabled on the PC. For that high end, CPU or GPU is required with multi cores for processing the data in parallel mode. However, this processing has some drawbacks that are loop holes for simulations, one such kind is that sometimes the algorithm is not designed in a way to parallel the process efficiently, which in turn gives ambiguous simulation output and convergence. For avoiding this, it is necessary for the user to know the correct working of the algorithm. Moreover, the field programmable gate array (FPGA) is good outbreak technology that is being implemented for paralleling the process [52, 53], nevertheless, again this technology requires new stuff and bits coding to be learned before operating or using this module for rapidly solving the simulation.

6. Conclusion

The chapter has brought about marvellous information and the literature for new geeks for conducting a nanofluid simulation. However, this chapter acts as a guide for a newbie for initialising the nanofluid simulation.

Nomenclature

Words	**Abbreviation**
Molecular dynamics simulation	MDS
Non-equilibrium molecular dynamics	NEMD
Equilibrium molecular dynamics	EMD
Embedded atom method	EAM
Condensed-phase optimized molecular potentials for atomistic simulation studies	COMPASS
Brownian dynamics	BD
Canonical	**NVT**
Grand canonical	ΔPT
Isobaric and isothermal	NPT
Micro canonical	NVE
Normal pressure hydrocephalus	NPH
Copper oxide	CuO
Titanium oxide	TiO_2
Cerium oxide	CeO_2
Poly ethylene oxide	PEO
Leonard Jones	L-J
Copper	**Cu**
Argon	**Ar**
Kelvin	**K**
Discrete particle dynamics	DPD
Large-scale atomic/molecular massively parallel simulator	LAMMPS
Optimized potential for liquid simulation	sOPLS
Protein Data Bank	PDB
File format output from material studio	CAR and.COR

Visual molecular dynamic	VMD
Radial distribution functions	RDF
Mean square displacement	MSD
Stress autocorrelation function	SACF
Heat autocorrelation function	HACF
Velocity autocorrelation function	VACF
Central processing unit	CPU
Graphic processing unit	GPU
Random access memory	RAM
Hard disk drive	HDD
Graphical user interface	GUI
Field programmable graphic array	FPGA

Author details

Adil Loyai

Address all correspondence to: loya_adil@yahoo.com

PAF-Karachi Institute of Economics and Technology, Karachi, Pakistan

References

[1] Cheung DL. Molecular dynamics study of nanoparticle stability at liquid interfaces: effect of nanoparticle-solvent interaction and capillary waves. Journal of Chemical Physics. 2011;135:054704.

[2] Ta TD, Tieu AK, Zhu H, Zhu Q, Kosasih PB, Zhang J, et al. Tribological behavior of aqueous copolymer lubricant in mixed lubrication regime. ACS Applied Materials & Interfaces. 2016;8:5641–52.

[3] Loya A, Stair JL, Ren G. Simulation and experimental study of rheological properties of CeO_2–water nanofluid. International Nano Letters. 2015;5:1–7.

[4] Loya A, Ren G. Molecular dynamics simulation study of rheological properties of CuO–water nanofluid. Journal of Materials Science. 2015;50:4075–82.

[5] Eastman J, Choi U, Li S, Thompson L, Lee S. Enhanced thermal conductivity through the development of nanofluids. MRS proceedings. Cambridge University Press, Cambridge; 1996. p. 3.

[6] Choi US. Enhancing thermal conductivity of fluids with nanoparticles. ASME FED. 1995;231:99–106.

[7] Karthik R, Harish Nagarajan R, Raja B, Damodharan P. Thermal conductivity of CuO–DI water nanofluids using 3-ω measurement technique in a suspended micro-wire. Experimental Thermal and Fluid Science. 2012;40:1–9.

[8] Sankar N, Mathew N, Sobhan C. Molecular dynamics modeling of thermal conductivity enhancement in metal nanoparticle suspensions. International Communications in Heat and Mass Transfer. 2008;35:867–72.

[9] Sarkar S, Selvam RP. Molecular dynamics simulation of effective thermal conductivity and study of enhanced thermal transport mechanism in nanofluids. Journal of Applied Physics. 2007;102:074302.

[10] Esmaeilzadeh P, Fakhroueian Z, Beigi M, Akbar A. Synthesis of biopolymeric α-lactalbumin protein nanoparticles and nanospheres as green nanofluids using in drug delivery and food technology. Journal of Nano Research: Trans Tech Publ. 2011;16:89–96.

[11] Rapoport L, Leshchinsky V, Lvovsky M, Lapsker I, Volovik Y, Feldman Y, et al. Superior tribological properties of powder materials with solid lubricant nanoparticles. Wear. 2003;255:799–800.

[12] Peng DX, Kang YA, Chen SK, Shu FC, Chang YP. Dispersion and tribological properties of liquid paraffin with added aluminum nanoparticles. Industrial Lubrication and Tribology. 2010;62:341–48.

[13] Rajabpour A, Akizi FY, Heyhat MM, Gordiz K. Molecular dynamics simulation of the specific heat capacity of water-Cu nanofluids. International Nano Letters. 2013;3:1–6.

[14] Mohebbi A. Prediction of specific heat and thermal conductivity of nanofluids by a combined equilibrium and non-equilibrium molecular dynamics simulation. Journal of Molecular Liquids. 2012;175:51–8.

[15] Li L, Zhang YW, Ma HB, Yang M. Molecular dynamics simulation of effect of liquid layering around the nanoparticle on the enhanced thermal conductivity of nanofluids. Journal of Nanoparticle Research. 2010;12:811–21.

[16] Lu W-Q, Fan Q-M. Study for the particle's scale effect on some thermophysical properties of nanofluids by a simplified molecular dynamics method. Engineering Analysis with Boundary Elements. 2008;32:282–9.

[17] Mingxiang L, Lenore LD. Molecular dynamics simulations of surfactant and nanoparticle self-assembly at liquid–liquid interfaces. Journal of Physics: Condensed Matter. 2007;19:375109.

[18] Rudyak VY, Krasnolutskii SL. Dependence of the viscosity of nanofluids on nanoparticle size and material. Physics Letters A. 2014;378:1845–9.

[19] Sheikholeslami M, Abelman S. Two-phase simulation of nanofluid flow and heat transfer in an annulus in the presence of an axial magnetic field. IEEE Transactions on Nanotechnology. 2015;14:561–9.

[20] Göktepe S, Atalık K, Ertürk H. Comparison of single and two-phase models for nanofluid convection at the entrance of a uniformly heated tube. International Journal of Thermal Sciences. 2014;80:83–92.

[21] Akbari M, Galanis N, Behzadmehr A. Comparative analysis of single and two-phase models for CFD studies of nanofluid heat transfer. International Journal of Thermal Sciences. 2011;50:1343–54.

[22] Zachariah MR, Carrier MJ, Blaisten-Barojas E. Properties of silicon nanoparticles: a molecular dynamics study. Journal of Physical Chemistry. 1996;100:14856–64.

[23] Ermak DLM. Computer-simulation of charged-particles in solution .1. Technique and equilibrium properties. Journal of Chemical Physics. 1975;62:4189–96.

[24] Ermak DL, McCammon JA. Brownian dynamics with hydrodynamic interactions. The Journal of chemical physics. 1978 Aug 15;69(4):1352-60..

[25] Kruyt HR. Colloid science. Elsevier, Amsterdam, 1952.

[26] Loya A, Stair JL, Ren G. The approach of using molecular dynamics for nanofluid simulations. International Journal of Engineering Research and Technology. 2014;3:1236–47.

[27] Loya A. Large scale dynamic molecular modelling of metal oxide nanoparticles in engineering and biological fluids. 2015.

[28] Loya A, Stair JL, Jafri AR, Yang K, Ren G. A molecular dynamic investigation of viscosity and diffusion coefficient of nanoclusters in hydrocarbon fluids. Computational Materials Science. 2015;99:242–6.

[29] Malevanets A, Kapral R. Mesoscopic model for solvent dynamics. Journal of Chemical Physics. 1999;110:8605–13.

[30] F. M. Lodge J, M. Heyes D. Transient colloidal gels by Brownian dynamics computer simulation. Physical Chemistry Chemical Physics. 1999;1:2119–30.

[31] Starr FW, SchrØder TB, Glotzer SC. Molecular dynamics simulation of a polymer melt with a nanoscopic particle. Macromolecules. 2002;35:4481–92.

[32] Barbier D, Brown D, Grillet A-C, Neyertz S. Interface between end-functionalized PEO oligomers and a silica nanoparticle studied by molecular dynamics simulations. Macromolecules. 2004;37:4695–710.

[33] Li L, Zhang Y, Ma H, Yang M. An investigation of molecular layering at the liquid-solid interface in nanofluids by molecular dynamics simulation. Physics Letters A. 2008;372:4541–4.

[34] Sun C, Lu W-Q, Liu J, Bai B. Molecular dynamics simulation of nanofluid's effective thermal conductivity in high-shear-rate Couette flow. International Journal of Heat and Mass Transfer. 2011;54:2560–7.

[35] Lin YS, Hsiao PY, Chieng CC. Roles of nanolayer and particle size on thermophysical characteristics of ethylene glycol-based copper nanofluids. Applied Physics Letters. 2011;98(15):153105.

[36] Kang H, Zhang Y, Yang M, Li L. Nonequilibrium molecular dynamics simulation of coupling between nanoparticles and base-fluid in a nanofluid. Physics Letters A. 2012;376:521–4.

[37] Loya A, Stair JL, Ren G. The study of simulating metaloxide nanoparticles in aqueous fluid. International Journal of Engineering Research & Technology. 2014;3:1954–60.

[38] Cooke DJ, Elliott JA. Atomistic simulations of calcite nanoparticles and their interaction with water. Journal of Chemical Physics. 2007;127:104706–9.

[39] Hoang VV. Diffusion in simulated SiO_2 nanoparticles. Nano. 2007;02:301–3.

[40] Kowsari MH, Alavi S, Ashrafizaadeh M, Najafi B. Molecular dynamics simulation of imidazolium-based ionic liquids. I. Dynamics and diffusion coefficient. Journal of Chemical Physics. 2008;129:224508–13.

[41] Spaeth JR, Kevrekidis IG, Panagiotopoulos AZ. A comparison of implicit- and explicit-solvent simulations of self-assembly in block copolymer and solute systems. Journal of Chemical Physics. 2011;134:164902–13.

[42] Symeonidis V, Em Karniadakis G, Caswell B. Dissipative particle dynamics simulations of polymer chains: scaling laws and shearing response compared to DNA experiments. Physical Review Letters. 2005;95:076001.

[43] Gao L, Shillcock J, Lipowsky R. Improved dissipative particle dynamics simulations of lipid bilayers. Journal of Chemical Physics. 2007;126:015101–8.

[44] Hoogerbrugge PJ, Koelman JMVA. Simulating microscopic hydrodynamic phenomena with dissipative particle dynamics. Europhysics Letters. 1992;19:155–60.

[45] Koelman JMVA, Hoogerbrugge PJ. Dynamic simulations of hard-sphere suspensions under steady shear. Europhysics Letters. 1993;21(3):363–368.

[46] Español P, Warren P. Statistical mechanics of dissipative particle dynamics. EPL (Europhysics Letters). 1995;30:191.

[47] Cheng S, Grest GS. Structure and diffusion of nanoparticle monolayers floating at liquid/vapor interfaces: a molecular dynamics study. Journal of Chemical Physics. 2012;136:214702.

[48] Bunte SW, Sun H. Molecular modeling of energetic materials: the parameterization and validation of nitrate esters in the COMPASS force field. Journal of Physical Chemistry B. 2000;104:2477–89.

[49] Zhao L, Liu L, Sun H. Semi-ionic model for metal oxides and their interfaces with organic molecules. Journal of Physical Chemistry C Nanomater Interfaces. 2007;111:10610–7.

[50] Plimpton S. Fast parallel algorithms for short-range molecular dynamics. Journal of Computational Physics. 1995;117:1–19.

[51] Sergis A, Hardalupas Y. Molecular dynamic simulations of a simplified nanofluid. Computational Methods in Science and Technology. 2014;20:14.

[52] Sukhwani B, Chiu M, Khan MA, Herbordt MC. Effective floating point applications on FPGAs: Examples from molecular modeling. HPEC'09: Proceedings of the Workshop on High Performance Embedded Computing. 2009.

[53] Waidyasooriya HM, Hariyama M, Kasahara K. Architecture of an FPGA accelerator for molecular dynamics simulation using OpenCL. 2016 IEEE/ACIS 15th International Conference on Computer and Information Science (ICIS). IEEE; 2016. pp. 1–5.

8

Cattanneo-Christov Heat Flux Model Study for Water-Based CNT Suspended Nanofluid Past a Stretching Surface

Noreen Sher Akbar, C. M. Khalique and
Zafar Hayat Khan

Abstract

This chapter discusses the magnetic field effects on the flow of Cattanneo-Christov heat flux model for water-based CNT suspended nanofluid over a stretching sheet. According to the authors, knowledge idea of Cattanneo-Christov heat flux model for water-based CNT suspended nanofluid is not explored so far for stretching sheet. The flow equations are modeled for the first time in the literature transformed into ordinary differential equations using similarity transformations. The numerical solutions are computed using shooting technique and compared with the literature for the special case of pure fluid flow and found to be in good agreement. Graphical results are presented to illustrate the effects of various fluid flow parameters on velocity, heat transfer, Nusselt number, Sherwood number, and skin friction coefficient for different types of nanoparticles.

Keywords: boundary layer flow, nanofluid, stretching sheet, Cattanneo-Christov heat flux model, numerical solution

1. Introduction

From recent few decades, heat transfer enhancement of the nanofluid has turned out to be a topic of main interest for the researchers and scientists. The word "nanofluid" was derived by Choi [1]. He defines a liquid suspension comprising ultrafine particles whose diameter is less than 50 nm. Xuan and Roetzel [2] investigated the mechanism of heat transfer enhancement of the nanofluid. According to them, the nanofluid is a solid-liquid mixture in which metallic or

$$q+\lambda_2\left(\frac{\partial q}{\partial t}+V.\nabla.q-q.\nabla V+(\nabla.V)q\right)=-K_{nf}\nabla T, \tag{4}$$

where λ_2 is the thermal relaxation time. Eliminating q from Eqs. (3) and (4) gives as follows:

$$\left(u\frac{\partial T}{\partial x}+v\frac{\partial T}{\partial y}\right)+\lambda_2\left(\begin{array}{l} u\frac{\partial u}{\partial x}\frac{\partial T}{\partial x}+v\frac{\partial v}{\partial y}\frac{\partial T}{\partial y}+u\frac{\partial v}{\partial x}\frac{\partial T}{\partial y}+v\frac{\partial u}{\partial y}\frac{\partial T}{\partial x} \\ +2uv\frac{\partial^2 T}{\partial x\partial y}+u^2\frac{\partial^2 T}{\partial x^2}+v^2\frac{\partial^2 T}{\partial y^2} \end{array}\right) = \frac{K_{nf}}{\rho_{nf}\left(c_p\right)_{nf}}\frac{\partial^2 T}{\partial y^2}, \tag{5}$$

Further, ρ_{nf} is the effective density, μ_{nf} is the effective dynamic viscosity, $(\rho c_p)_{nf}$ is the heat capacitance, α_{nf} is the effective thermal diffusibility, and k_{nf} is the effective thermal conductivity of the nanofluid, which are defined as follows:

$$\begin{gathered} \rho_{nf}=(1-\phi)\rho_f+\phi\,\rho_{CNT},\mu_{nf}=\frac{\mu_f}{(1-\phi)^{2.5}},\alpha_{nf}=\frac{k_{nf}}{\left(\rho_{cp}\right)_{nf}}, \\ \left(\rho_{cp}\right)_{nf}=(1-\phi)\left(\rho_{cp}\right)_f+\phi\left(\rho_{cp}\right)_{CNT}, \\ k_{nf}=k_f\left(\frac{(1-\phi)+\frac{2\varphi k_{CNT}}{k_{CNT}-k_f}\log\left(\frac{k_{CNT}+k_f}{2k_f}\right)}{(1-\phi)+\frac{2\varphi k_f}{k_{CNT}-k_f}\log\left(\frac{k_{CNT}+k_f}{2k_f}\right)}\right), \end{gathered} \tag{6}$$

where μ_f is the viscosity of base fluid, ϕ is the nanoparticles fraction, $(\rho C_p)_f$ is the effective heat capacity of a fluid, $(\rho C_p)_{CNT}$ is the effective heat capacity of a carbon nanotubes, k_f and k_{CNT} are the thermal conductivities of the base fluid and carbon nanotubes, respectively, ρ_f and ρ_{CNT} are the thermal conductivities of the base fluid and carbon nanotubes, respectively.

Corresponding boundary conditions are as follows:

$$u=ax+N\nu_f\frac{\partial u}{\partial y},v=0,T=T_w,\ \text{at}\ y=0, \tag{7a}$$

$$u\to 0,v\to 0,T\to T_\infty,\ \text{as}\ y\to\infty.zd \tag{7b}$$

where T, T_w and N are the ambient, wall fluid temperature, and slip parameter, respectively.

Introducing the following similarity transformations, we have

$$\eta=\sqrt{\frac{a}{\nu_f}}y,\ u=axf'(\eta),\ v=-\sqrt{a\nu_f}f(\eta),\theta=\frac{T-T_\infty}{T_f-T_\infty} \tag{8}$$

Making use of Eqs. (6, 8) in Eqs. (1–5), we have

$$f'''+(1-\phi)^{2.5}\left[\left(1-\phi+\phi\frac{\rho_{CNT}}{\rho_f}\right)\left\{ff''-f'^2\right\}-M^2f'\right]=0, \tag{9}$$

$$\left(\frac{k_{nf}}{k_f}\right)\theta''+\Pr\left(1-\phi+\phi\frac{(\rho c_p)_{CNT}}{(\rho c_p)_f}\right)\left[(f\theta')-\gamma\left(ff'\theta'+f^2\theta''\right)\right]=0, \tag{10}$$

$$f(0)=0,f'(0)=1,f'(\infty)=0,\theta(0)=1,\ \theta(\infty)=0, \tag{11}$$

where $\Pr = \frac{(\mu c_p)_f}{k_f}$ is the Prandtl number, $\gamma = a\lambda_2$ is the non-dimensional thermal relaxation time, and $\beta = N\sqrt{a\nu_f}$ is the slip parameter.

The quantity of practical interest, in this study, is the skin friction coefficient c_f and Nusselt number Nu_x, which is defined as follows:

$$c_f=\frac{\mu_{nf}}{\rho_f U_w{}^2}\left(\frac{\partial u}{\partial y}\right)_{y=0},Nu_x=\frac{-xK_{nf}}{k_f(T_f-T_\infty)}\left(\frac{\partial T}{\partial y}\right)_{y=0} \tag{12}$$

where q_w is the heat flux and K_{nf} is the effective thermal conductivity. Using variables (8), we obtain:

$$\mathrm{Re}_x^{1/2}c_f=\frac{f''(0)}{(1-\phi)^{2.5}},\ \mathrm{Re}_x^{-1/2}Nu_x=-\frac{k_{nf}}{k_f}\theta'(0). \tag{13}$$

3. Numerical scheme

The nonlinear ordinary differential equations (9)–(10) subject to the boundary conditions (11) have been solved numerically using an efficient Runge-Kutta fourth-order method along with shooting technique. The asymptotic boundary conditions given by Eq. (11) were replaced by using a value of 15 for the similarity variable η_{max}. The choice of η_{max} = 15 and the step size

$\Delta\eta = 0.001$ ensured that all numerical solutions approached the asymptotic values correctly. For validating of the proposed scheme, a comparison for the Nusselt number with the literature [4, 8, 9] has been shown in **Table 2** for both active and passive control of ϕ in the special case when. Therefore, we are confident that the applied numerical scheme is very accurate.

4. Results and discussion

In this section, the graphical explanation of the numerical results for velocity, temperature, skin friction coefficients, Nusselt number, and stream lines is expressed with respect to certain changes in the physical parameters through illustrations (**Figures 2–7**). A comparative study for pure water, SWCNT and MWCNT, is also depicted through **Tables 1–5**.

(**Figure 2a** and **b**) represents the changes in the fluid velocity profiles with respect to different values of solid nanoparticle volume fraction. **Figure 2(a)** shows the variation in solid volume fraction of nanoparticles with respect to Hartmann number M. As Hartmann number is the ratio of electromagnetic forces to the viscous forces. It is observed that when Hartmann number increases, electromagnetic forces will be dominant to the viscous forces that give declines in the velocity field (see **Figure 2(a)**). **Figure 2(b)** shows the variation in solid nanoparticle volume fraction with slip parameter β on velocity profile. It is analyzed that with an increase in slip parameter, velocity profile decreases. Further with an increase in solid nanoparticle volume fraction, velocity profile increases, and boundary layer thickness also increases with the increase in Hartmann number M, slip parameter, and solid nanoparticle volume fraction.

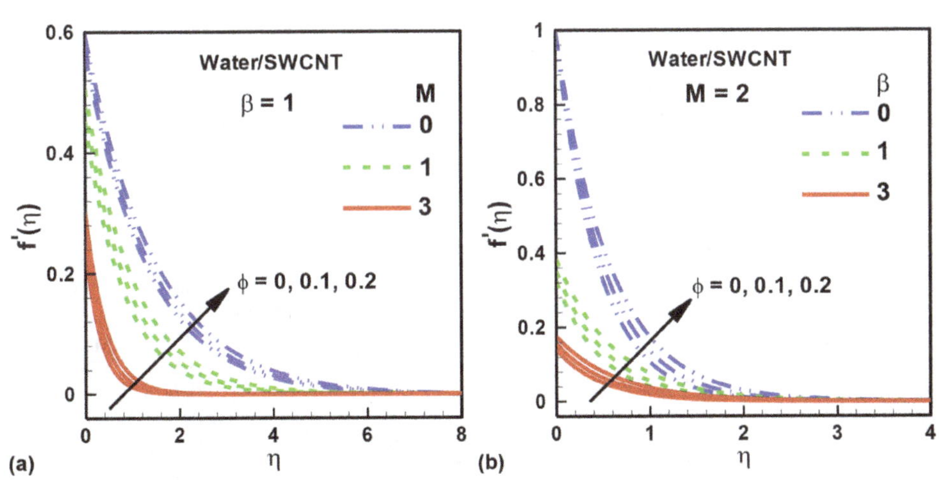

Figure 2. Velocity profile for different values of solid nanoparticle volume fraction. (a) Shows the variation with Hartmann number M. (b) Shows the variation with slip parameter β.

Temperature profile for different values of solid nanoparticle volume fraction with the variation in thermal relaxation time γ, Hartmann number M, and slip parameter β is presented in **Figure 3(a–c)**. Temperature profile decreases with the rise in thermal relaxation time, but thermal boundary layer increases with an increase in thermal relaxation time (see **Figure 3(a)**).

Figure 3(b) depicts that with an increase in electromagnetic forces as compared to the viscous forces, temperature profile and thermal boundary layer increase rapidly. Temperature profile and thermal boundary layer also increase rapidly with the rise in slip parameter (see **Figure 3(c)**). Moreover, temperature profile and thermal boundary layer increase with an increase in solid nanoparticle volume fraction.

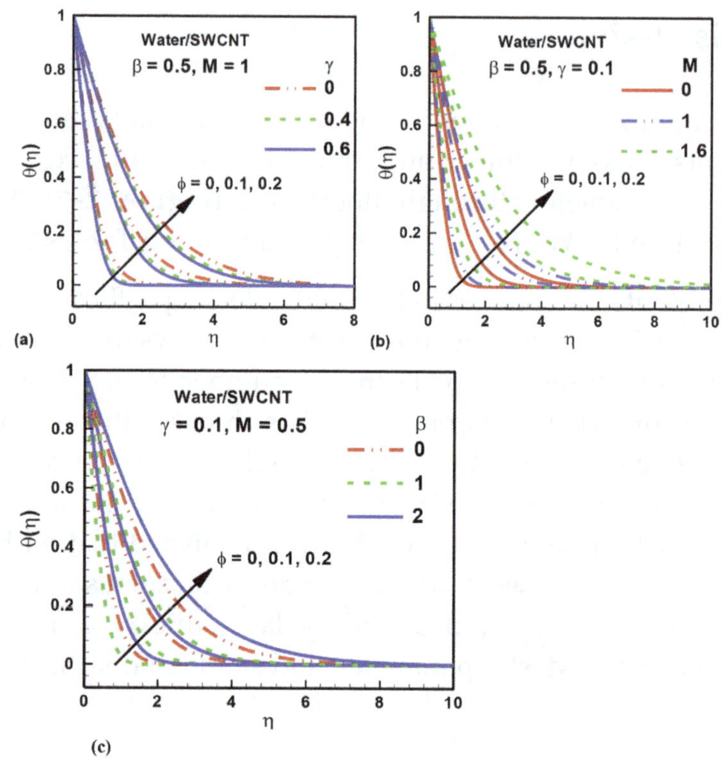

Figure 3. Temperature profile for different values of solid nanoparticle volume fraction. (a). Shows the variation with thermal relaxation time γ. (b) Shows the variation with Hartmann number M. (c) Shows the variation with slip parameter β.

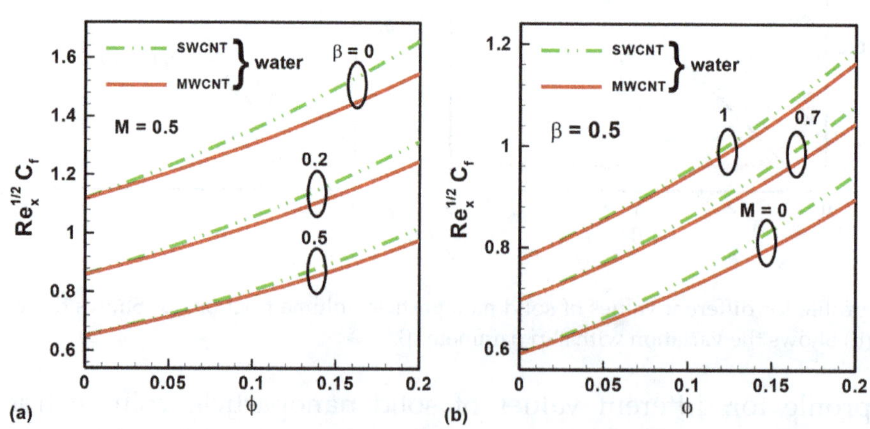

Figure 4. Skin friction coefficient for SWCNT and MWCNT. (a) Shows the variation with slip parameter β. (b) Shows the variation with Hartmann number M.

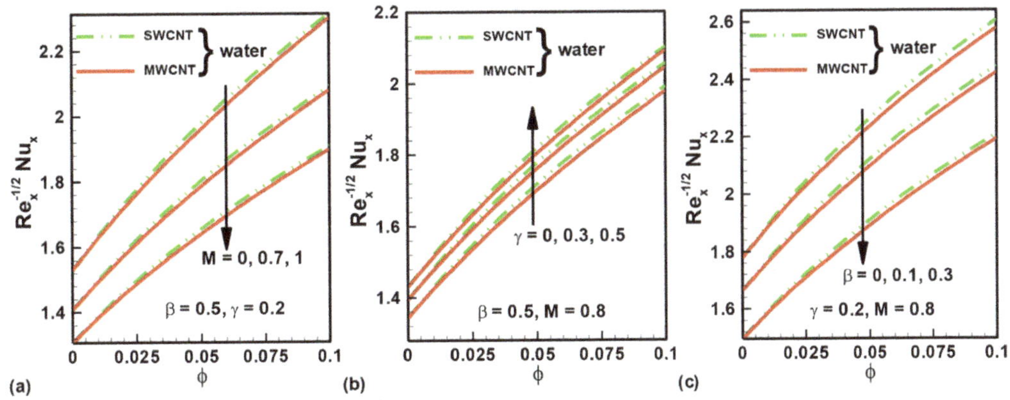

Figure 5. Nusselt number for SWCNT and MWCNT. (a) Shows the variation with Hartmann number M. (b) Shows the variation with thermal relaxation time γ. (c) Shows the variation with slip parameter β.

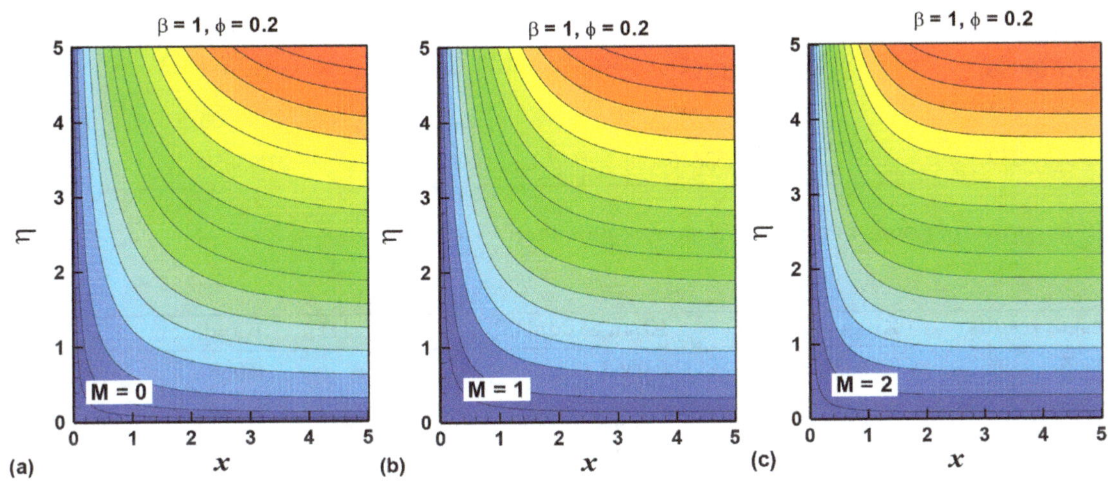

Figure 6(a-c). Streamlines for different values of Hartmann number M other parameters are $\beta = 0.4$, $\gamma = 0.3$.

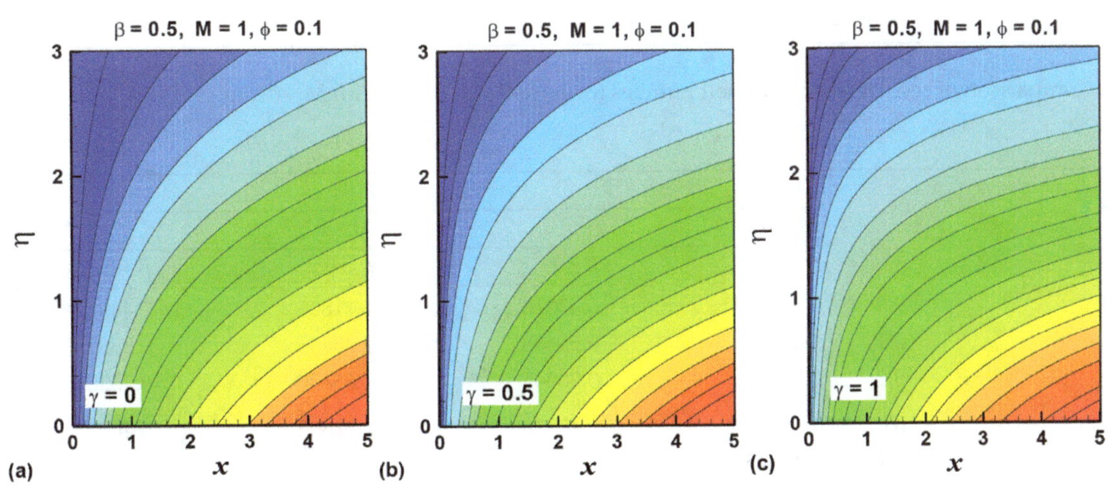

Figure 7(a-c). Isotherms for different values of thermal relaxation time γ other parameters are $M = 2$, $\beta = 0.2$.

Physical properties	Base fluid	Nanoparticles	
	Water	SWCNT	MWCNT
ρ (kg/m^3)	997	2600	1600
c_p (J/kg K)	4179	425	796
k (W/m K)	0.613	6600	3000

Table 1. Thermophysical properties of different base fluid and CNTs.

M	Present results	Salahuddin et al. [18]	Noreen et al. [19]
0.0	1	1	1
0.5	−1.11803	−1.11801	−1.11803
1	−1.41421	−1.41418	−1.41421
5	−2.44949	−2.44942	−2.44949
10	−3.31663	−3.31656	−3.31663
100	−10.04988	−10.04981	−10.04988
500	−22.38303	−22.38393	−22.38303
1000	−31.63859	−31.63846	−31.63859

Table 2. Comparison of results for the skin friction for pure fluid ($\phi = 0$).

Pr	Present results	Khan et al. [20]	Khan & Pop. [5]	Wang [11]	Kandasamy et al. [12]
0.07	0.0663	0.0663	0.0663	0.0656	0.0661
0.20	0.1691	0.1691	0.1691	0.1691	0.1691
0.70	0.4539	0.4539	0.4539	0.4539	0.4542
2	0.9114	0.9114	0.9113	0.9114	0.9114
7	1.8954	1.8954	1.8954	1.8954	1.8952
20	3.3539	3.3539	3.3539	3.3539	–
70	6.4622	6.4622	6.4621	6.4622	–

Table 3. Comparison of results for the Nusselt number for pure fluid ($\phi = 0$) with $M = 0$ and $\gamma = 0$.

ϕ		Cf			
		$M = 0$		$M = 0.5$	
		$\beta = 0$	$\beta = 1$	$\beta = 0$	$\beta = 1$
SWCNT	0.0	1	0.43016	1.11803	0.46912
	0.1	1.22904	0.54480	1.35495	0.58833
	0.2	1.51940	0.70245	1.65690	0.75286
MWCNT	0.0	1	0.43016	1.11803	0.46912
	0.1	1.17475	0.53298	1.30590	0.57942
	0.2	1.39935	0.67377	1.54755	0.73047

Table 4. Skin friction coefficient for different values of Hartmann number M and slip parameter β.

ϕ		Nusselt number							
		$\gamma = 0$		$\gamma = 0$		$\gamma = 0.1$		$\gamma = 0.1$	
		$M = 0$		$M = 0.5$		$M = 1$		$M = 2$	
		$\beta = 0$	$\beta = 1$	$\beta = 0$	$\beta = 1$	$\beta = 0$	$\beta = 1$	$\beta = 0$	$\beta = 1$
SWCNT	0.0	1.77095	1.33685	1.74513	1.25641	1.71287	1.08479	1.52688	0.70076
	0.1	2.66543	2.03231	2.60424	1.89346	2.48958	1.58280	2.07111	0.92785
	0.2	3.22397	2.49289	3.13585	2.31928	2.95535	1.92489	2.37985	1.11466
MWCNT	0.0	1.77095	1.33685	1.74513	1.25641	1.71287	1.08479	1.52688	0.70076
	0.1	2.63339	2.02350	2.57329	1.88386	2.46327	1.57551	2.05686	0.92736

Table 5. Nusselt number for different values of Hartmann number M and thermal relaxation time γ.

Variation in skin friction coefficient for SWCNT and MWCNT with slip parameter β and Hartmann number M is presented in **Figure 4**(**a** and **b**). It is seen that with the augment in M, electromagnetic strength is elevated in contrast to thick strength, skin friction coefficient rises for SWCNT as well as for MWCNT, but with the increase in slip parameter, skin friction coefficient decreases for both SWCNT and MWCNT. It is also seen that density and thermal conductivity of SWCNT are greater as compared to the MWCNT; therefore, the skin friction coefficient for SWCNT is greater as compared to the MWCNT.

Nusselt number for SWCNT and MWCNT shows the variation in Hartmann number M, thermal relaxation time γ, and slip parameter β. It is observed that the higher values of thermal relaxation time γ raise the Nusselt number for SWCNT as well as for MWCNT, and it is also analyzed that Nusselt number gives the larger values for SWCNT than MWCNT (see **Figure 5(a–c)**). Increasing values of Hartmann number M and slip parameter β decrease the Nusselt number for both SWCNT and MWCNT. But due to high density and thermal conductivity of SWCNT, Nusselt number for SWCNT is higher than MWCNT.

Streamlines and Isotherms are presented in **Figures 6(a–c)** and **7(a–c)**, respectively. It is analyzed from **Figures 6** and **7** that for increasing Hartmann number M and thermal relaxation time γ, streamlines and Isotherms are going close to origin.

5. Conclusions

This chapter discussed the magnetic field effects on the flow of Cattanneo-Christov heat flux model for water-based CNT suspended nanofluid over a stretching sheet. Key points of the performed analysis are as follows:

1. It is observed that when Hartmann number increases, electromagnetic forces will be dominant to the viscous forces that give declines in the velocity.
2. It is analyzed that with an increase in slip parameter, velocity profile decreases. Further, with an increase in solid nanoparticle volume fraction, velocity profile increases.
3. Boundary layer thickness also increases with the increase in Hartmann number M, slip parameter β, and solid nanoparticle volume fraction φ.

4. Temperature profile decreases with the rise in thermal relaxation time, but thermal boundary layer increases with an increase in thermal relaxation time.

5. It depicts that with an increase in electromagnetic forces as compared to the viscous forces, temperature profile and thermal boundary layer increase rapidly.

6. Temperature profile and thermal boundary layer also increase rapidly with the rise in slip parameter.

7. Temperature profile and thermal boundary layer increase with an increase in solid nanoparticle volume fraction.

8. Skin friction coefficient increases for SWCNT as well for MWCNT, but with the increase in slip parameter, skin friction coefficient decreases for both SWCNT and MWCNT.

9. It is also seen that thickness and thermal conductivity of SWCNT are better as compared to the MWCNT; consequently, the skin friction coefficient for SWCNT is better as compared to MWCNT.

10. It is pragmatic that the superior values of thermal relaxation time γ hoist the Nusselt number for SWCNT as well as for MWCNT, and it is also analyzed that Nusselt number gives the well-built principles for SWCNT as evaluated to MWCNT.

11. It is analyzed from **Figures 6** and **7** that for rising Hartmann number M and thermal relaxation time γ, streamlines and isotherms are departing shut to source.

Nomenclature

μ_f: Viscosity of base fluid

$(\rho C_p)_f$: Effective heat capacity of a fluid

$(\rho C_p)_{CNT}$: Effective heat capacity of a carbon nanotubes

k_f: Thermal conductivities of the base fluid

k_{CNT}: Thermal conductivities of the carbon nanotubes

B_0: Magnitude of magnetic field strength

T: Local fluid temperature

T_∞: Ambient temperature

u, v: Velocity components along x and y directions

P: Pressure

ϕ: Nanoparticle volume fraction

q: Heat flux

U_w, V_w along the x-axis and y-axis: Thermal Grashof number

g: Acceleration due to gravity

λ_2: Thermal relaxation time

B_C: Solutal Grashof number

v: Kinematic viscosity of the fluid

P_r: Prandtl number

M: Hartmann number

$(\rho c_p)_{nf}$: Heat capacitance

ρ_{nf}: Effective density

Re_x: Local Reynolds number

η: Similarity variable (transformed coordinate)

Nu_x: Local Nusselt number

$(\rho c)_p$: Effective heat capacity of the nanoparticle material

θ: Dimensionless temperature

γ: Non-dimensional thermal relaxation time

α_{nf}: Effective thermal diffusibility

x, y: Coordinate along and normal to the sheet

$(\rho c)_f$: Heat capacity of the fluid

μ_{nf}: Effective dynamic viscosity

k_{nf}: Effective thermal conductivity of the nanofluid

Author details

Noreen Sher Akbar[1*], C. M. Khalique[2] and Zafar Hayat Khan[3]

*Address all correspondence to: noreensher1@gmail.com

1 DBS&H CEME, National University of Sciences and Technology, Islamabad, Pakistan

2 International Institute for Symmetry Analysis and Mathematical Modelling, Department of Mathematical Sciences, South Africa

3 Department of Mathematics, University of Malakand, Dir (Lower), Khyber Pakhtunkhwa, Pakistan

References

[1] S.U.S. Choi, Enhancing thermal conductivity of fluids with nanoparticles, in: D.A. Siginer, H.P. Wang (Eds.), Developments and Applications of Non-Newtonian Flows, ASME, New York, vol. 66 (1995) 99–105.

[2] Y. Xuan, W. Roetzel, Conceptions for heat transfer correlation of nanofluids, Int. J. Heat Mass Transfer. 43 (2000) 3701–3707.

[3] K. Khanafer, K. Vafai, M. Lightstone, Buoyancy-driven heat transfer enhancement in a two-dimensional enclosure utilizing nanofluids, Int. J. Heat Mass Transfer. 46 (2003) 3639–3653.

[4] A.V. Kuznetsov, D.A. Nield, Natural convective boundary-layer flow of a nanofluid past a vertical plate, Int. J. Therm. Sci. 49 (2010) 243–247.

[5] W.A. Khan, I. Pop, Boundary-layer flow of a nanofluid past a stretching sheet, Int. J. Heat Mass Tranfer 53 (2010) 2477–2483.

[6] A. Ebaid, H.A. El-arabawy, Y. Nader, New exact solutions for boundary-layer flow of a nanofluid past a stretching sheet, Int. J. Differential Equ. 2013 (Article ID 865464) 8.

[7] A. Ebaid, E.H. Aly, Exact analytical solution of the peristaltic nanofluids flow in an asymmetric channel with flexible walls: Application to cancer treatment, Comput. Math. Methods Med. 2013 (Article ID 825376) 8.

[8] A. Ebaid and A.M. Wazwaz, On the generalized Exp-function method and its application to boundary layer flow at nano-scale, J. Comput. Theor. Nanosci. 11(1) (2014) 178–184.

[9] A. Ebaid, E.H. Aly, N.Y. Abdelazem, Analytical and numerical investigations for the flow and heat transfer of nanofluids over a stretching sheet with partial slip boundary condition, J. Appl. Math. Inf. Sci. 8(4) (2014) 1639-1645.

[10] A. Ebaid, H.K. Al-Jeaid, H. Al-Aly, Notes on the Perturbation Solutions of the Boundary Layer Flow of Nanofluids Past a Stretching Sheet, Appl. Math. Sci. 7(122) (2013) 6077–6085.

[11] C.Y. Wang, Free convection on a vertical stretching surface, ZAMM J. Appl. Math. Mech. 69(11) (1989) 418–420.

[12] R. Kandasamy, P. Loganathan, P. Puvi Arasu, Scaling group transformation for MHD boundary-layer flow of a nanofluid past a vertical stretching surface in the presence of suction/injection, Nuclear Eng. Design 241(6) (2011) 2053–2059.

[13] B.J. Fourier, Theorie Analytique De Lachaleur, Jacques Gabay, Paris, 1822.

[14] C. Cattaneo, Sulla conduzionedelcalore, AttiSemin. Mat. Fis. Univ. Modena Reggio Emilia 3 (1948) 83–101.

[15] C.I. Christov, On frame indifferent formulation of the Maxwell–Cattaneo model of finite-speed heat conduction, Mech. Res. Commun. 36 (2009) 481–486.

[16] V. Tibullo, V. Zampoli, A uniqueness result for the Cattaneo–Christov heat conduction model applied to incompressible fluids, Mech. Res. Commun. 38 (2011) 77.

[17] M. Mustafa, Cattaneo–Christov heat flux model for rotating flow and heat transfer of upper convected Maxwell fluid, AIP Adv. 5 (2015) 047109.

[18] T. Salahuddin, M.Y. Malik, A. Hussain, S. Bilal, M. Awais, MHD flow of Cattanneo–Christov heat flux model for Williamson fluid over a stretching sheet with variable thickness: Using numerical approach, J. Magnet. Magnet. Mater. 401 (2016) 991–997.

[19] N.S. Akbar, E. Abdelhalim, Z.H. Khan, Numerical analysis of magnetic field effects on Eyring-Powell fluid flow towards a stretching sheet, J. Magnet. Magnet. Mater. 382 (2015) 355–358.

[20] Z.H. Khan, W.A. Khan, R.J. Culham, Estimation of boundary-layer flow of a nanofluid past a stretching sheet: A revised model, J. Hydrodyn. (2015).

[21] N.S. Akbar, Z.H. Khan, S. Nadeem, W.A. Khan, Double-diffusive natural convective boundary-layer flow of a nanofluid over a stretching sheet with magnetic field, Int. J. Numer. Methods Heat Fluid Flow 26(1) (2016) 108–121.

[22] M. Sheikholeslami, D.D. Ganji, M. Gorji-Bandpy, S. Soleimani, Magnetic field effect on nanofluid flow and heat transfer using KKL model, J. Taiwan Inst. Chem. Eng. 45 (2014) 795–807.

[23] H. Togun, G. Ahmadi, T. Abdulrazzaq, A.J. Shkarah, S.N. Kazi, A. Badarudin, M.R. Safaei, Thermal performance of nanofluid in ducts with double forward-facing steps, J. Taiwan Inst. Chem. Eng. 47 (2015) 28–42.

[24] M. Sheikholeslami, R. Ellahi, H.R. Ashorynejad, G. Domairry, Effects of heat transfer in flow of nanofluids over a permeable stretching wall in a porous medium, J. Comput. Theor. Nanosci. 11(2) (2014) 486–496.

[25] M. Sheikholeslami, D.D. Ganji, M.Y. Javed, R. Ellahi, Effect of thermal radiation on magnetohydrodynamics nanofluid flow and heat transfer by means of two phase model, J. Magnet. Magnet. Mater. 374 (2015) 36–43.

[26] M. Sheikholeslami, M. Gorji-Bandpy, Free convection of ferrofluid in a cavity heated from below in the presence of an external magnetic field, Powder Technol. 256 (2014) 490–498.

[27] M. Sheikholeslami, M. Gorji-Bandpy, D.D. Ganji, Lattice Boltzmann method for MHD natural convection heat transfer using nanofluid, Powder Technol. 254 (2014) 82–93.

[28] M. Sheikholeslami, M. Gorji-Bandpy, D.D. Ganji, Numerical investigation of MHD effects on Al_2O_3-water nanofluid flow and heat transfer in a semi-annulus enclosure using LBM, Energy 60 (2013) 501–510.

[29] M. Sheikholeslami, R. Ellahi, M. Hassan, S. Soleimani, A study of natural convection heat transfer in a nanofluid filled enclosure with elliptic inner cylinder, Int. J. Numer. Methods Heat Fluid Flow 24 (2014) 1906–1927.

[30] M. Sheikholeslami, M. Gorji, M. Bandpy, R. Ellahi, M. Hassan, S. Soleimani, Effects of MHD on Cu-water nanofluid flow and heat transfer by means of CVFEM, J. Magnet. Magnet. Mater. 349 (2014) 188–200.

[31] M. Sheikholeslami, M. Bandpy, M.G.R. Ellahi, R. Zeeshan, A simulation of MHD CuO-water nanofluid flow and convective heat transfer considering Lorentz forces, J. Magnet. Magnet. Mater. 369 (2014) 69–80.

[32] R. Ellahi, M. Gulzar, M. Sheikholeslami, Effects of heat transfer on peristaltic motion of Oldroyd fluid in the presence of inclined magnetic field, J. Magnet. Magnet. Mater. 372 (2014) 97–106.

9

Thermophysical Properties of Metal Oxides Nanofluids

Zafar Said and Rahman Saidur

Abstract

Thermophysical properties of TiO_2, Al_2O_3 and SiO_2 nanofluids are experimentally investigated and compared with published data. Density has been measured over a range of 25–40°C for nanoparticle volumetric concentration of 0.05–4%. Viscosity experiments were carried out over a wide temperature range, from 25 to 80°C, to determine their applicability in such ranges. Nanofluids with particle volume fraction ranging from 0.02 to 0.03% and 1–4 kg/min were examined for the convective heat transfer and pumping power. The heat transfer coefficient of the nanofluid rises with rising mass flow rate, as well as rising volume concentration of metal oxide nanofluids; however, increasing the volume fraction results in increasing the density and viscosity of nanofluid, leading to a slight increase in friction factor which can be neglected. Addition of surfactants results in part of the increment in viscosity as well. An empirical formula for density is proposed, which also contributes to the novelty of this paper.

Keywords: nanofluids, surfactants, density, viscosity, pressure drop, heat transfer coefficient

1. Introduction

Thermophysical properties of nanofluids are significant to enhance the heat transfer behaviour. It is tremendously significant in controlling industrial and energy saving prospects. Great interest has been shown by the industry in nanofluids. Unlike conventional particle-fluid suspension (millimetre- and micrometre-sized particles), nanoparticles have great ability to enhance the thermal transport properties. In the last decade, due to the ability of improving thermal properties, nanofluids have gained prominent attention. Based on the broad research, it has been recognized that the suspension of metallic particles in a base fluid significantly increases the thermal conductivity of the mixture [1], therefore improving the heat transfer capability. Such observations have inspired the industrial as well as the science community to discover the thermophysical properties of nanofluids, such as density, viscosity, thermal conductivity and heat capacity. Obtaining the viscosity of nanofluids is of importance for establishing sufficient pumping power. Besides, the convective heat transfer coefficients, Prandtl and Reynolds numbers, are also reliant on viscosity.

The available literature has suggested that the nanofluids tend to enhance the heat transfer performance, most importantly the heat transfer coefficient, thermal conductivity and viscosity, which in turn affects the pumping power. Das et al. [2] experimentally showed that thermal conductivity of a nanofluid can be augmented up to fourfold by increasing temperature. Xuan and Li [3] also found that heat transfer coefficient could be enhanced by the use of nanofluids, particularly when increasing the flow or nanoparticle concentration.

Different results on the density of nanofluids have been reported by the researchers, most of which are for a particular temperature or for a particular nanofluid [4–12]. There is no generalized method or model to obtain the density of different nanofluids theoretically. Density of the nanofluids strongly depends on nanoparticle material and increases with the increase in volume concentration. Base fluid also plays a significant role in the density of the nanofluids, whereas the other parameters such as nanoparticles shape, size, zeta potential and additives do not affect the density of the nanofluids. For engineering applications, larger density is more preferable [13]. The equation for the density of two phase mixtures for particles of micrometre size is available in the literature for slurry flows [14]. Densities of solids are greater than that of the liquids, with the increase in the concentration of nanoparticles in the fluid, and the density of the nanofluid is found to increase. Density of the nanofluids is proportional to the volume ratio of nanoparticles (solid) and base fluid (liquid) in a system. Pak and Cho [15] conducted an experiment at one temperature (25°C) for Al_2O_3 and TiO_2 nanofluids of up to 4 vol.%.

Studies on viscosity have been reported by numerous researchers. Several researchers [16–19] observed a Newtonian behaviour in TiO_2-ethylene glycol, Al_2O_3-water, single wall carbon nanohorns (SWCNH)-water and TiO_2-water nanofluids, respectively. With the increase in particle concentration, increase in the viscosity is noted. Timofeeva et al. [20] observed that viscosity decreases with the increase in particle size; however, Pastoriza-Gallego et al. [21] observed a different behaviour. A non-Newtonian behaviour was found by other researchers. Some authors have also developed models to describe the rheological behaviour of nanofluids. Koo [22] introduced a model to predict the thermal conductivity and viscosity of nanofluid in terms of nanoparticle size, concentration and density. Masoumi et al. [23] also presented a model to calculate the effective viscosity by considering the Brownian motion and the relative viscosity between the fluid and particles.

Pumping power is deserted in more than a few studies although it directly effects the usefulness of the fluid in applications. This is due to the reason that heat transfer coefficients could be enhanced by merely increasing the flow velocity of the fluid, which results in additional pumping power due to the surged pressure losses. Any effort to improve heat transfer results in improved pressure losses. Also with the addition of nanoparticles to the base fluid, an antagonism between heat transfer growth and enhanced pressure losses is present. Hence, the real measure of efficiency of a heat transfer fluid is not only the convective heat transfer coefficient alone, but also the pressure losses need to be taken into account for calculations as well [24]. Heat transfer of a nanofluid flow which is squeezed between parallel plates is investigated analytically using homotopy perturbation method (HPM). It was reported from the findings that the Nusselt number has a direct relationship with the nanoparticle volume fraction, the squeeze number and Ecket number when two plates are separated, but an inverse effect is noted when the plates are squeezed [25]. In another study, heat and mass transfer characteristics of unsteady nanofluid flow between two parallel plates are investigated considering thermal radiation. Ordinary differential equations are solved numerically using the fourth-order Runge-Kutta method. Results indicated that the radiation parameter increased with the concentration boundary layer thickness. It was also reported that the Eckert number, Schmidt number, squeeze parameter and radiation parameter have direct relationship with Nusselt number [26].

For nanofluids to be applied for practical applications, it is essential to study the effect of these nanofluids on the flow features and their effect on the pumping power and pressure drop, in addition to heat transfer performance enhancement. Calculations are carried out for different volume fractions of nanoparticles and for changing mass flow rate. The effectiveness of nanofluids is investigated by comparing the required pumping power of oxides nanofluids and the base fluid. The reason these oxides are considered for this study is that they are easy to produce and cheaper compared to the CNTs and graphene.

On the basis of the inclusive literature review, the main objectives of this study are to examine the effects of nanofluids on the density, viscosity, the pressure drop and the convection heat transfer characteristics. The nanofluids contained Al_2O_3, SiO_2, and TiO_2 nanoparticles in water as a base fluid. The analyses were done for several nanofluids and then associated them with the water as the base fluid. The potential outcomes and their details were also described.

2. Methodology

2.1. Materials

Deionized water was used as the base fluid. The nanoparticles (TiO_2, P25 ≥ 99.5% trace metals basis), (SiO_2, 99.5% trace metals basis) and (Al_2O_3, 99.8% trace metals basis) were purchased from Sigma-Aldrich. Sodium dodecyl sulphate (SDS, 92.5–100.5%, Sigma-Aldrich), polyvinylpyrrolidone (PVP, Sigma-Aldrich), poly(ethylene glycol) 400 (PEG, Sigma-Aldrich) and hexadecytrimethyl-ammonium bromide (HTAB, ≥ 98%) were used as surfactants (**Figures 1–4**).

2.2. Nanofluids preparation

Ultra sound sonication was used to homogenize the suspensions. The particles can be mixed into many different liquids at preferable concentrations. Probe-type sonicators break particle agglomerates faster and more thoroughly than bath sonicators, and thus, it was chosen for our experiments. The nanoparticles were dispersed mechanically in distilled water at a concentration of 0.05, 0.5, 1, 2 and 4% by volume for density and 0.05–0.5% by volume for viscosity.

Figure 1. SEM images of (a) TiO_2, (b) SiO_2 and (c) Al_2O_3 nanoparticles.

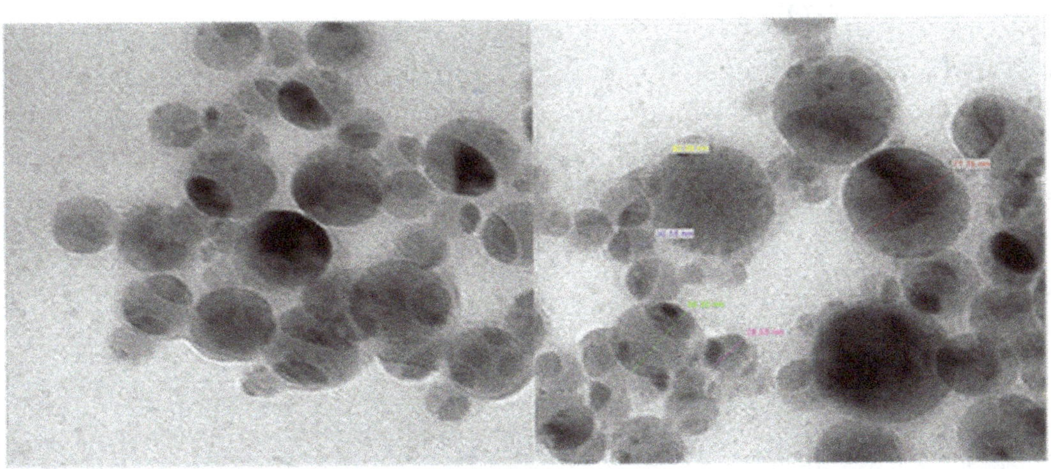

Figure 2. TEM images of silica (SiO_2) nanoparticles ~10 to 20 nm and with surfactant (0.14% PEG).

Aforementioned surfactants with 1:2, 1:3 and 1:5 nanoparticle to surfactant ratios (by volume) were used to prepare nanofluids for this research.

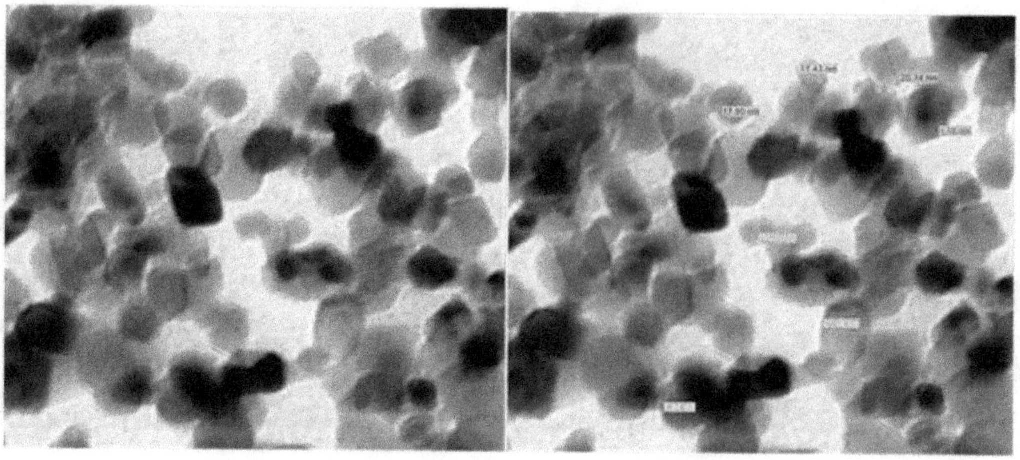

Figure 3. TEM images of titania (TiO_2) nanoparticles ~21 nm and the surfactant (0.1 %vol. PEG).

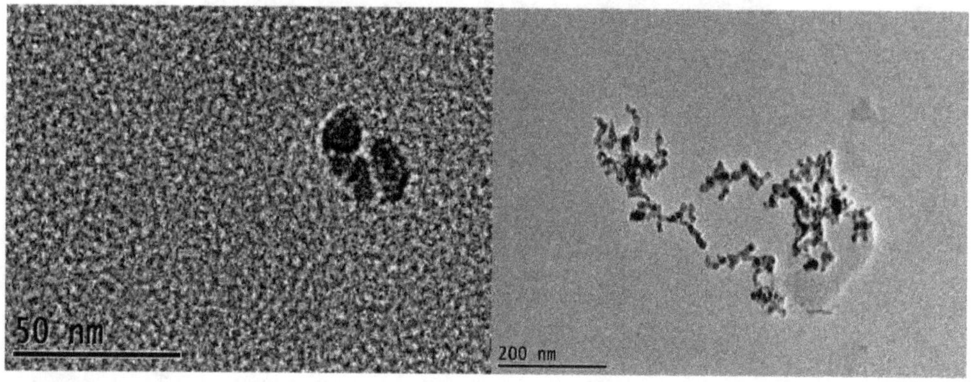

Figure 4. TEM images of alumina (Al_2O_3) nanoparticles, average 13 nm and the surfactant (0.1%vol. HTAB).

2.3. Nanofluid characterization

In order to characterize the prepared nanofluids, particle size, dynamic viscosity and density were measured as functions of temperature and particle volumetric fraction. Field emission scanning electron microscopy (FESEM) and transmission electron microscopy (TEM) were used to obtain the morphological characterization of the nanoparticles with SIGMA Zeiss instrument (Carl Zeiss SMT Ltd., UK). The Density Metre DA-130N from Kyoto Electronics Shinjuku-ku, Tokyo, Japan was used to measure the density of the nanofluids.

2.4. Sedimentation

Different nanofluids of 10 ml volume were used to investigate the sedimentation rate, and data were collected for 1 month from the date of preparation. **Table 1** shows the detailed description of our investigation. It was also found that 0.1 %vol. HTAB for TiO_2/DW and 0.1 %vol. PVP for Al_2O_3/DW and $TiSiO_4$/DW work as the best surfactants for stability.

From **Table 1**, it is noticed that all the nanoparticles, dispersed in DW, showed stability for longer periods of time, except for SiO_2.

Nanofluid	Surfactant	Sedimentation after 30 days (%)	Sedimentation rate (ml/day)
TiO_2/DW	0.1 %vol. PVP	95	0.317
	0.1 %vol. PEG	44	0.147
	0.15 %vol. PEG	42	0.140
	0.25 %vol. PEG	35	0.117
	0.1 %vol. HTAB	31	0.103
Al_2O_3/DW	0.1 %vol. PVP	10	0.033
	0.1 %vol. PEG	20	0.066
	0.15 %vol. PEG	27	0.090
	0.25 %vol. PEG	21	0.070
	0.1 %vol. HTAB	20	0.066
SiO_2/DW	0.1 %vol. PVP	99	9.9
	0.1 %vol. PEG	99	9.9
	0.15 %vol. PEG	99	9.9
	0.25 %vol. PEG	99	9.9
	0.1 %vol. HTAB	99	9.9

Table 1. Sedimentation rate of different nanofluids.

2.5. Density of nanofluid

Experimental data on density measurements are not sufficient for various nanofluids at varying temperatures in the literature. Therefore, we carried out comprehensive measurements to obtain density and provide data as well as to verify the applicability of Eq. (1) (which is also called as mixing theory) [15] for various nanofluids. **Figure 5** shows that density varies with temperature. Therefore, density equation should include temperature as a variable, whereas the mixing theory does not consider effect of temperature.

$$\rho_{nf} = \left(\frac{m}{V}\right)_{nf} = \frac{m_f + m_p}{V_f + V_p} = \frac{\rho_f V_f + \rho_p V_p}{V_f + V_p} = (1-\phi_p)\rho_{bf} + \phi_p \rho_p, \quad (1)$$

where $\phi_p = \frac{V_p}{V_f + V_p}$ is the volume fraction of the nanoparticles.

2.6. Theoretical models

In this section, the existing theoretical models and correlations for the viscosity of nanofluid suspensions are presented. Each model is used for specific circumstances.

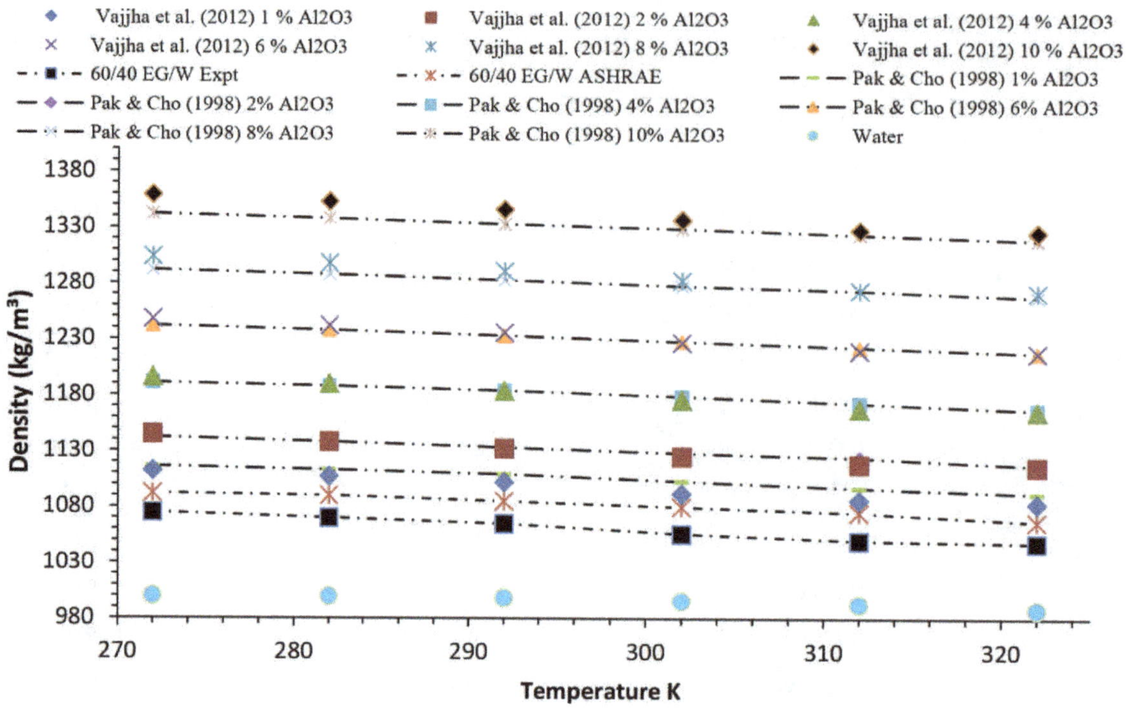

Figure 5. Densities of particle volumetric concentrations as a function of temperature [15, 27].

2.6.1. *Implemented models for viscosity and density*

In this part, the correlations that we have implemented for comparison between the experimental data and the predicted data are indicated.

$$\frac{\mu_{eff}}{\mu_f} = \frac{1}{1\text{–}34.87(d_p/d_f)^{-0.3}\varphi^{1.03}}, \quad \textit{where}$$

$$d_f = 0.1\left(\frac{6M}{N\pi\rho_{bf}}\right)^{1/3} \tag{2}$$

It may be noted that once the base fluid is designated, the dimensionless effective viscosity of the nanofluid $\frac{\mu_{eff}}{\mu_f}$ increases with the decrease in particle diameter and increase in volume concentration.

2.6.2. *Pumping power*

The system counted is a forced flow nature. A pump is needed to mingle nanofluids throughout the system. The pump would require electrical energy. It is crucial to comprehend the entire energy needed by the pump to sustain a constant flow across the collector. The pumping power is analysed as follows [28]. The pressure drop throughout the collector is specified by Δp, which is determined from the subsequent equation [29]

$$\Delta p = f\frac{\rho V^2}{2}\frac{\Delta l}{d} + K\frac{\rho V^2}{2}, \tag{3}$$

where K is the loss coefficient because of entrance effects, exit effects, bends, elbows, valves, etc. V is the mean flow velocity of nanofluids in the system and is given by

$$V = \frac{\dot{m}}{\rho_{nf}\pi D_H^2/4}, \tag{4}$$

where D_H represents the hydraulic diameter. In present analysis consider D_H = pipe diameter (d). ρ_{nf} *was* calculated from Eq. (1). The frictional factor, f, for laminar and turbulent flow, correspondingly, is as follows [30]

$$f = \frac{64}{Re} \quad \text{for laminar flow}$$

$$f = \frac{0.079}{Re^{1/4}} \quad \text{for turbulent flow.}$$

The Reynolds number is composed as

$$\mathrm{Re} = \frac{\rho V D_H}{\mu}. \tag{5}$$

Now, the pumping power can be calculated using Eq. (20)

$$\text{Pumping power} = \left(\frac{\dot{m}}{\rho_{nf}}\right)\Delta p. \tag{6}$$

2.6.3. Convective heat transfer

Elementary forced convective heat transfer model utters the interactions in the middle of fluid flow and convective heat transfer in terms of correlations between the dimensionless Reynolds, Prandtl and Nusselt numbers (Nu, Re and Pr, respectively). For the regular instances of laminar flows inside a pipe of diameter d, the theoretically developed relations are given as follows [31]:

$$h_{nf} = \frac{q}{T_W - T_f} \tag{7}$$

$$Nu_{nf} = \frac{h_{nf} d}{k_{nf}}. \tag{8}$$

The Nusselt number for the laminar flow throughout a circular pipe is the function of the Reynolds and Prandtl numbers and can be resolved by employing Eq. (8) [32]. Nusselt number,

Parameters of collector	Value
Type	Black paint flat plate
Glazing	Single glass
Agent fluids	Water, ethylene glycol and water mixture and Al_2O_3 nanofluids
Absorption area, A_p	1.51 m^2
Wind speed	20 m/s
Collector tilt, β_o	20°
Ambient temperature, T_a	300 K
Apparent sun temperature, T_s	4350 K
Optical efficiency, η_o	0.84
Glass thickness, t	4 mm
Insulation thermal conductivity, k_i	0.06 W/mK
Incident solar energy per unit area of the absorber plate, I_T	1000 W/m^2
Inner diameter of pipes, d	0.01 m

Table 2. Environmental and analysis conditions for the flat plate solar collector.

$$Nu = 0.000972Re^{1.17}Pr^{1/3} \text{ for Re} < 2000, \tag{9}$$

where Pr and k_{nf} can be expressed as,

$$Pr = \frac{C_{p,nf}\mu_{nf}}{k_{nf}} \tag{10}$$

$$\frac{k_{nf}}{k_f} = \frac{k_p + (SH-1)k_f - (SH-1)\varphi(k_f - k_p)}{k_p + (SH-1)k_f + \varphi(k_f - k_p)}. \tag{11}$$

In Eq. (12), SH is the shape factor, which is given to be three for the spherical shape of nanoparticle [33] (**Table 2**).

3. Results and discussions

3.1. Size distribution of the nanoparticles

SEM nano-graphs of (a) TiO_2, (b) SiO_2 and (c) Al_2O_3 nanoparticles are presented in **Figure 1**. **Figure 2** presents the TEM images of silica (SiO_2) nanoparticles with surfactant (0.14% PEG). **Figure 3** presents the TEM images of titania (TiO_2) nanoparticles ~21 nm and the surfactant (0.1 %vol. PEG). **Figure 4** presents the TEM images of alumina (Al_2O_3) nanoparticles, average 13 nm and the surfactant (0.1%vol. HTAB).

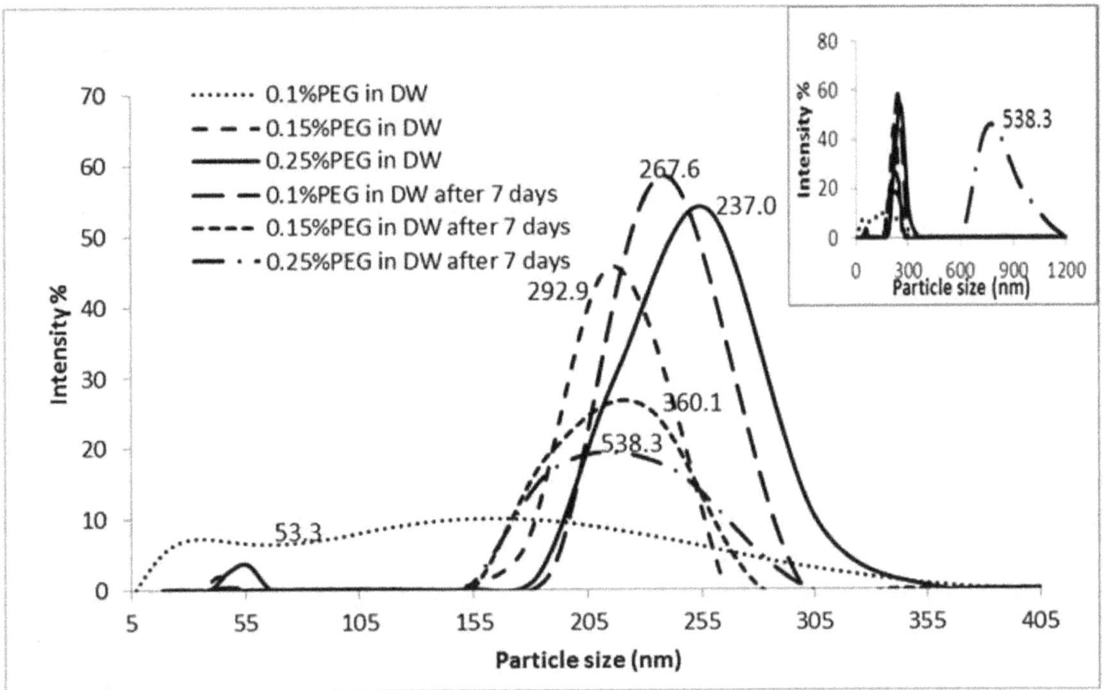

Figure 6. Size distribution of the nanoparticles in 0.05 %vol. water-based TiO_2 nanofluids with surfactants. Inset is the full size distribution.

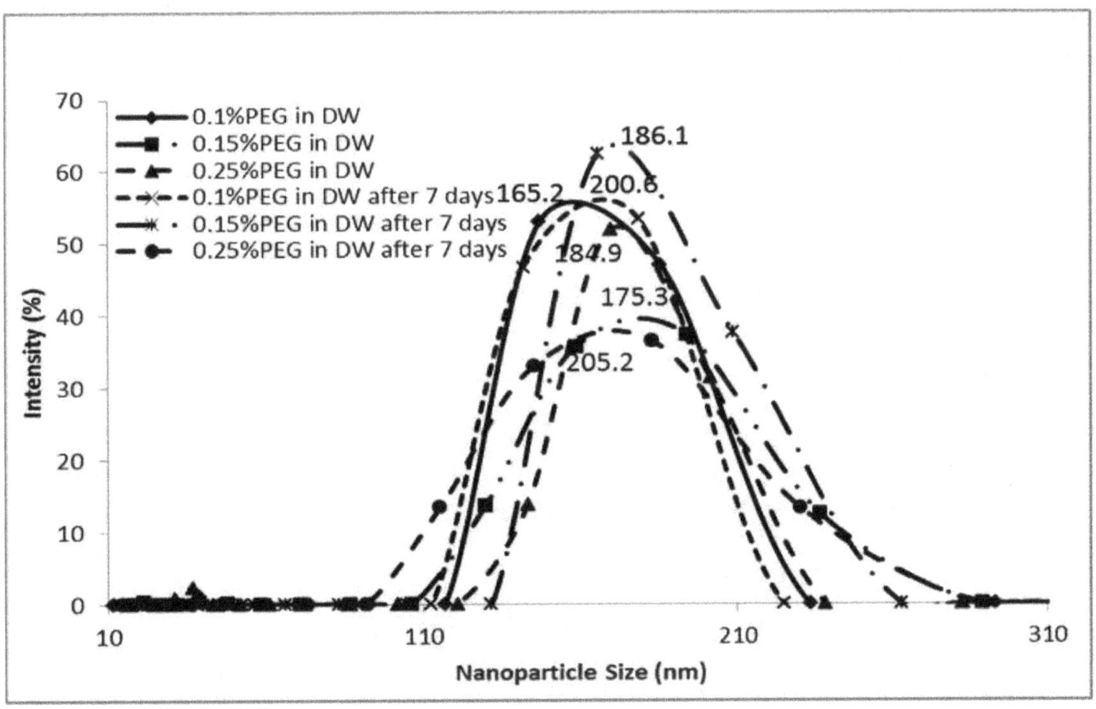

Figure 7. Size distribution of the nanoparticles in 0.05 %vol. water-based Al_2O_3 nanofluids with surfactants.

Sizes of the nanoparticles in all the prepared nanofluids were measured using Zetasizer3000HSa (Malvern), and results for the most stable nanofluids are shown in **Figures 6** and **7**.

The numbers placed at the apexes describe the average particle size as obtained from the machine. It should be noted that the Malvern Nanosizer measures hydrodynamic properties based on the Strokes-Einstein equation, which is expected to be slightly larger than the actual size. Incorporating the findings from those in **Figures 6** and **7**, it is found that the visual sedimentation rate of TiO_2/water nanofluid decreases with increasing surfactant (PEG), whereas particle size increases with the increase in surfactant. For Al_2O_3/water nanofluid, the sedimentation rate for PEG in different concentrations is approximately similar and the size distribution of the particles is approximately also in the same region. The stability of the stable nanofluids as shown in **Figures 6** and **7** was obtained for more than 1 week.

3.2. Density of nanofluids

First, a benchmark test for the density of the base fluid is presented showing excellent agreement with the data presented in the handbook of the American Society of Heating, Refrigerating and Air-Conditioning Engineers (ASHRAE). Next, density measurements of the TiO_2, Al_2O_3 and SiO_2 nanofluids over a temperature range of 25–40°C for several particle volume concentrations are presented. These measured results were compared with a widely used theoretical equation, and good agreements between the theoretical equation and measurements were obtained for the TiO_2, Al_2O_3 and SiO_2 nanofluids.

We found the density decreasing with the increase in temperature. Eq. (1) is used to make the comparison with our experimental data. **Figures 8** and **9** show the experimental values at different temperatures and the theoretical value obtained from Pak and Cho [15]. The trend lines generated from the experimental values clearly certify the linear relationship of density with concentration at a particular temperature. It also validates that the density increases with the increase in concentration. The pattern of the change in density is analysed, and the

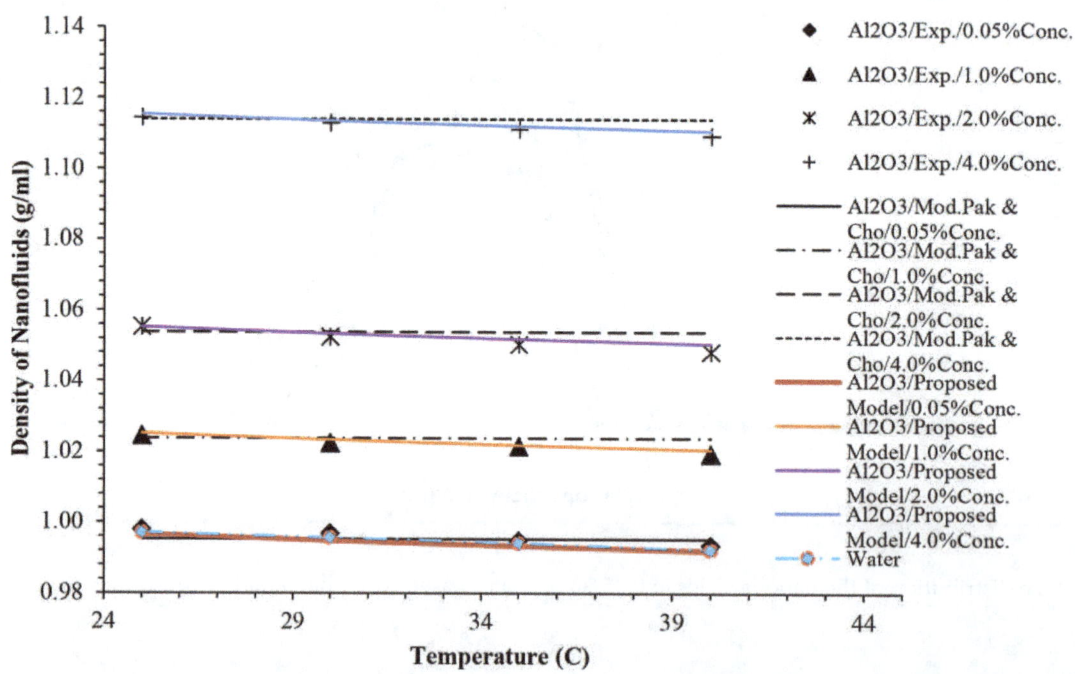

Figure 8. Density vs. temperature graph of Al_2O_3–water nanofluid at different concentrations.

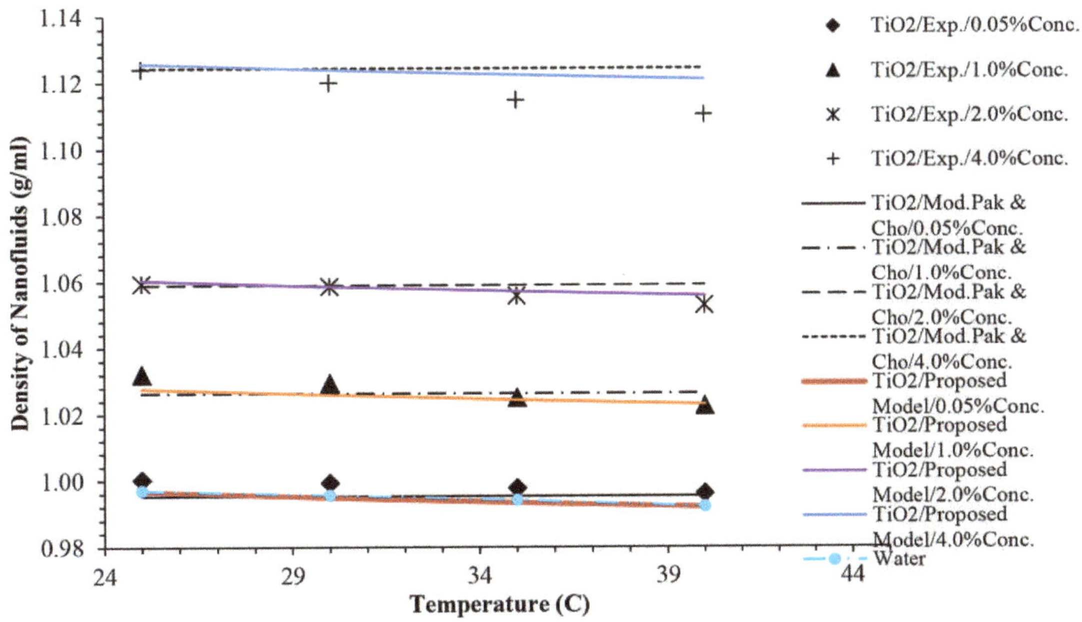

Figure 9. Density vs. temperature graph of TiO_2–water nanofluid at different concentrations.

proposed empirical formula is presented below. The effective density of the nanofluids is measured using authors' proposed formula derived based on Einstein's theory. This model considers the changing temperature, volume fraction as well as the size of the nanoparticles. No such models have been presented previously, which also justifies the novelty of this proposed formula.

$$\rho_{nf} = (1-\phi_p)\rho_{bf} + \phi_p\rho_p + \left(a - \frac{\ln(T)}{100}\right), \text{where } a = 0.03358. \quad (12)$$

Moreover, as we examine the average absolute percentage deviation, we notice a systematic increase in percentage deviation with the concentration, as shown in **Figures 8–10**. Further examination shows a gradual increase in percentage deviation with the temperature. Therefore, the proposed equation may have limitations for some nanofluids. The proposed formula is found valid for up to 2 %vol. concentration. The maximum relative difference between the experimental and theoretical values is 0.3% for Al_2O_3, 0.44% for TiO_2 and 0.28% for SiO_2. **Figures 8–10** represent the comparative graph for experimental data, Pak and Cho model data and proposed model data (solid line).

3.3. Viscosity

Viscosity of water-based nanofluids with and without surfactant was experimentally obtained. Viscosity of nanofluid was measured using Brookfield viscometer (DV-II + Pro Programmable Viscometer), which was connected with a temperature controlled bath. To verify the accuracy of our equipment and experimental procedure, viscosity of the ethylene glycol and water mixture (60:40 by mass) was measured and compared with the data from the American Society of Heating, Refrigerating and Air-Conditioning Engineers (ASHRAE) handbook [34].

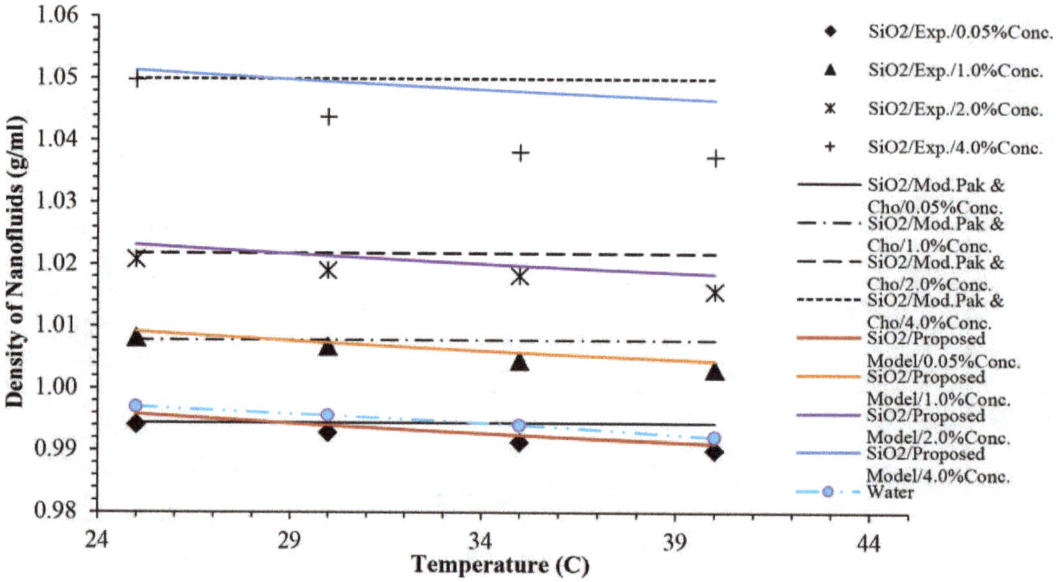

Figure 10. Density vs. temperature graph of SiO_2–water nanofluid at different concentrations.

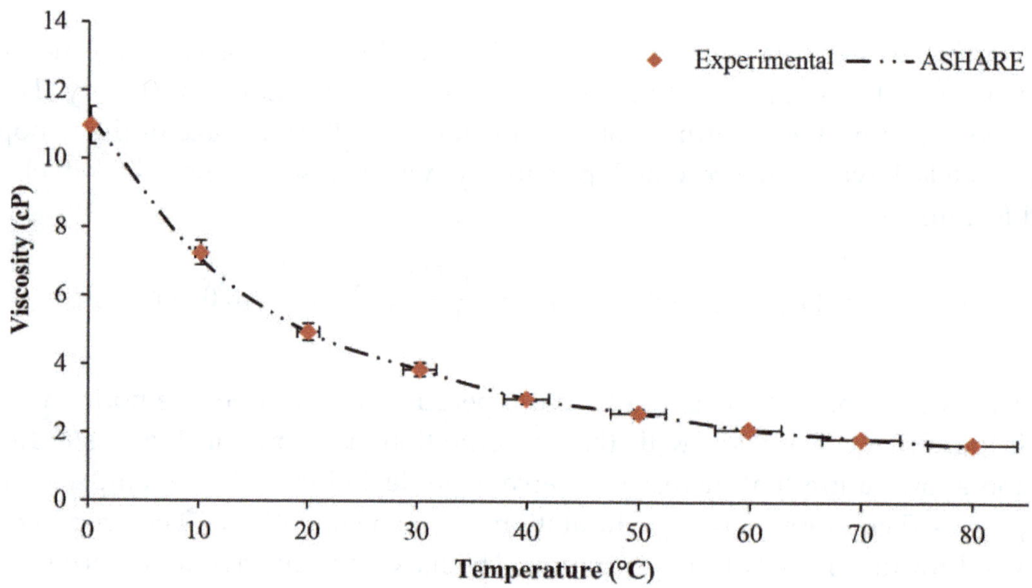

Figure 11. Comparison of ASHRAE viscosity values of 60:40 ethylene glycol and water mixture (by mass) and experimental data. 1 cP (centipoise) = 1 mPa s.

Before conducting the density measurement, the equipment must be calibrated. Density measurements of a mixture of 60:40 EG/W by mass were first conducted to confirm the accuracy of our apparatus and the calibration procedure. The results of these measurements and the data from ASHRAE are presented in **Figure 11** over a temperature range of 0–80°C. Excellent agreement is observed between the current measurements and the ASHRAE data. **Figure 11** shows that the experimental values of viscosity for the ethylene glycol mixture and the ASHRAE data match fairly with a maximum alteration of ±2.0%. The deviations at low shear rates were due to machine error.

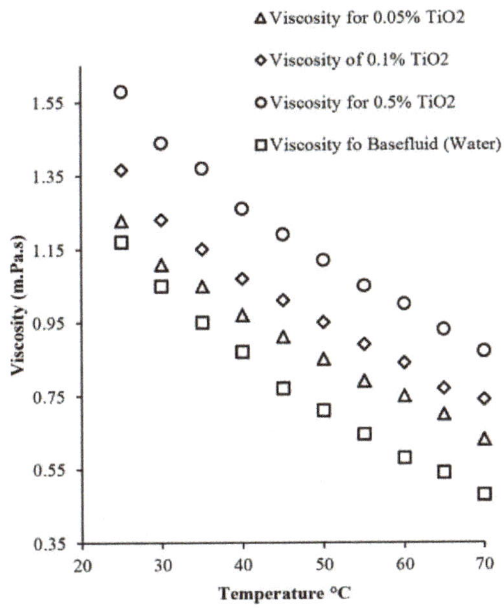

Figure 12. Viscosity of TiO_2 at different temperatures and concentrations.

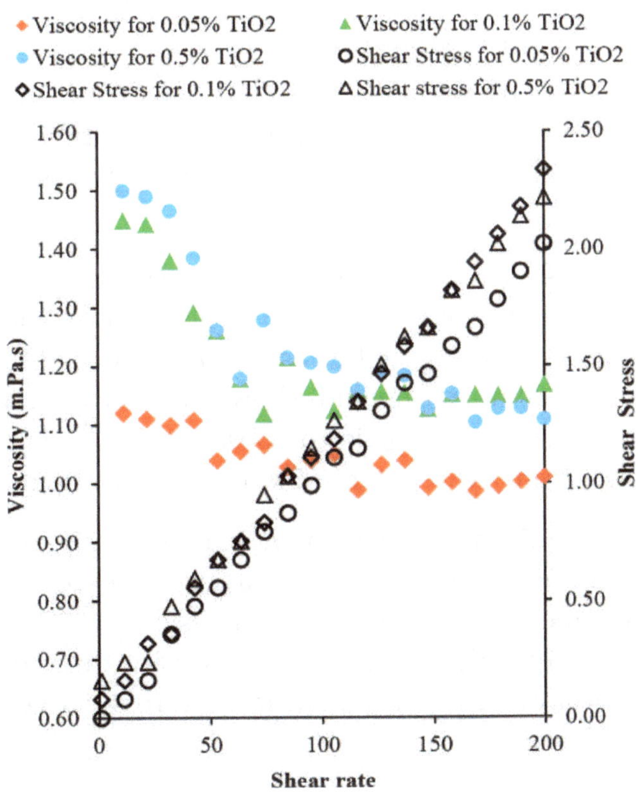

Figure 13. Viscosity of TiO_2 at different shear rates and concentrations.

Viscosity of TiO_2-H_2O with different volume concentration and changing temperature is presented in **Figure 12**. It is observed that the viscosity reduces with the increase in temperature and rises with the increase in volume fraction. From **Figure 13** at 0.5 vol.%, titanium

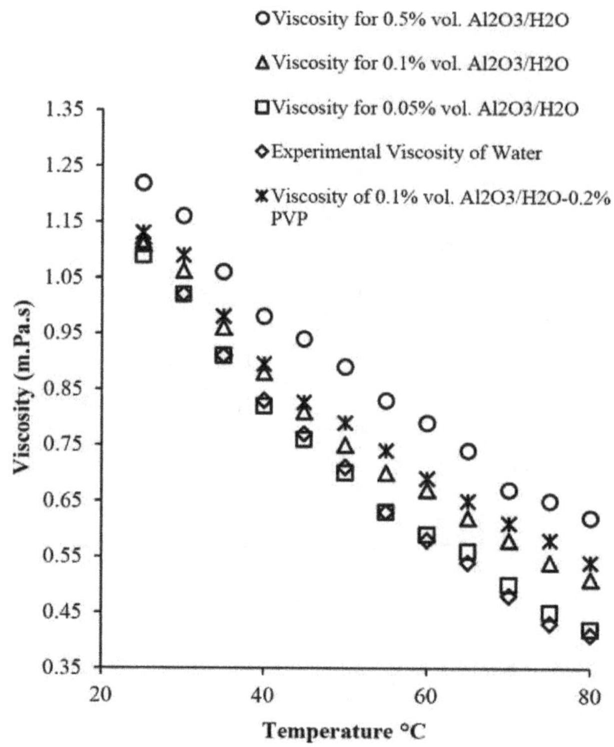

Figure 14. Viscosity of Al_2O_3 at different temperatures and concentrations.

behaves in a non-Newtonian way after 40°C. For the whole temperature range, this nanofluid demonstrates a Newtonian behaviour at 0.05 vol.%, while a non-Newtonian nature is found for 0.1 vol.%.

From **Figure 14**, the effect of surfactant is observed to increase viscosity. Addition of surfactant augments the viscosity of the nanofluid. Similarly, viscosity tends to increase in **Figure 15**, for alumina with the rise in the volume concentration as well as addition of surfactants. As opposed to thermal conductivity, viscosity of these nanofluids showed a decreasing trend with the increase in temperature. Furthermore, a nonlinear relation is observed between the viscosity of alumina nanofluid and particle concentration except for 0.05 vol.% and temperatures lower than 40°C.

As it is observed from **Figure 15**, in the case of nanoparticles with surfactants, viscosity tends to decrease due to augmenting the temperature. This can be explained by the change in the shape of micelles. Worm shapes change to spherical or vesicles which lead to the destruction of network structure and consequently a decrease in viscosity through temperature rise [35]. After the breakdown in network occurs, the attractive forces between the particles become dominant to produce aggregates.

Viscosity of SiO_2 at different temperatures and concentrations is illustrated in **Figure 16**. It is witnessed that the rise in temperature results in reduced viscosity and addition of volume fraction also results in higher viscosity. **Figure 17** presents the changes in the viscosity for SiO_2 nanofluid. In this case, the viscosity keeps increasing up to the concentration of 0.1 vol.%, but remains unchanged afterwards. This is because SiO_2 nanofluid is not stable, and due to the high rate of aggregation, sedimentation of nanoparticles takes place and consequently no

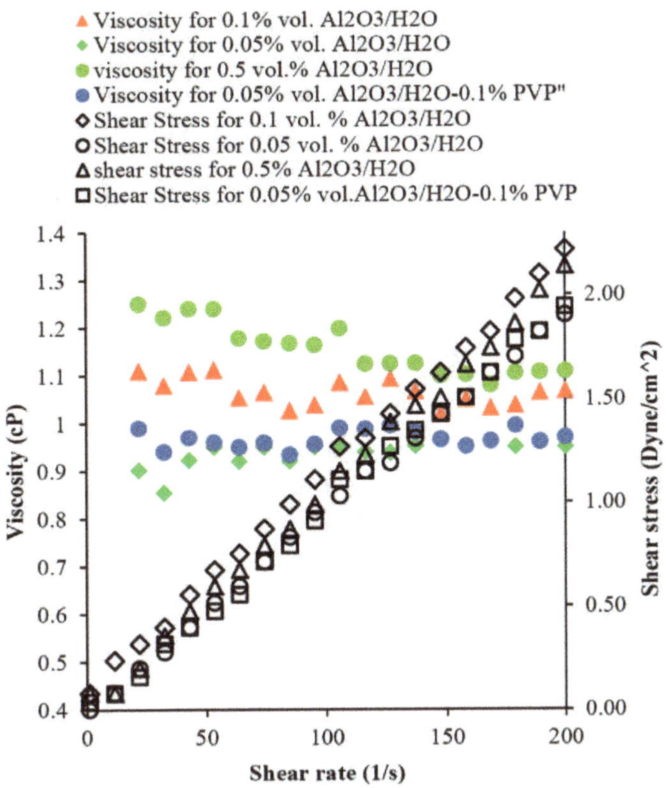

Figure 15. Viscosity of Al_2O_3 at different shear rate and concentrations.

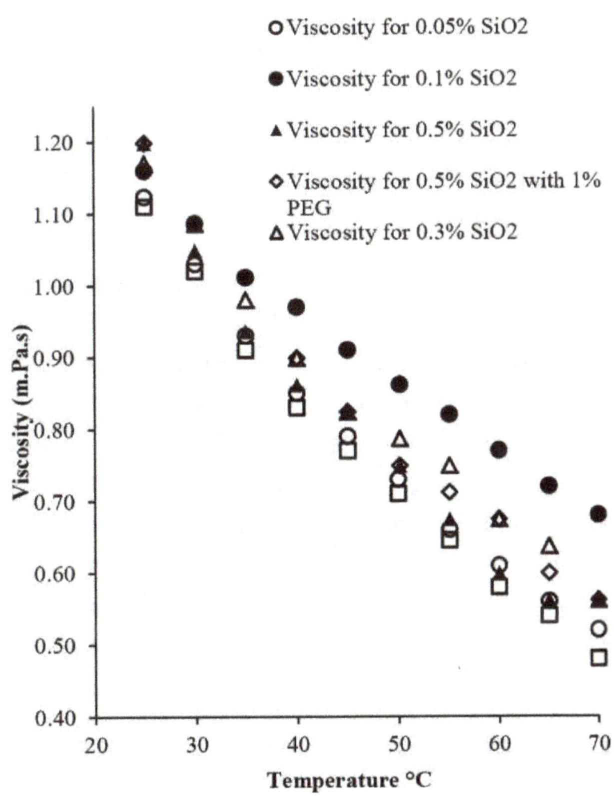

Figure 16. Viscosity of SiO_2 at different temperatures and concentrations.

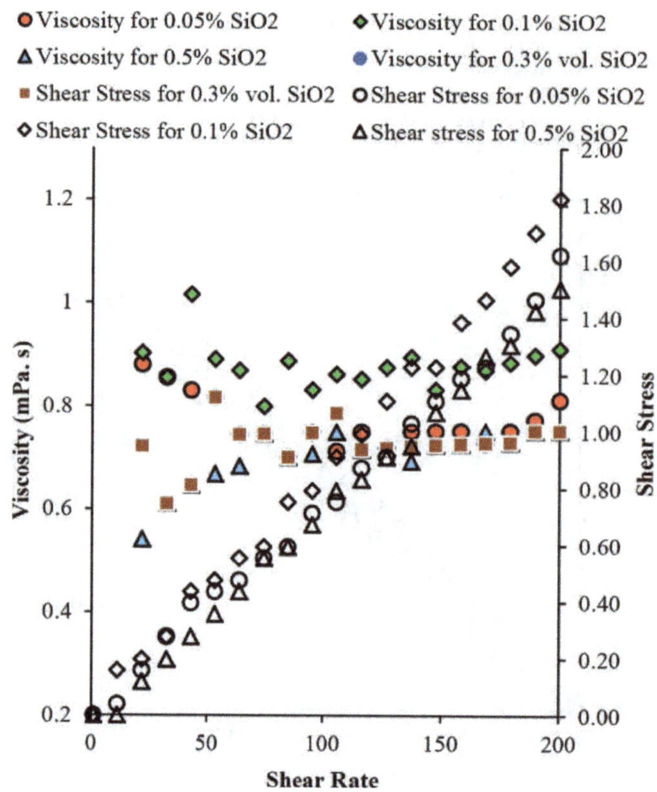

Figure 17. Viscosity of SiO_2 at different shear rate and concentrations.

increment is observed. Adding surfactant causes minor increments in the viscosity of the nanofluid, with no significant changes in the stability of the colloid even after a period of half an hour. Newtonian behaviour is found for this nanofluid at 0.05 vol.% for temperatures over 50°C, but non-Newtonian characteristics are shown beyond this point. A similar trend occurred for 0.5 vol.% and at 60°C. A non-Newtonian behaviour for the entire range of temperature is observed for 0.1 vol.%.

In the previous articles, it was claimed that the viscosity of nanofluids is mainly dependent on the concentration of nanoparticles, properties of base fluid and temperature. However, some also reported about the size of the nanoparticles. From the results of current study, it is observed that for the same concentration (0.5%vol.), viscosities of the nanofluids are different, and TiO_2 results in higher viscosity followed by Al_2O_3 and SiO_2 nanofluids for 0.5 vol.% concentration. Therefore, it can be said that viscosity depends on nanoparticles' properties such as the size and density. The available models are not appropriate to forecast or measure accurate viscosity of nanofluids as they are not related to temperature variation and nanoparticles' properties. Therefore, the results suggest the requirement of providing a more generalized viscosity model.

3.4. Pumping power and convective heat transfer

The resulting pressure losses are analysed in detail in this section. **Figure 18** presents the pumping power and pressure drop with respect to volume fraction and volume flow rate for

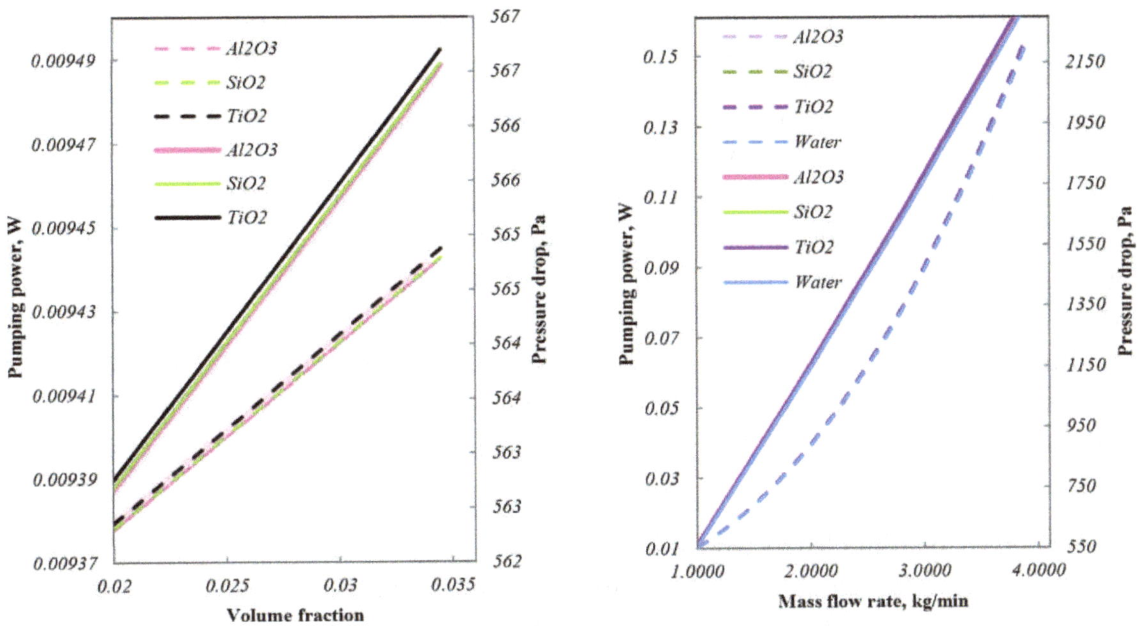

Figure 18. Effect of volume fraction and mass flow rate on pumping power (solid line) and pressure drop (dotted line).

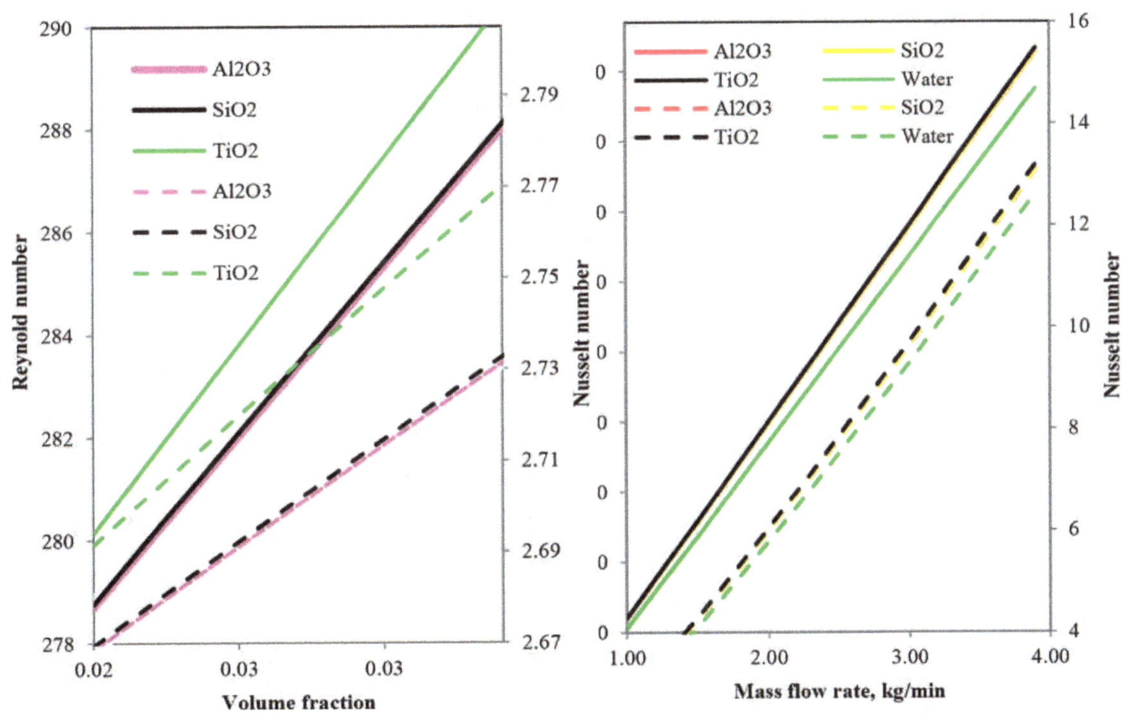

Figure 19. Effect of volume fraction and mass flow rate on the Reynolds number and Nusselt number.

the laminar flow, respectively. These two parameters are calculated using Eqs. (4) to (7) and **Table 1**.

The results show that the friction factor enhances with the increase in volume fraction and flow rate. It is observed from the figures that the friction factors of the nanofluids are almost the same as that of the base fluid under same nanoparticle volume concentrations, therefore resulting in little supplementary pumping power for the process [36–38].

Effect of volume fraction and mass flow rate is presented in **Figure 19**. These parameters are calculated using Eqs. (6) and (9). With the rising volume fraction of the nanoparticles suspended in water, it is observed that the both the Reynolds number and Nusselt number increase slightly in comparison with water. Reynolds number and Nusselt numbers of the nanofluid are greater compared to water, and the numbers are rising with the growing volume fraction as well as with the rising mass flow rate. This shows that increasing the molecular thermal diffusion due to increasing the nanoparticles volume fraction is the foremost purpose of the heat transfer enhancement for an exact Reynolds number. As a significance, the increase in the Nusselt number is found with a rise in Re. An alike trend was also witnessed by Maïga et al. [39].

The experimental results clearly show that the nanoparticles suspended in water enhance the convective heat transfer coefficient, although the volume fraction of nanoparticles is very low ranging from 0.01 to 0.3 vol.%. The convective heat transfer coefficient of water-based Al_2O_3 nanofluids increases with volume fraction of Al_2O_3 nanoparticles as shown in **Figure 20**.

The overall heat transfer coefficient of water rises with the rising mass flow rate, with a maximum value of 804 W/m^2 compared to that of water, which is 732 W/m^2K with the highest flow rate investigated. Heat transfer coefficient is improved with the addition of volume fraction as well as with the mass flow rate. Heat transfer coefficient is directly proportional to heat rate.

From the above results presented in the figures, it can be concluded that the heat transfer coefficient enhanced up to 15% with the suspended nanoparticles in the base fluid. The

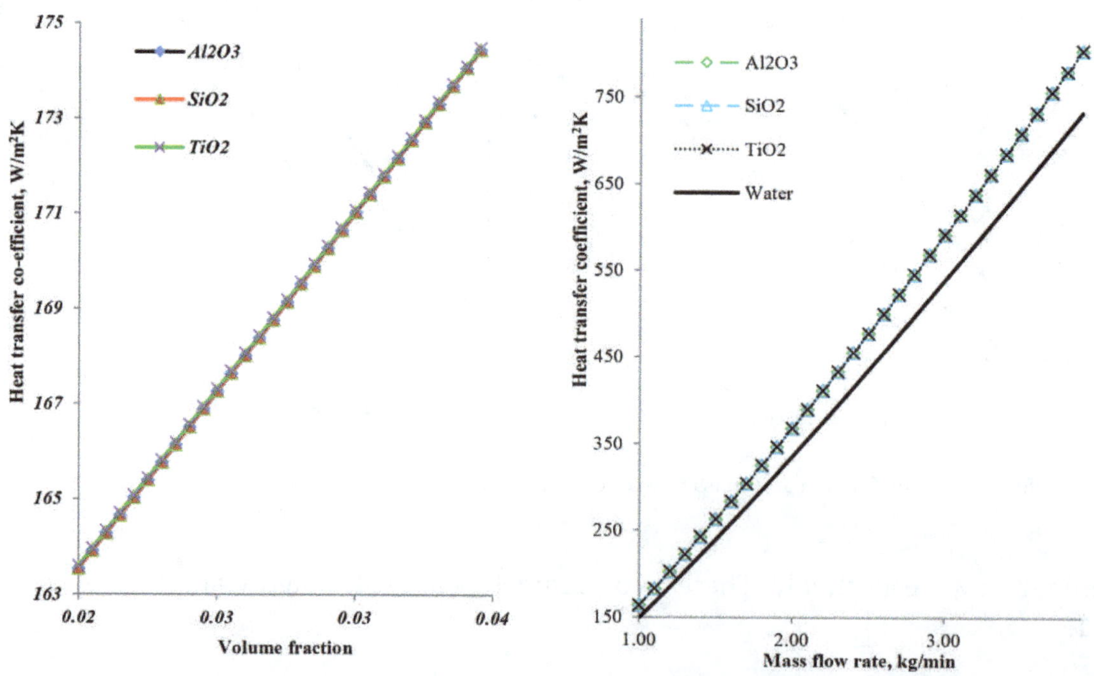

Figure 20. Heat transfer coefficient with respect to changing volume fraction and mass flow rate.

patterns shown by the oxide nanofluids are due to the fact that the addition of nanoparticles tends to enhance the thermal conductivity, density and viscosity of the base fluid. The enhancement in Reynolds number and Nusselt number was observed to be 8.4, 7.6 and 7.5% for TiO_2, SiO_2 and Al_2O_3, respectively. Nusselt number was improved by 6.8, 5.5 and 5.4%, for TiO_2, SiO_2 and Al_2O_3, respectively. This enhancement results in heat transfer performance. Enhancement in pumping power was observed to be 1.5%.

4. Future recommendations

The prospect of using nanofluids in different applications is related to their thermal and flow properties. Measurement of these properties will be performed in near future in order to access the total improvement in energy efficiency where these fluids are being used. Despite some undesirable changes such as rising viscosity or decreasing specific heat, we can consider nanofluids as good thermal fluids. However, researches to obtain a comprehensive formula to determine other physical properties of nanofluid must be continued.

5. Conclusion

The effect of volume fraction, temperature and mass flow rate was investigated on density, viscosity, pumping power and convective heat transfer of nanofluids in this article. Stability of nanofluids was obtained using different surfactants. Important conclusions have been attained and summarized as follows:

1. The density is found decreasing with the increase in temperature. An empirical model is proposed to describe the behaviour. The maximum deviation between the experimental values and the proposed model is 0.3% for Al_2O_3, 0.44% for TiO_2 and 0.3% for SiO_2 which is very small and well below the minimum acceptable limit (1%).

2. Viscosity of our nanofluids increases dramatically with the increase in particle concentration. Addition of surfactants results in part of the increment.

3. Friction factor rises with the rising volume fraction. This is due to the increasing density and viscosity with the addition of nanoparticles. Since this effect is slightly higher compared to base fluid, therefore little penalty in pressure drop and pumping power occurs.

4. At a particle volume concentration of 3%, the use of oxide nanofluid gives significantly higher heat transfer characteristics. For example, at the particle volume concentration of 3%, the overall heat transfer coefficient is 804 W/m^2 compared to that of water, which is 732 W/m^2 for a mass flow rate of 1 kg/min, so the overall heat transfer coefficient of the oxides nanofluid is 15% greater than that of distilled water as a base fluid.

5. The enhancement in Reynolds number Nusselt number was observed to be 8.4, 7.6 and 7.5% for TiO_2, SiO_2 and Al_2O_3, respectively. Nusselt number was improved by 6.8, 5.5

and 5.4%, for TiO_2, SiO_2 and Al_2O_3, respectively, which is greater than that of distilled water.

Acknowledgements

The authors would like to acknowledge the financial support from the Ministry of higher education (MOHE), project no: FP019-2011A & High Impact Research Grant (HIRG) project no: UM.C/HIR/MOHE/ENG/40.

Nomenclature

d	diameter
f	friction factor
FESEM	field emission scanning electron microscopy
k	thermal conductivity, W/mK
Re	Reynolds number
SEM	scanning electron microscopy
T	temperature, °C
TEM	transmission electron microscopy
Δp	pressure drop, Pa

Greek symbols

ρ	density, kg m^{-3}
ϕ	volumetric fraction
μ	volumetric fraction
η	intrinsic viscosity
δ	thickness of nano-layer

Subscripts

bf	base fluid
nf	nanofluid
p	particle

Author details

Zafar Said[1,2]* and Rahman Saidur[2,3]

*Address all correspondence to: zaffar.ks@gmail.com; zsaid@masdar.ac.ae

1 Department of Sustainable and Renewable Energy Engineering (SREE), University of Sharjah, Sharjah, United Arab Emirates

2 Department of Mechanical Engineering, Faculty of Engineering, University of Malaya, Kuala Lumpur, Malaysia

3 Centre of Research Excellence in Renewable Energy (CoRE-RE), King Fahd University of Petroleum and Minerals (KFUPM), Dhahran, Saudi Arabia

References

[1] Eastman, J., S. Choi, S. Li, W. Yu, and L. Thompson, Anomalously increased effective thermal conductivities of ethylene glycol-based nanofluids containing copper nanoparticles. Applied Physics Letters. **78**(6)(2001): p. 718–720.

[2] Das, S.K., N. Putra, P. Thiesen, and W. Roetzel, Temperature dependence of thermal conductivity enhancement for nanofluids. Journal of Heat Transfer. **125**(2003): p. 567.

[3] Xuan, Y. and Q. Li, Investigation on convective heat transfer and flow features of nanofluids. Journal of Heat Transfer. **125**(2003): p. 151.

[4] Reddy, M., V.V. Rao, B. Reddy, S.N. Sarada, and L. Ramesh, Thermal conductivity measurements of ethylene glycol water based TiO_2 nanofluids. Nanoscience and Nanotechnology Letters. **4**(1)(2012): p. 105–109.

[5] Wamkam, C.T., M.K. Opoku, H. Hong, and P. Smith, Effects of ph on heat transfer nanofluids containing ZrO_2 and TiO_2 nanoparticles. Journal of Applied Physics. **109** (2011): p. 024305.

[6] Xie, H., W. Yu, and W. Chen, Mgo nanofluids: higher thermal conductivity and lower viscosity among ethylene glycol-based nanofluids containing oxide nanoparticles. Journal of Experimental Nanoscience. **5**(5)(2010): p. 463–472.

[7] Turgut, A., I. Tavman, M. Chirtoc, H. Schuchmann, C. Sauter, and S. Tavman, Thermal conductivity and viscosity measurements of water-based TiO_2 nanofluids. International Journal of Thermophysics. **30**(4)(2009): p. 1213–1226.

[8] Li, Y., J.e. Zhou, S. Tung, E. Schneider, and S. Xi, A review on development of nanofluid preparation and characterization. Powder Technology. **196**(2)(2009): p. 89–101.

[9] Duangthongsuk, W. and S. Wongwises, Measurement of temperature-dependent thermal conductivity and viscosity of TiO_2-water nanofluids. Experimental Thermal and Fluid Science. **33**(4)(2009): p. 706–714.

[10] Ghadimi, A., R. Saidur, and H. Metselaar, A review of nanofluid stability properties and characterization in stationary conditions. International Journal of Heat and Mass Transfer. **54**(17)(2011): p. 4051–4068.

[11] Ramesh, G. and N.K. Prabhu, Review of thermophysical properties, wetting and heat transfer characteristics of nanofluids and their applicability in industrial quench heat treatment. Nanoscale Research Letters. **6**(1)(2011): p. 1–15.

[12] Trisaksri, V. and S. Wongwises, Critical review of heat transfer characteristics of nanofluids. Renewable and Sustainable Energy Reviews. **11**(3)(2007): p. 512–523.

[13] Timofeeva EV, Yu W, France DM, Singh D, Routbort JL. Nanofluids for heat transfer: an engineering approach. Nanoscale research letters. 2011 Feb 28;6(1):1.

[14] Cheremisinoff, N. P. "Encyclopedia of fluid mechanics. Volume 1-Flow phenomena and measurement." Encyclopedia of Fluid Mechanics. Volume 1. Vol. 1. 1986.

[15] Pak, B.C. and Y.I. Cho, Hydrodynamic and heat transfer study of dispersed fluids with submicron metallic oxide particles. Experimental Heat Transfer an International Journal. **11**(2)(1998): p. 151–170.

[16] Bobbo, S., L. Fedele, A. Benetti, L. Colla, M. Fabrizio, C. Pagura, and S. Barison, Viscosity of water based swcnh and TiO_2 nanofluids. Experimental Thermal and Fluid Science. **36** (2012): p. 65–71.

[17] Chandrasekar, M., S. Suresh, and A. Chandra Bose, Experimental investigations and theoretical determination of thermal conductivity and viscosity of Al_2O_3 water nanofluid. Experimental Thermal and Fluid Science. **34**(2)(2010): p. 210–216.

[18] Chen, H., Y. Ding, Y. He, and C. Tan, Rheological behaviour of ethylene glycol based titania nanofluids. Chemical Physics Letters. **444**(4)(2007): p. 333–337.

[19] Fedele, L., L. Colla, and S. Bobbo, Viscosity and thermal conductivity measurements of water-based nanofluids containing titanium oxide nanoparticles. International Journal of Refrigeration. **35**(5)(2012): p. 1359–1366.

[20] Timofeeva, E.V., A.N. Gavrilov, J.M. McCloskey, Y.V. Tolmachev, S. Sprunt, L.M. Lopatina, and J.V. Selinger, Thermal conductivity and particle agglomeration in alumina nanofluids: experiment and theory. Physical Review E. **76**(6)(2007): p. 061203.

[21] Pastoriza-Gallego, M., C. Casanova, J. Legido, and M. Piñeiro, Cuo in water nanofluid: influence of particle size and polydispersity on volumetric behaviour and viscosity. Fluid Phase Equilibria. **300**(1)(2011): p. 188–196.

[22] Koo, J., Computational nanofluid flow and heat transfer analyses applied to micro-systems. (2005).

[23] Masoumi, N., N. Sohrabi, and A. Behzadmehr, A new model for calculating the effective viscosity of nanofluids. Journal of Physics D: Applied Physics. **42**(5)(2009): p. 055501.

[24] Meriläinen, A., A. Seppälä, K. Saari, J. Seitsonen, J. Ruokolainen, S. Puisto, N. Rostedt, and T. Ala-Nissila, Influence of particle size and shape on turbulent heat transfer characteristics and pressure losses in water-based nanofluids. International Journal of Heat and Mass Transfer. **61**(2013): p. 439–448.

[25] Sheikholeslami, M. and D. Ganji, Heat transfer of Cu-water nanofluid flow between parallel plates. Powder Technology. **235**(2013): p. 873–879.

[26] Sheikholeslami, M. and D.D. Ganji, Unsteady nanofluid flow and heat transfer in presence of magnetic field considering thermal radiation. Journal of the Brazilian Society of Mechanical Sciences and Engineering. **37**(3)(2015): p. 895–902.

[27] Vajjha, R.S. and D.K. Das, A review and analysis on influence of temperature and concentration of nanofluids on thermophysical properties, heat transfer and pumping power. International Journal of Heat and Mass Transfer. **55**(15)(2012): p. 4063–4078.

[28] Garg, H.P. and R.K. Agarwal, Some aspects of a pv/t collector/forced circulation flat plate solar water heater with solar cells. Energy Conversion and Management. **36**(2)(1995): p. 87–99.

[29] White, F.M., Fluid mechanics. 5th edition. Boston: McGraw-Hill Book Company. (2003).

[30] Kahani, M., S. Zeinali Heris, and S. M. Mousavi. "Effects of curvature ratio and coil pitch spacing on heat transfer performance of Al2O3/water nanofluid laminar flow through helical coils." Journal of Dispersion Science and Technology 34.12 (2013): 1704-1712

[31] Li, Q., Y. Xuan, and J. Wang, Investigation on convective heat transfer and flow features of nanofluids. Journal of Heat Transfer. **125**(2003): p. 151–155.

[32] Owhaib, W. and B. Palm, Experimental investigation of single-phase convective heat transfer in circular microchannels. Experimental Thermal and Fluid Science. **28**(2)(2004): p. 105–110.

[33] Li, Feng-Chen, et al. "Experimental study on the characteristics of thermal conductivity and shear viscosity of viscoelastic-fluid-based nanofluids containing multiwalled carbon nanotubes." Thermochimica Acta 556 (2013): 47-53.

[34] Handbook, A., 1985 Fundamentals. American Society of Heating, Refrigerating, and Air Conditioning Engineers, Inc., Atlanta, Georgia. (1985).

[35] Mingzheng, Z., X. Guodong, L. Jian, C. Lei, and Z. Lijun, Analysis of factors influencing thermal conductivity and viscosity in different kinds of surfactant solutions. Experimental Thermal and Fluid Science. **36**(2012): p. 22–29.

[36] Mahian, O., A. Kianifar, S.A. Kalogirou, I. Pop, and S. Wongwises, A review of the applications of nanofluids in solar energy. International Journal of Heat and Mass Transfer. **57**(2)(2013): p. 582–594.

[37] Gherasim, I., G. Roy, C.T. Nguyen, and D. Vo-Ngoc, Heat transfer enhancement and pumping power in confined radial flows using nanoparticle suspensions (nanofluids). International Journal of Thermal Sciences. **50**(3)(2011): p. 369–377.

[38] Hwang, K.S., S.P. Jang, and S.U. Choi, Flow and convective heat transfer characteristics of water-based Al_2O_3 nanofluids in fully developed laminar flow regime. International Journal of Heat and Mass Transfer. **52**(1)(2009): p. 193–199.

[39] Maïga, S.E.B., S.J. Palm, C.T. Nguyen, G. Roy, and N. Galanis, Heat transfer enhancement by using nanofluids in forced convection flows. International Journal of Heat and Fluid Flow. **26**(4)(2005): p. 530–546.

Permissions

All chapters in this book were first published in NHMTEP, by InTech Open; hereby published with permission under the Creative Commons Attribution License or equivalent. Every chapter published in this book has been scrutinized by our experts. Their significance has been extensively debated. The topics covered herein carry significant findings which will fuel the growth of the discipline. They may even be implemented as practical applications or may be referred to as a beginning point for another development.

The contributors of this book come from diverse backgrounds, making this book a truly international effort. This book will bring forth new frontiers with its revolutionizing research information and detailed analysis of the nascent developments around the world.

We would like to thank all the contributing authors for lending their expertise to make the book truly unique. They have played a crucial role in the development of this book. Without their invaluable contributions this book wouldn't have been possible. They have made vital efforts to compile up to date information on the varied aspects of this subject to make this book a valuable addition to the collection of many professionals and students.

This book was conceptualized with the vision of imparting up-to-date information and advanced data in this field. To ensure the same, a matchless editorial board was set up. Every individual on the board went through rigorous rounds of assessment to prove their worth. After which they invested a large part of their time researching and compiling the most relevant data for our readers.

The editorial board has been involved in producing this book since its inception. They have spent rigorous hours researching and exploring the diverse topics which have resulted in the successful publishing of this book. They have passed on their knowledge of decades through this book. To expedite this challenging task, the publisher supported the team at every step. A small team of assistant editors was also appointed to further simplify the editing procedure and attain best results for the readers.

Apart from the editorial board, the designing team has also invested a significant amount of their time in understanding the subject and creating the most relevant covers. They scrutinized every image to scout for the most suitable representation of the subject and create an appropriate cover for the book.

The publishing team has been an ardent support to the editorial, designing and production team. Their endless efforts to recruit the best for this project, has resulted in the accomplishment of this book. They are a veteran in the field of academics and their pool of knowledge is as vast as their experience in printing. Their expertise and guidance has proved useful at every step. Their uncompromising quality standards have made this book an exceptional effort. Their encouragement from time to time has been an inspiration for everyone.

The publisher and the editorial board hope that this book will prove to be a valuable piece of knowledge for researchers, students, practitioners and scholars across the globe.

List of Contributors

Monir Noroozi and Azmi Zakaria
Physics Department, Faculty of Science, Universiti Putra Malaysia, Malaysia

Wenzheng Cui, Zhaojie Shen and Jianguo Yang
School of Automotive Engineering, Harbin Institute of Technology, Weihai, China

Shaohua Wu
School of Energy Science and Engineering, Harbin Institute of Technology, Harbin, China

Lucian Pîslaru-Dănescu and Gabriela Telipan
National Institute for Electrical Engineering ICPE-CA, Bucharest, Romania

Floriana D. Stoian, Sorin Holotescu and Oana Maria Marinică
Politehnica University of Timisoara, Timisoara, Romania

Seval Genc
Engineering Faculty, Metallurgical and Materials Engineering, Marmara University, Istanbul, Turkey

Stanford Shateyi
Department of Mathematics, University of Venda, Thohoyandou, South Africa

Aaiza Gul and Sharidan Shafie
Department of Mathematical Sciences, Faculty of Science, Universiti Teknologi Malaysia, Skudai, Malaysia

Ilyas Khan
Basic Engineering Sciences Department, College of Engineering Majmaah University, Majmaah, Saudi Arabia

Adil Loyai
PAF-Karachi Institute of Economics and Technology, Karachi, Pakistan

Noreen Sher Akbar
DBS&H CEME, National University of Sciences and Technology, Islamabad, Pakistan

C. M. Khalique
International Institute for Symmetry Analysis and Mathematical Modelling, Department of Mathematical Sciences, South Africa

Zafar Hayat Khan
Department of Mathematics, University of Malakand, Dir (Lower), Khyber Pakhtunkhwa, Pakistan

Zafar Said
Department of Sustainable and Renewable Energy Engineering (SREE), University of Sharjah, Sharjah, United Arab Emirates
Department of Mechanical Engineering, Faculty of Engineering, University of Malaya, Kuala Lumpur, Malaysia

Rahman Saidur
Department of Mechanical Engineering, Faculty of Engineering, University of Malaya, Kuala Lumpur, Malaysia
Centre of Research Excellence in Renewable Energy (CoRE-RE), King Fahd University of Petroleum and Minerals (KFUPM), Dhahran, Saudi Arabia

Index

www.ingramcontent.com/pod-product-compliance
Lightning Source LLC
LaVergne TN
LVHW060559210926
848485LV00005B/1345